Georg-Wilhelm Oetjen

Freeze-Drying

Other titles of interest:

Common Fragrance and Flavor Materials
3rd edition, 278 pages, ISBN: 3-527-28850-5
Flavourings

Georg-Wilhelm Oetjen

Freeze-Drying

Weinheim · New York · Chichester · Brisbane · Singapore · Toronto

Dr. Georg-Wilhelm Oetjen
Tondernstraße 7
D-23556 Lübeck

The cover illustration shows the course of a main drying observed with a cryomicroscope. The photographs are taken after 1.5, 3.0, 4.5, and 6 minutes, respectively.

Library of Congress Card No.: applied for

A CIP catalogue record or this book is available from the British Library.

Deutsche Bibliothek Cataloguing-in-Publication Data:
Georg-Wilhelm Oetjen:
Freeze-Drying. / Georg-Wilhelm Oetjen. – Weinheim ; New York ; Chichester ; Brisbane ; Singapore ; Toronto :
Wiley-VCH, 1999
ISBN 3-527-29571-2

Printed on acid-free and chlorine-free paper.

Composition: Text- und Software-Service Manuela Treindl, D-93059 Regensburg
Printing: Strauss Offsetdruck, D-69509 Mörlenbach
Bookbinding: Großbuchbinderei J. Schäffer, D-67269 Grünstadt
Printed in the Federal Republic of Germany

Preface

This book is dedicated to my esteemed teacher, the late Prof. Dr. K. H. Hellwege, from whom I learned to tackle a problem from as many sides as I could imagine, with wide open curiosity.

One of humanity's oldest methods of preservation is the drying of food and herbs. Freeze-drying was first carried out by Altmann, who freeze-dried organ pieces in 1890, as Dr. K. H. Neumann wrote in his book 'Grundriß der Gefriertrocknung' in 1954. In 1932 Gersh designed an effective vacuum plant for the freeze-drying of histological preparations with the help of the diffusion pump, just invented by Gaede at that time.

Sawyer, Lloyd and Kitchen successfully freeze-dried yellow fiber viruses in 1929.

Industrial freeze-drying began with the production of preserved blood plasma and penicillin, as shown by E. W. Flosdorf in his book 'Freeze-drying' in 1949.

Vacuum technology and penicillin were also my introduction to freeze-drying. After my studies of physics at the university in Göttingen, I worked in the development department of E. Leybold's Nachf. where I had to build a freeze-drying plant for penicillin. From then on, I was engaged in vacuum process technology for almost 25 years, including the time from 1952 as Managing Director of Leybold Hochvakuum Anlagen GmbH. Freeze-drying has always fascinated me as being the most complex vacuum process. Mechanical and chemical engineering, chemistry and biology, sterility and regulatory issues are all part of the freeze-drying process.

I intended to write this book many years ago, but only after my retirement as a member of the managing board of Drägerwerke AG did I have the time to do so.

It was at this time that Mr. Wolfgang Suwelack, Managing Partner of Dr. Otto Suwelack Nachf. GmbH & Co., asked me to work for him as a consultant in freeze-drying and I have to thank him for the permission to use some of the results achieved in the last years.

Furthermore, I am grateful to Dipl.-Ing. P. Haseley, Managing Director of AMSCO Finn-Aqua GmbH, now Steris GmbH, for the permission to use results, drawings and photographs of his company. Several companies and publishing houses have granted permission to use drawings and photographs to which they own the © copyright. I am grateful to all of them, because they have made it possible to present freeze-drying under many aspects.

I have tried to show the interconnection between the property of the product, the goal to make it stable and the necessary processes to achieve this. The problems of the different process steps are discussed with examples and the parameters are described which influence each step. I have avoided following the many theoretical attempts describing one or more of the freeze-drying steps, but have restricted myself to a few equations which permit the calculation of process and product data with sufficient accuracy, or at least, allow an estimate, if some data is mentioned.

The freezing of a product is a very important step. The structure in the frozen product decides whether the product can be freeze-dried at all and under which conditions it can be done. For this reason, the consequences of the freezing rate, layer thickness of the product and excipients are discussed in some detail. The second main point is the measurement and control of the two drying phases; the main and secondary drying and the third concentrates on the residual moisture content, its measurement and the consequences during storage of the dry product. There will be critical opinions that some of the processes are unilaterally represented. My aim was to show the limits and the advantages of certain procedures to enable the reader to decide whether the ideas of the quoted authors, or my own can be applied to his tasks.

The approx. 220 references in the 1997 (German) edition are supplemented by approx. 50 new ones.

Contents

1 Foundations and Process Engineering

Freeze drying or lyophilization is a drying process, in which the solvent and/or the medium of suspension is crystallized at low temperatures and thereafter sublimated from the solid state directly into the vapor phase.

Freeze drying is mostly done with water as solvent. Fig. 1.1 sows the phase diagram of water and the area in which this transfer from solid to vapor is possible. This step is difficult, even for pure water. If the product contains two or more components in true solutions or suspensions, the situation can become so complicated that simplified model substances have to be used. Such complex systems occur ubiquitously in biological substances.

The second step, the drying, transforms the ice or water in an amorphous phase into vapor.

Due to the low vapor pressure of the ice, the vapor volumes become large, as can be seen in Fig. 1.2.

The most important goal of freeze-drying is to produce a substance with good shelf stability and which is unchanged after reconstitution with water, though this depends also very much on the third step of the process: the packing and conditions of storage.

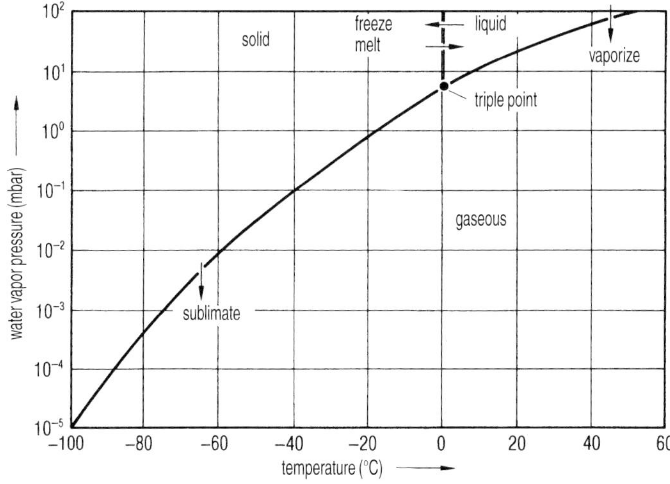

Fig. 1.1. Phase diagram of water.

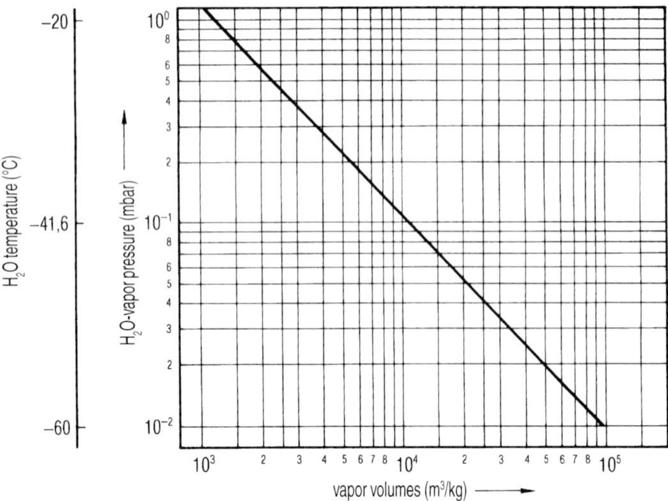

Fig. 1.2. Specific volume of water vapor as a function of the water vapor pressure. The temperature of the vapor in this diagram is that of the ice.

1.1 Freezing

To freeze a substance, it must be cooled to such a temperature at which the water and the solids are fully crystallized, or at which areas of crystallized ice and solids are enclosed in zones in which amorphous concentrated solids and water remain in mechanically solid state (see Section 1.1.2). In the zone of freezing, the ice crystals are growing first, thus concentrating the remaining solution, which can vary the pH value. In many substances a eutectic temperature can be determined, but in many others this value does not exist. The crystallization depends on several factors which influence each other: cooling velocity, initial concentration, end temperature of cooling, and the time at this temperature. In several products no crystallization takes place and the product remains in an amorphous, glass-like phase, or a mixture of both occurs.

1.1.1 Amount of Heat, Heat Conductivity, Heat Transfer and Cooling Velocity

For pure water, the melting heat to be withdrawn for freezing (Q_{tot}) can be calculated by Eq. (1), if the starting and the desired final temperatures are known.

$$Q_{tot} = c_w (T_1 - T_0) + Q_e + c_{ice} (T_0 - T_2) \quad \text{(kJ/kg)} \tag{1}$$

where

c_w = specific heat capacity of water;
Q_e = melting heat of ice;
c_{ice} = specific heat capacity of ice;
T_0 = freezing temperature of ice;
T_1 = initial temperature of water; and
T_2 = final temperature of ice.

The temperature dependence of c_w between +20 °C and 0 °C and c_{ice} between 0 °C and –50 °C has to be adopted by average values.

For solutions and suspensions the solid content has to be recognized. This is reflected in Eq. (2)

$$Q_{tot} = [(c_w\, x_w + c_{f'}\, x_f)\,(T_1 - T_0)] + x_w\, Q_e + [(c_e\, x_w + c_f\, x_{w'})\,(T - T_0)] \tag{2}$$

where

x_w = part of water above 0 °C and
c_f = specific heat of solids,
per example, for animal products ≈ 1.47 kJ/kg °C
for plant products ≈ 1.34 kJ/kg °C
or for some solids:
carbohydrates ≈ 1.42 kJ/kg °C
proteins ≈ 1.55 kJ/kg °C
fats ≈ 1.7 kJ/kg °C
salts ≈ 0.8 kJ/kg °C
x_f = part of solids
$x_{w'}$ = part of ice, which freezes until temperature T_2 is reached. If not all water is frozen at T_2, an additional term has to be introduced, which reflects the cooling of the unfrozen water.

Table 1.1: Percentage of water frozen out at various temperatures for some foods (Part of Table 1 in [1.1] and [1.2]).

Product	Frozen out water at °C (% of the total water)				UFW
	–10	–15	–20	–30	(% of total water)
Lean beef	82	85	87	88	12
Haddock	84	87	89	91	9
Whole eggs, liquid	89	91	92	93	7
Yolk	85	86	87	87	13
Egg white	91	93	94	94	6
Yeast	80	85	88	89	11
Fruit juice	85	90	93	96	(3)
Peas	80	86	89	92	(7)

Table 1.2: Enthalpy of meat, fish and egg products (Part of Table 3 in [1.1] and [1.2]).

Product	Water content (weight %)	Enthalpy kJ/kg at a temperature of °C					
		–30	–20	–10	0 +	5 +	20
beef, 8 % fat	74.0	19.2	41.5	72.4	298.5	314.8	368.4
cod	80.3	2.1	41.9	74.1	322.8	341.2	381.0
egg white	86.5	18.4	38.5	64.5	351.3	370.5	427.1
whole egg	74.0	18.4	38.9	66.2	308.1	328.2	386.9

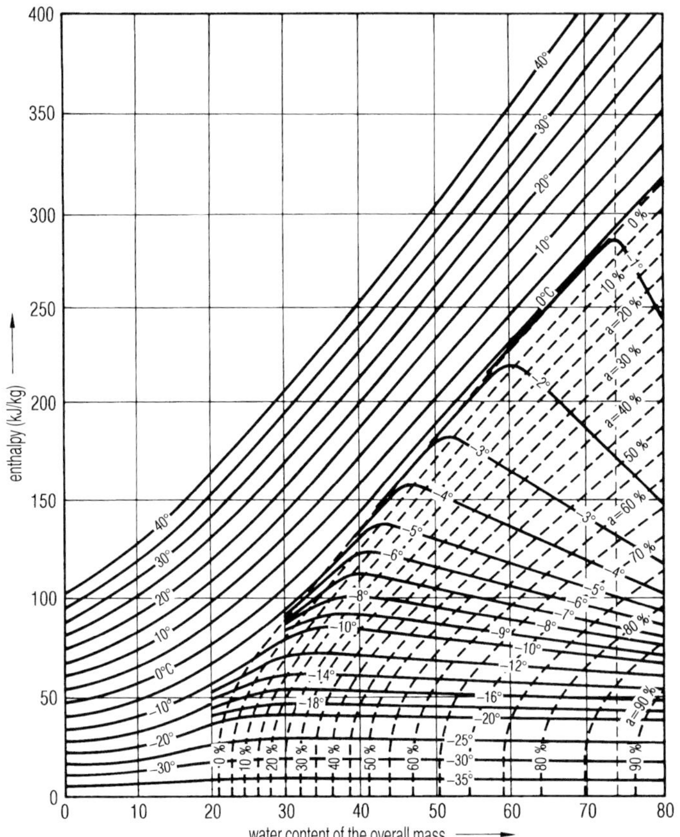

Fig. 1.3. Enthalpy of lean beef meat as a function of its water content (0 kJ/kg at –40 °C).
The temperatures at the beginning of cooling the and the desired end temperatures for freezing are plotted as parameters. The dotted lines indicate the percentage of water frozen at the end temperatures (Fig 1 from [1.1] and [1.2]).
Example: Beef meat has 74 % water. At +10 °C the enthalpy is approx. 325 kJ/kg, at –20 °C approx. 40 kJ/kg. 285 kJ/kg have therefore to be removed and 83 % of the water frozen. The maximum possible (88 %) (see Table 1.1) is reached at approx. –30 °C.

Table 1.1 shows the non-freezable water (UFW) in various foods. The reasons and the consequences are described in Sections 1.1.3 and 1.1.4. In comparing these data with other publications, e. g. [1.3], smaller values may be found. This can depend not only on the different raw materials and the history of the probe until measurement, but also the methods of measurement.

For meat with less than 4 % fat content, Riedel [1.1] has published an enthalpy diagram (shown in Fig. 1.3). For some other foods Table 1.2 shows enthalpy data at various temperatures. At –40 °C the enthalpy is set at 0 kJ/kg.

The transport of the calculated energy from the freezing zone of the product to the cooling medium can be described in a simplified way by the following steps:

The product is an infinite plate, which is cooled from one site only, and the energy flows only perpendicular to its infinite expansion. The crystallization energy flows from the crystallization zone, through the already frozen ice, through the container bottom to a shelf and into the cooling brine.

The freezing time (t_e) is approximately given by Eq. 3 [1.4]

$$t_e = \Delta J/\Delta T\ \rho_g\ (d^2/2\ \lambda_g + d/K_{su}) \tag{3}$$

$$t_e = \Delta J/\Delta T\ \rho_g\ (w + u) \tag{3a}$$

where

t_e = freezing time;
ΔJ = enthalpy difference between the the initial freezing point and the final temperature;
ΔT = difference of temperature between the freezing point and the cooling medium;
d = thickness of the product parallel to direction of prevailing heat transfer
ρ_g = density of the frozen product
λ_g = thermal conductivity of the frozen product; and
K_{su} = surface heat transfer coefficient between cooling medium and the freezing zone.

The thermal conductivity of ice and of dried products is relatively well known, but the surface heat transfer coefficient, K_{su} during freezing and the total heat transfer coefficient K_{tot} during freeze-drying vary largely as described in the various chapters. Table 1.3 gives a survey of some data of interest in freeze-drying.

The influence of the variables in Eq. (3) can be studied by an example: A slice of lean beef with a thickness which is small compared with its horizontal dimensions is to be frozen to –20 °C. The influences of the border of the slice are neglected. The thickness of the slice is $d = 2$ cm. As can be seen in Fig. 1.3, the enthalpy difference for beef with 74 % water is approximately 240 kJ/kg. If the freezing process starts between 0 and –3 °C and is mostly finished at –20 °C, the cooling medium has a temperature of –43 °C and an average $\lambda = 1.38 \cdot 10^{-2}$ J/ °C cm s is used when the slice is in contact with a liquid, having a similar behavior as water at 20 °C, $K_{su} = 4.61 \cdot 10^{-2}$ J/ °C cm^2 s can be used for the calculation. The freezing time

$$t_{e\ d20}^{fl} = 5.4\ (0.725 \cdot 10^2 + 0.43 \cdot 10^2) \approx 12 \text{ min} \tag{4}$$

Table 1.3: Surface heat transfer coefficient, total heat transfer coefficient and thermal conductivity.

K_{su} from gases to a solid surface (KJ/m^2 h° C): free convection	17–21
laminar flow 2 m/s	50
laminar flow 5 m/s	100
K_{su} between the shelf of a freeze-drying plant and a product in vials or trays during freezing (kJ/m^2 h° C)	200–400
K_{su} between a liquid and a solid surface (kJ/m^2 h° C) oil in tubes, laminar	160–250
LN_2 by drops on the product[1]	900
from liquids similar to water [2]	1600
from water at 1bar, temperature difference < 7 °C[3]	3600
K_{tot} between the shelf of a freeze-drying plant and the sublimation front in the product contained in vials or trays under vacuum[4] (kJ/m^2 h° C)	62–124
λ thermal conductivity (kJ//m h° C)	
λ_g frozen product (ice)	4.0[5]–6.3[6]
λ_{tr} dry product[7]	0.059 to 0.29

1 Rinfret, A. P.: Factors affecting the erythrocyte during rapid freezing and thawing Annals of the New York Academy of Sciences, Vol. 85, Art. 2, p. 576–594, 1960
2 from [1.2]
3 Wärmeatlas, 5. Auflage, Bild 38, Seite A 26, VDI-Verlag, Düsseldorf, 1988
4 from Table 1.9 and 1.10
5 from [1.2]
6 from [1.50]
7 from [1.54], [1.55] and [1.56]

As shown in Eqs. (3) and (3a) the thickness d has a mayor influence if the conductivity term w, which includes d^2, is large compared with the transfer term u, which includes only d.

In Eq. (4) $w : u = 1.7 : 1$, showing, that the influence of the conductivity is almost double compared with influence of the transfer. Assuming that d is only 0.2 cm, the freezing time falls to

$$t_{e\ d2}^{fl} = 5.4\ (0.725 + 4.35) \approx 28\ \text{s} \tag{5}$$

In this case $w : u = 1 : 6$, and the transfer term is overwhelming. The freezing time is neither reduced by d^2 nor by d, since the importance of w and u has changed. An increase in d by a factor of 3, to 6 cm prolongs the freezing time

$$t_{e\ d60}^{fl} \approx 70\ \text{min} \tag{6}$$

$w : u = 5 : 1$, the freezing time depends mostly on the heat conductivity of the material.

The freezing of a slice of beef in direct contact with a model liquid has been used to demonstrate the influence of the two terms *w* and *u*. To freeze a product for freeze-drying, two methods are mainly used: (i) freezing of the product in trays or in vials on cooled surfaces; or (ii) in a flow of cold air. If these methods do not result in a sufficient freezing rate, LN_2 in direct contact with the vials is used (see Fig. 2.2) or droplets of the product are sprayed into LN_2 (see Section 2.1.4).

The heat transfer coefficient K_{su} in air varies strongly with the gas velocity, surface conditions of the product, and the geometry of the installation. In practical operations it will be difficult to achieve K_{su} values of 1.7–2.5 · 10^{-3} J/cm^2 s °C, or ≈ 75 kJ/m^2 h °C, and in many applications only half of this value (or less) may be possible. However, even with this high K_{su} the above discussed slice of beef (2 cm thick) has a freezing time

$$t_{e}^{lu}{}_{d20} = 5.4\,(0.72 \cdot 10^2 + 9.5 \cdot 10^2) \approx 92 \text{ min} \tag{7}$$

compared with 12 min when cooled by a liquid, since the K_{su} of a gas is ≤ 10 % that of a liquid.

The time to reach a desired temperature level can expressed as freezing rate v_t, the change in temperature per unit time, e. g. °C min^{-1}. Thus, the values of Eqs. (4) to (7) are approximately:

(4) v_f = 1.7 °C min^{-1} (5) v_f = 43 °C min^{-1}
(6) v_f = 0.3 °C min^{-1} (7) v_f = 0.2 °C min^{-1}

These data are calculated by using 0 °C as the start and –20 °C as the end temperature to show the relative data. The exact calculation requires more information, as given below.

Figure 1.4 is the cooling curve of vials filled with a solution of 4 % solid content and 27 mm filling height. From curve, the v_f can be estimated:

0 °C	to –10 °C ≈ 0.15 °C min^{-1}
0 °C	to –14 °C ≈ 0.18 °C min^{-1}
–14 °C	to –30 °C ≈ 0.73 °C min^{-1}
0 °C	to –30 °C ≈ 0.3 °C min^{-1}

During the freezing of the main part of the water v_f is only 25 % compared with the value after most of the water is crystallized. To take the average value between 0 and –30 °C can therefore be misleading: The intention to freeze at a rate of 0.3 °C/min has not occurred during an important part of the operation. The difference between 0.15 °C min^{-1} and 0.7 °C min^{-1} influences the structure of the product. How important the change is has to be checked from case to case, but one should be aware of this possibility.

With Eq. (3), it is also possible to estimate K_{su}. The uncertainties are the differences between the freezing of the product around the temperature sensor and in the undisturbed product, the position of the sensors, the correlation between time and temperature and occasionally also the actual amount of frozen water. From Fig. 1.4, the estimated K_{su} is

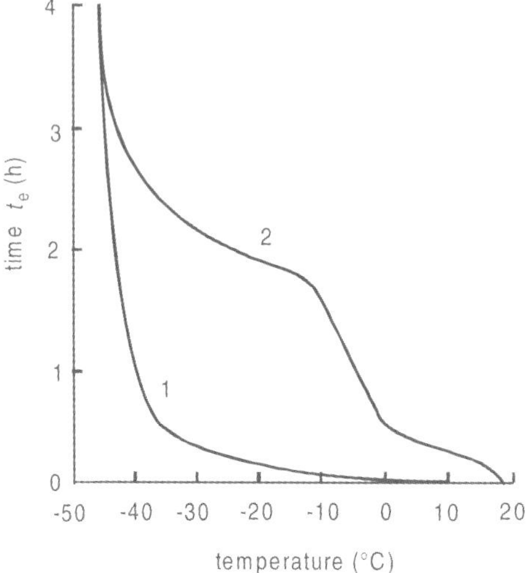

Fig. 1.4. Temperatures during freezing as a function of time [97].
1 = shelf temperature; 2 = product temperatures in a product with d = 2,7 cm, solid content ≈ 4 %.

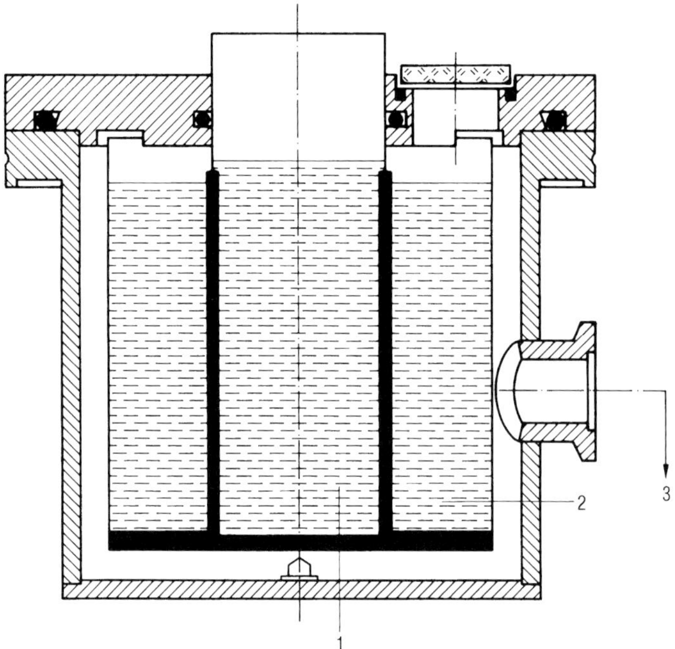

Fig. 1.5. Apparatus to produce a mixture of liquid and solid nitrogen.
1 Inner container with LN_2; 2 external container with LN_2 connected to a vacuum pump; 3 the container 2 is evacuated to approx. 124 mbar and kept at that pressure. The evaporating nitrogen reduces the temperature in 2 and thereby also in 1, since the two containers are in close thermal contact. A temperature of –210 °C is reached in container after approx. 5 min (from [1.99]).

Table 1.4: Comparison of cooling rates, measured in the same installation, with comparable vials and comparable *d*.

Run	Time from 0 °C to –10 °C/(min)	Cooling rate (°C/min)	Time from –10 °C to –30 °C (min)	Cooling rate (°C/min)
1	34 ± 5	0.29 ± 15 %	13 ± 5	1.5 ± 38 %
2	25	0.4	17	1,2
3	23 ± 1,5	0.4 ± 6 %	15 ± 2	1.3 ± 13 %
4 [1]	19 ± 2,5	0.5 ± 13 %	21 ± 3,5	0.95 ± 17 %
5 [2]	79 ± 7	0.13 ± 9 %	38 ± 5	0.5 ± 13 %

1 During the cooling phase –10 °C to –30 °C is $\Delta T \approx$ *13 °C instead of* $\approx$ *20 °C* in run 1–3; taking this into account, the value of 0.95 corresponds to 1.4 in run 1–3.

2 The shelf temperature has been constantly lowered with ≈ 10 °C/30 min. Therefore ΔT is only ≈ 8 °C during the freezing phase, compared with ≈ 30 °C in run 1–3 0.13 °C/min corresponds therefore to ≈ 0.48 °C in run 1–3. The same applies to the 0.5 °C/min during the cooling phase, making it comparable with 1.3 °C/min.

approximately 480 kJ/m^2 h° C with a possible error of ±10 % and maximum error of ±20 %. Such high values can only be expected if the vials are carefully selected for their uniformity, especially with respect to a very even and flat bottom. Otherwise, the K_{su} can be much smaller e. g. ≤ 200 kJ/m^2 h °C. If the vials are placed in trays and these are loaded on the shelves, the K_{su} will be reduced, very likely to ≤ 50 %.

Equation (3) can be used to estimate the influence of the variation of the layer thickness and the shelf temperature, if the K_{su} values are measured for one type of vials on one type of well manufactured shelves (see Section 1.2).

Table 1.4 shows a comparison of cooling rates [1.5]. The percentage indicates the maximum differences between the measurements with three temperature sensors in three vials. In run 2, only one sensor has been used. The expected v_f at a shelf temperature of –45 °C and d = 10 mm are in the order of 0,5 °C/min to 1.5 °C/min. To increase v_f the following possibilities can be used: (I) a reduction of *d;* (II) reduce the shelf temperature; (III) precooling of the vials e. g. to –80 °C and filling the precooled product, e. g. +4 °C into the cold vials; (IV) cooling of the vials directly by LN_2; and (V) dropping the product into LN_2. With precooled vials v_t can be in the order of 10 to 20 °C/min and with direct cooling by LN_2 40 to 60 °C/min and more is possible. With droplet freezing up to 1000 °C/min can be achieved.

For laboratory work different cooling liquids can be used as shown in Table 1.5. However these substances are not easy to use, they boil and are partially explosive. The cooling method shown in Fig. 1.5 can be helpful. LN_2 is evaporated under vacuum, freezing a part of the N_2 as a solid. In this mixture the solid melts, if energy is produced from cooling and crystallization. Thereby the forming of gaseous N_2 is greatly reduced, which otherwise limits the heat transfer, Fig. 1.6 shows the relative cooling rates for different forms of N_2.

Table 1.5: Physical data of cooling liquids.

Medium	Boiling point T_s (°C)	c_p of liquid at T_s (kJ/kg °C)	λ of liquid at T_s (kJ/m · h °C)	heat of vaporization at T_s (kJ/kg)
Helium (He^4)	–268.9	4.41	0.098	20.5
Nitrogen	–195.8	2.05	0.506	197.6
Propane	–42.3	2.19		426.2
n-Pentane (Fig. 2 from [1.98])	+36.1	2.2		234.1

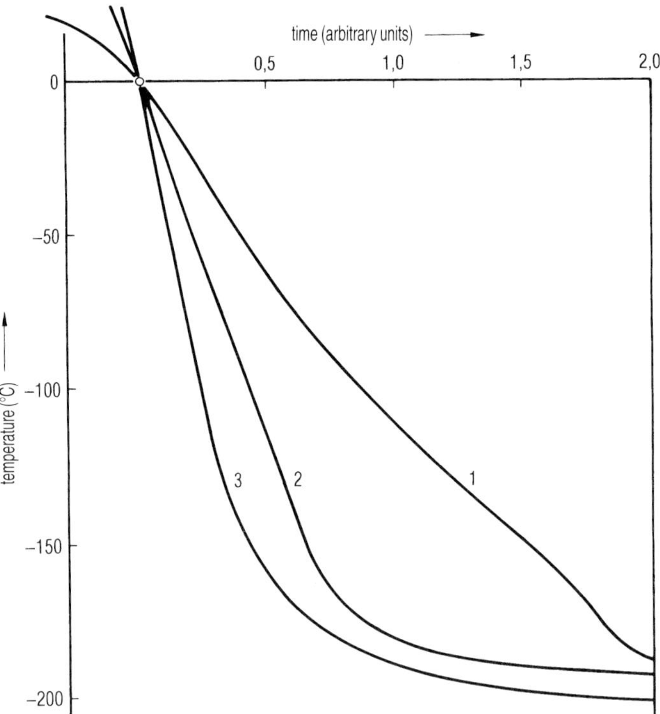

Fig. 1.6. Relative cooling rate of a small sample in different forms of N_2.
(The plot for LN_2 depends mostly on the successful removal of the nitrogen gas.) Melting solid nitrogen reduces the formation of gaseous N_2, since the crystallization energy melts the solid nitrogen and does not evaporate the liquid. (*Note*: Theoretically cooling in solid N_2 would be the fastest method, but liquid N_2 will be formed and the heat transfer is not stable.)
1 LN_2; 2 LN_2 + solid N_2; 3 melting of solid N_2 (from [1.100]).

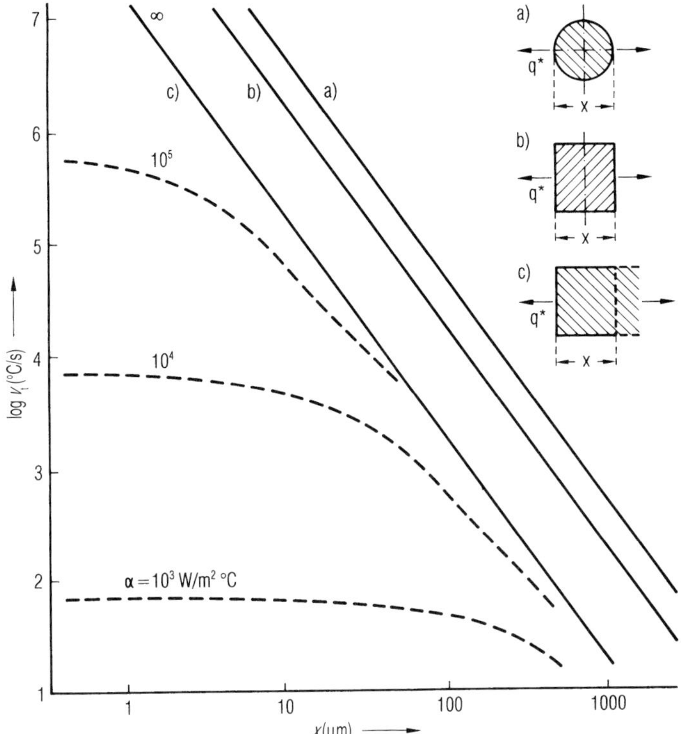

Fig. 1.7. Maximum theoretical cooling rate for different geometric configurations (a, b, c) of water by cooling with LN_2, if α is assumed = ∞. The dotted lines are calculated for three values of $\alpha = 10^3$, 10^4, 10^5 W/m^2 s (Fig. 2 from [1.6]).

Riehle [1.6] has calculated the theoretically possible cooling velocities for small objects between 1 and 10^{-3} mm as shown in Fig. 1.7. These calculations are made for a substance consisting of water only and K_{su} is assumed to be infinitely large for the geometric dimensions shown in a) a sphere, b) a square cylinder of infinite length and c) a plate of infinite length and the thickness X, cooled only from one side. For the plate (c) v_t is also calculated for three limited K_{su} : 10^3, 10^4 and 10^5 W/m^2 s (Chain lines). The purpose of this calculation is to show, that freezing rates of 10^3 to 10^4 °C/s ($6 \cdot 10^4$ to $6 \cdot 10^5$ °C/min) cannot be achieved. However these rates are necessary to reduce the velocity of crystal growth in pure water sufficiently to obtain water in a glass-like phase with irregular particle size $< 10^{-8}$ m.

Riehle shows, that such freezing rates can only be reached for layers < 0.1 mm under a pressure of 1.5–2.5 kbar.

A different way obtaining short cooling and freezing times is to evaporate a part of the water in the product under vacuum. The evaporation energy of water at 0 °C is approximately $2.5 \cdot 10^3$ kJ/kg. To cool 1 kg of beef from 0 °C to –20 °C, 240 kJ have to be removed, which corresponds to ≈ 0,1kg of water to be evaporated, or 15 % of the water in the beef.

This quick evaporation will produce foam or bubbles in the product. This is unacceptable in most cases, since the original structure is changed and that part of the product which is vacuum dried will have different qualities than the freeze-dried part. Often the product frozen in this way can not be freeze-dried at all.

1.1.2 Structure of Ice, Solutions and Dispersions

The water molecule has a configuration as shown in Fig. 1.8 [1.7], having a pronounced dipole moment, which produces the liquid phase at relative high temperatures and ensures a structure in the envelope of molecules which surround ions [1.8]. However clusters are also in water without ions; these consist of approximately ten water molecules in a tetrahedral geometry surrounded by O–H–O groups. The clusters are not stabile units with always the same molecules, and they are constantly exchanging molecules with their surroundings, having an average life time of between 10^{-10} and 10^{-1} s. The number of clusters decreases as temperatures are lowered untill freezing is reached.

In water which is very well cleared of all foreign particles, the clusters begin to crystallize in the subcooled water at –39 °C, this is called homogeneous nucleation. Foreign, undissolved particles in water act as nuclei for the crystallization of ice, and this is called heterogeneous nucleation. In normal water there exist approximately 10^6 particles per cm^2, and these act as nuclei for crystallization. They become increasingly effective, if their structure equals that of water. If a nucleus has formed, it grows faster at the outside than at the inside, producing (depending on subcooling and cooling velocity) structures of ice stars (Fig. 1.9). During further freezing, branches grow at an angle of 60 °, well known as frost flowers. For a crystal of $1 \cdot 10^{-9}$ mm^3 $2.7 \cdot 10^{10}$ molecules have to be brought into position. It is difficult to visualize how such a crystal can be formed in a small fraction of a second, but it is obvious that the growth of such a crystal will be influenced or disturbed by many factors.

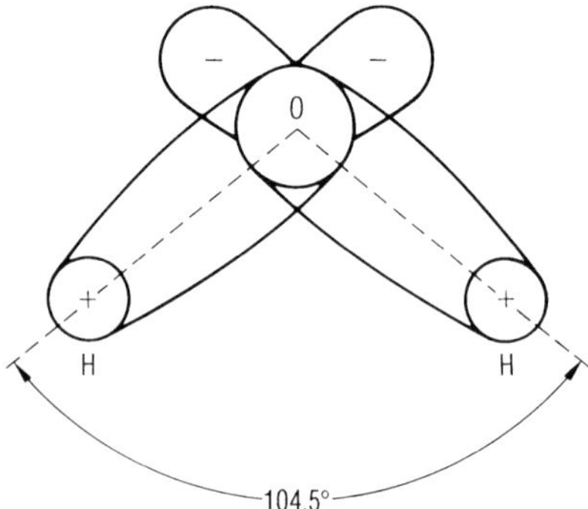

Fig. 1.8. Configuration of the electrical charges in a water molecule (Fig. 2 from [1.7]).

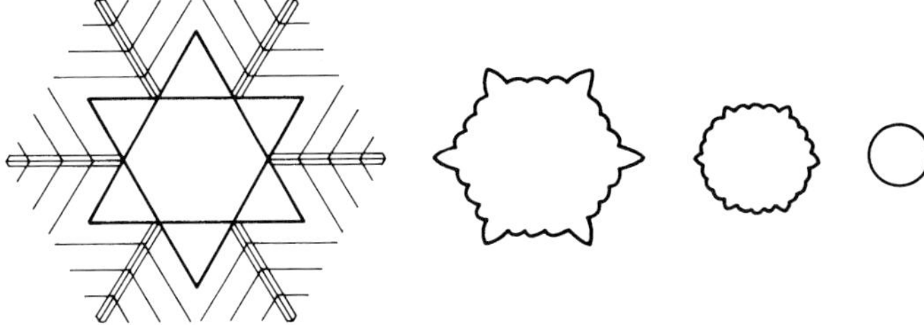

Fig. 1.9. Growth of ice crystals in water. The subcooling is increased from left to right (Fig. 10 from [1.7]).

In Fig 1.10, Riehle shows log J^* (J^* = nuclei per time and volume) as a function of the temperature of the phase transition water – ice different pressures of 1 and 2100 bar. At 2100 bar, J^* is comparable with J* at an approximately 35 °C higher temperature. Under pressure, water can be subcooled further, with a delayed formation of nuclei.

The growing of crystals is determined by the diffusion of molecules to the surface of the nucleus, the finding of a proper place, and the distribution of the freed energy to the surroundings. Under normal conditions (cooling speed $v_f < 10^2$ °C/s and subcooling $T_{sc} < 10$ °C) Eq. (8) can be used:

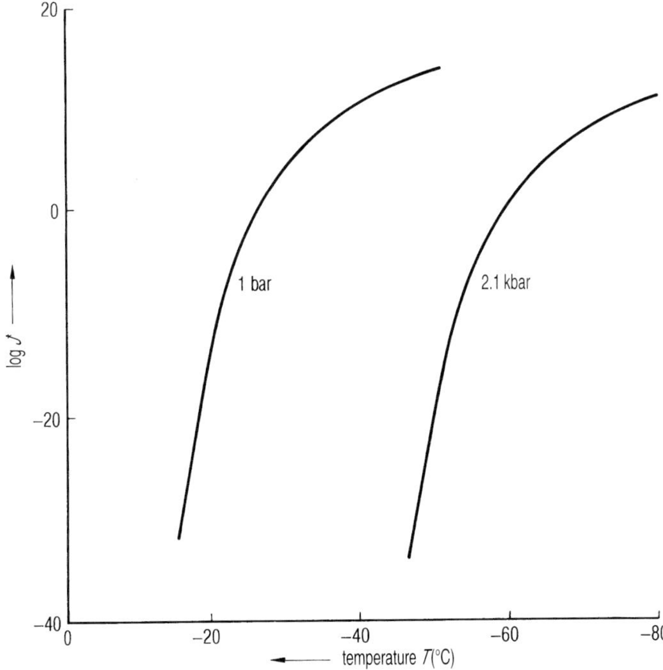

Fig. 1.10. Nucleation rate J^* (nuclei/volume time) as a function of the temperature of the phase transformation water ice (Fig. 4 from [1.6]).

$$v_k = \text{const. } T_{sc}^{\,n} \qquad (8)$$

where $n = 1$ if the energy transport and $n = 1.7$ if the surface reactions are decisive (Hillig and Turnbull, *J. Chem. Physiol.* 24, 914, 1956). If $T_{sc} > 10$ °C, the diffusion process has to be taken into account. Since v_k is furthermore dependent on the concentration, the calculation of v_k insecure.

To summarize it can be stated that:

To produce large crystals:

- the rate of nucleation should be small, therefore the subcooling should be small and the freezing should take place in a quasi-equilibrium situation between solution and crystals.
- the temperature should be as high as possible, since the crystals grow with the function $e^{-1/T}$.
- the time given for crystallization has to be increased, since v_K is inversely proportional to the size of the crystal.

To produce only very few or no crystals:

- freezing should take place under high pressure (Fig. 1.10)
- the freezing rate should be as high as possible, to produce a large degree of subcooling.

As can be seen from Fig. 1.11, the phase of Ih (hexagonal ice) has to be passed quickly to reach the zone of Ic (cubic ice). At or below –160 °C, Ic is glassy and unstructured. In pure water Ic can only be reach under high pressure. The addition of e. g. glycerol (10 %) and a freezing rate of $4 \cdot 10^3$ cm/s leads to vitrification of the solution.

Dowell and Rinfret [1.9] demonstrated that the phase at temperatures above –160 °C consist of small crystals with ≈ 400 A° in size and having cubic and pseudohexagonal structures.

Figure 1.12 shows the three phases of ice which exist under normal pressure as a function of temperature, indicating also the time it takes to change from one type to another. If water vapor is condensed on a cold surface in a very thin film, amorphous ice is formed and remains stable at –160 °C for a long time. As shown in Fig 1.12 the change from amorphous to cubic ice will take ≈ $5 \cdot 10^5$ min, or more than a year. The rate of change depends very much on the temperature: at –135 °C the same change takes only 1 min. This change is called devitrification. At –125 °C the change from cubic to hexagonal ice at takes ≈ 1000 h, while at –65 °C only hexagonal ice is stable.

To summarize, amorphous ice is stable below –160 °C, until –125 °C when cubic ice is formed irreversibly from the amorphous phase; above this temperature, hexagonal ice develops. Between –160 °C and –130 °C, cubic ice can be embedded in an amorphous surrounding. During warming It is likely that some amorphous some ice changes directly into the hexagonal form. Between –130 °C and –65 °C all three phases could be present, depending on time-temperature function. This behavior of pure water changes if water solutions, suspensions in water and mixtures with water are studied as be the case for virtually all products to be freeze dried.

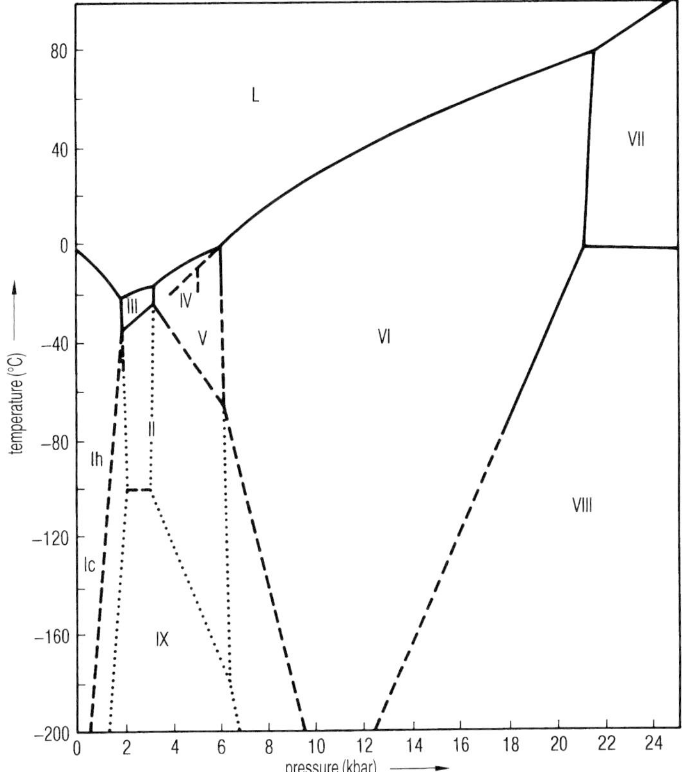

Fig. 1.11. Phase diagram of water.
L, liquid water; *Ih,* hexagonal ice; *Ic,* cubic ice; III to IX, crystal configurations of ice (Fig. 1. from [1.7])

The freezing process will be discussed with model substances, which will be used as cryo-protective agents, called CPAs. If a solution of water and glycerol is cooled quickly, a 10 % solution in a layer of $3 \cdot 10^{-3}$ mm and $v_f = 10^6$ °C/s can be vitrified ([1.6], page 218), but in a 5 % solution crystals of 1000 A° are formed. At high pressures (1.5–2.1 kbar), $4 \cdot 10^3$ °C/s are sufficient for a 10 % solution and $2 \cdot 10^4$ C/s for a 5 % solution to achieve vitrification. For these measurements the absence of foreign particles must be presumed in order to fully use the subcooling effect. Foreign particles could also come from containers, holding devices, etc.

Riehle has proved the existence of such vitrification by electron microscopy. With higher concentrations of glycerol vitrification is more simple. Luyet [1.10] shows diagrammatically (Fig. 1.13) how various phase changes take place at different glycerol concentrations. At 60 % glycerol devitrification takes place at ≈ –115 °C and increases with increasing glycerol concentration to ≈ –85 °C. Howevere such high concentrations of glycerol can normally not be used to freeze dry organic substances.

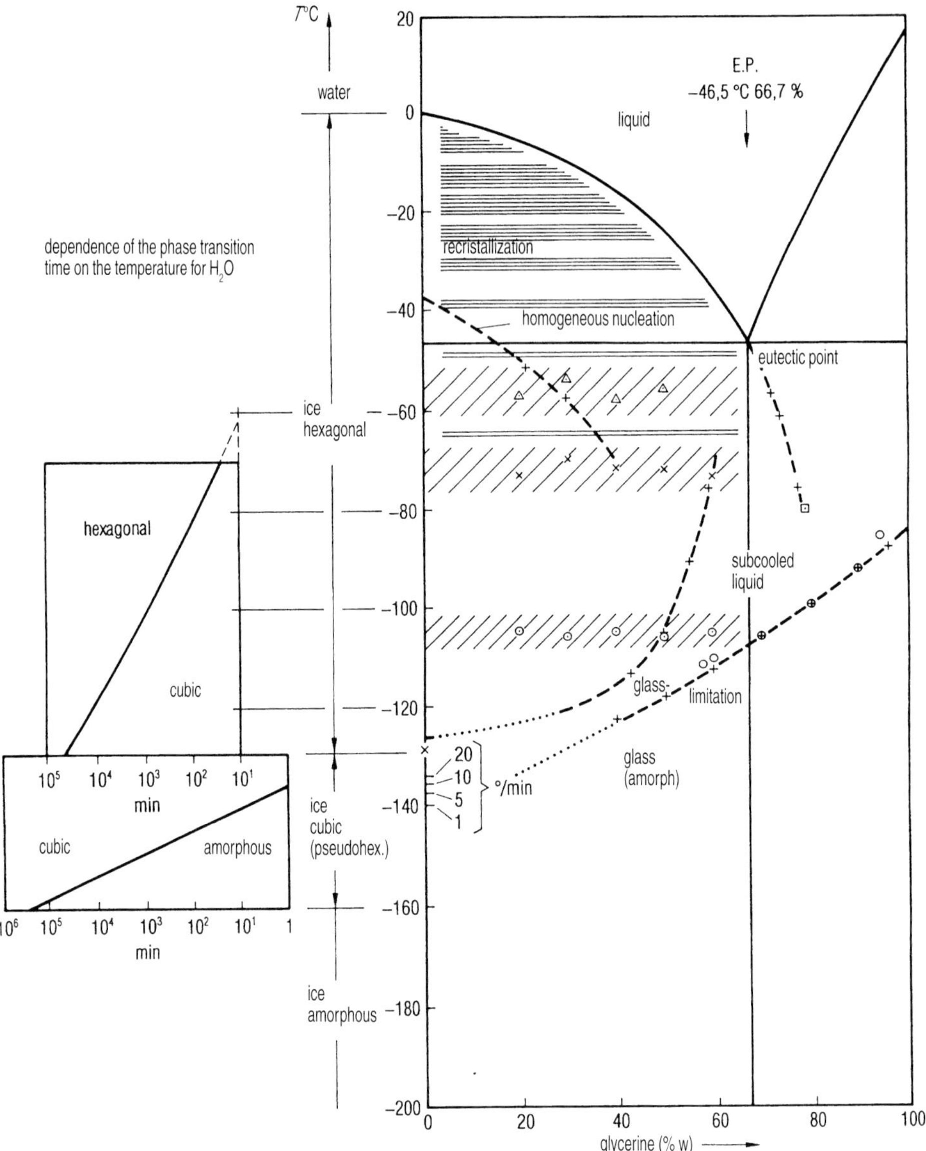

Fig. 1.12. Phase diagram of water – glycerine. On the left hand side the dependence of the phase transition time from the ice temperature is shown: At −140 °C, amorphous ice transforms into cubic ice in approx. 10 min (Fig. 8 from [1.98]).

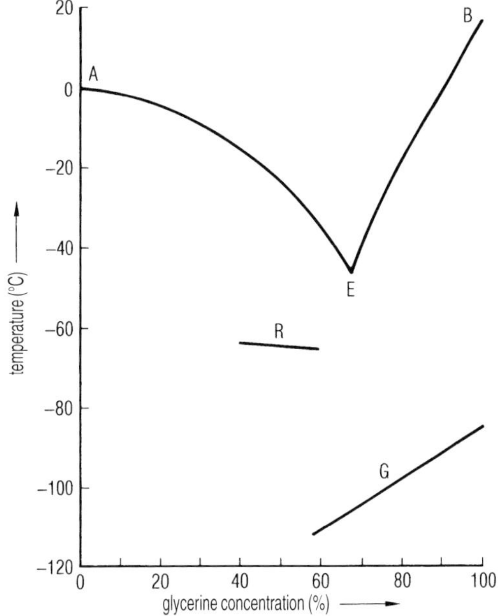

Fig. 1.13. Temperature as a function of the concentration water – glycerine mixture at which phase transformations occur (Fig. 14 from [1.10]). Definitions by Luyet: AE, Forming of small crystals or molecular groups; E, etectic point; EB, forming of clusters; R, irruptive recristallization; G, glass transition.

As shown for pure water, the phase transitions depend on the cooling rate, the end temperature of cooling, and the temperature and time of the treatment after cooling.

The rate of rewarming is especially critical. One has to differentiate between quasi-static situations, which are independent of time and all other dynamic states, in which the history of the present situation and the rate of the further changes play an important role.

Freezing processes can be divided in two categories: one type is so slow, so that they run under almost equilibrium conditions, others are too fast to approach the equilibrium situation. Figures 1.14.1 to 1.14.3 show the effect of the freezing rate on the structure of the dried product. In Fig. 1.14.1, milk has been frozen slowly (0.2–0.4 °C/min) in trays. In Fig. 1.14.2 mannit solution has been frozen in vials at a rate of approx. 1 °C/min, the arch at the bottom represents the vial bottom. In Fig. 1.14.3 γ-globulin has been frozen in LN_2 (approx. 10–15 °C/min). This shows only the upper part of the dry product. The cake has been frozen so quickly from the bottom and the walls, that the concentrated liquid has been pushed to the center, where it has been pressed to form a cone. The cake is cut and in the center of the cone a channel can be seen, in which high concentrated solution has been included, leaving a channel. Since the solids of this part are agglomerated to the surrounding areas, the structure of the channel is partially collapsed during drying.

The non equilibrium status can be seen during a slow cooling of a water-glycerine solution: Starting with a 20 % glycerol solution pure ice crystals will first be formed until at –46.5 °C when the glycerol concentration has reached 66.7 %. At this temperature, the eutectic should solidify. Howevere it is possible to reduce the temperature to –58 °C with a glycerol concentration of 73 %. A further decrease of the temperature does not crystallize

Fig. 1.14.1. Milk frozen slowly (0.2–0.4 °C/min) in a tray.

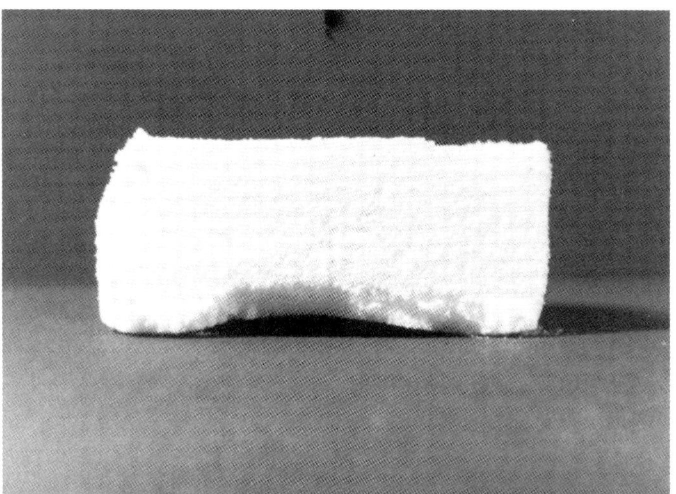

Fig. 1.14.2. Mannitol solution frozen at approx. 1 °C/min in a vial on precooled shelf.

any more water. The solution is so highly concentrated and highly viscous, and the mobility of the water molecules is so much reduced, that the remaining water is unfreezable (UFW) in an amorphous state between the glycerol and ice molecules.

Figure 1.15 ([1.101], page 286) shows diagramatically at a given starting concentration. which part will be ice, unfrozen water and glycerol at a freezing temperature actually used under equilibrium conditions. A solution of 20 % initial glycerol contains, when cooled to –50 °C, 70 % ice, 10 % unfreezable water (UFW) and 20 % glycerol. At –58 °C, the line marked UFW is effective, 72 % are glycerol and 27 % water no longer.

Fig. 1.14.3. Gammaglobulin solution frozen in a vial by LN_2 at approx. 10 °C/min. (Only the upper part of the product is shown; Fig. 1.14.1–1.14.3 by AMSCO Finn-Aqua, D-50354 Hürth.)

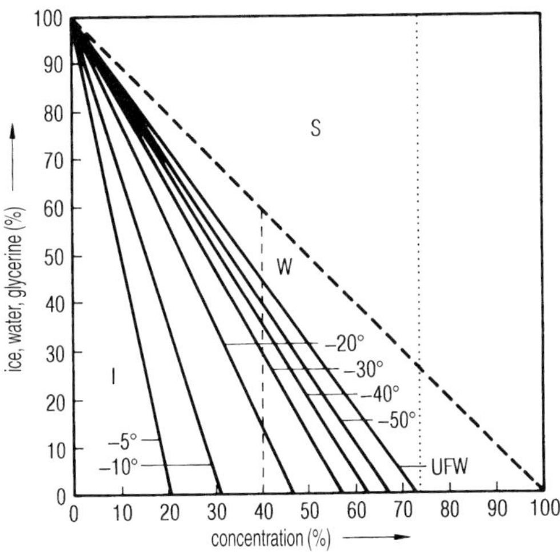

Fig. 1.15. Rate of ice, water and dissolved substance in the state of equilibrium of a glycerine water solution as a function of the initial glycerine concentration. Different freezing temperatures between –5 °C and –50 °C are plotted as parameters. A 40 % glycerine solution frozen at –30 °C contains in the state of equilibrium approx. 32 % ice, 30 % water and 38 % glycerin. The line marked as UFW represents the temperature at which the glycerine concentration becomes so high that no more water can be frozen (the water molecules become highly unmovable). The glycerine concentration is approx. 73 % and the UFW concentration 27 %. The diagram shows the equilibrium conditions, which may not exist during quick freezing (Fig. 1 from [1.101]).

The fact that a certain amount of water cannot crystallize in a high concentrated solution, and that the molecules cannot move any more to the existing crystals, is important during the freezing of biological substances. Table 1.1 shows this for some food products.

The combination of this knowledge and the results of quick-freezing processes provide a theoretical opportunity to freeze products into a solid, amorphous state. If the freezing velocity is smaller than required for vitrification, but large enough to avoid an equilibrium state, an amorphous mixture will result of hexagonal ice, concentrated solids and UFW.

1.1.3 Influence of Exipients: Growers of Structures and Cryoprotectant Agents (CPAs)

The freezing of complex organic solutions and suspensions is often difficult to predict theoretically. The methods to analyze the freezing process and the formed structure are described in Section 1.1.5. The freezing is influenced by several factors, which often act in opposing directions:

1. Freezing rate
   - slow: quasi in equilibrium
   - very fast: dynamically governed
2. Number and geometry of foreign particles, which influence the heterogeneous nucleation: the more their structure is similar to the ice structure, the better is their effectiveness as nuclei.
3. The degree of subcooling, which depends on the substance, but is strongly influenced by the two points above.
4. The rate of growth of the ice crystals, which depends on temperature and the viscosity of the solution; the latter increases strongly with the concentration of the solution.
5. That part of the water which is not frozen due to high freezing rate, forms highly viscous occlusions in between the ice crystals.
6. The crystallization of the solved substance(s) (or part of it) or the subcooling, and the delay of this crystallization, which depends again not only on the temperature but also very much on the viscosity of mixture.

By adding excipients it is possible to influence the cooling and solidification proceses.

Such additives are also necessary to form a structure, if the amount of solids is very small and could be carried out with the water vapor during drying. Further-more, more CPAs could be used to adjust the pHvalue or to act as a buffer to reduce changes in pH value. CPAs for some products are discussed in Chapter 3.

The cooling rate directly influences the size of the ice crystals, which can be measured after drying by the size of the pores in the product. Thijssen and Rulkens [1.11] give the size of the pores in chicken meat (Table 1.6). Figure 1.16 shows the average size of pores in

Table 1.6: Size and number of pores in chicken meat as a function of freezing rate.

Freezing rate (°C/min)	Size of pores (μ)	Rate of pores (%)
0,5	> 10	95
9	> 10	65
230	> 10	25
0,5	> 30	85
9	> 30	22
230	> 30	12
(1μ = 10^{-3} mm, 10^{-6} m)		

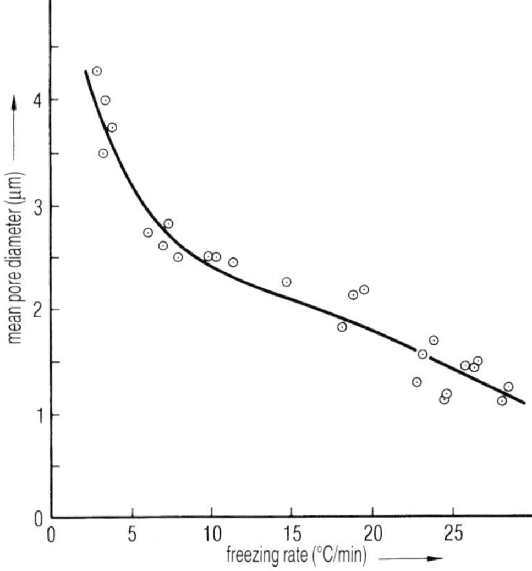

Fig. 1.16. Pore diameter as a function of freezing rate in a 20 % dextran solution (Fig. 3 from [1.11]).

a 20 % dextrose solution as a function of the freezing rate. The pore size influences the drying rate and the retention of aroma (see Section 1.2.5).

Reid et al. [1.12] described the effect of 1 % addition certain polymers on the heterogeneous nucleation rate: at –18 °C the rate was 30 times greater than in distilled, microfiltered water and at –15 °C, the factor was still 10 fold hogher. All added polymers (1 %) influenced the nucleation rate in a more or less temperature-dependent manner. However, the authors could not identify a connection between the polymer structure and nucleation rate. None the less it became clear that the growth of dendritic ice crystals depended on to factors: (i) the concentration of the solution (5 % to 30 % sucrose); and (ii) the rate at which the phase boundary water – ice crystals moved. However, the growth was found to be independent of the freezing rate. (Note of the author: the freezing rate influences the boundary rate).

The chances of a water molecule reaching dendritic ice decrease as sucrose concentration increases, and the distance between the points of the ice stars increases. The addition of polymers reinforces this effect.

Burke and Lindow [1.13] showed, that certain bacteria (e. g. *Pseudomonas syringae*) can act as nuclei for crystallization if their surface qualities and their geometric dimensions are close to those of ice. Rassmussen and Luyet [1.14] developed a connection for solutions of water with ethyleneglycol (EG), glycerol (GL) and polyvinylpyrrolidone (PVP) between the subcooling down to the heterogeneous and homogeneous nucleation of ice.

The heterogeneous and homogeneous temperatures of nucleation during cooling (5 °C/min) and the melting temperatures during rewarming may be measured by differential thermal analysis (DTA). Figure1.17.1 shows the resulting phase diagram for water – glycerol, and Fig. 1.17.2 for water – PVP. Glycerol reduces the temperature of homogeneous crystallization to a much more pronounced degree than PVP; the melting temperatures follow the same tendency. Figures 1.17.1 and 1.17.2 show also the temperatures of devitrification: 50 % PVP is sufficient to avoid crystallization at ≈ –68 °C, while 50 % glycerol reaches this effect only at ≈ –132 °C.

PVP decreases the temperatures of crystallization less than GL, the temperatures of devitrification being higher with PVP than with GL. With GL crystallization can be avoided until ≈ –70 °C, but PVP pushes devitrification in an amorphous product to higher temperatures.

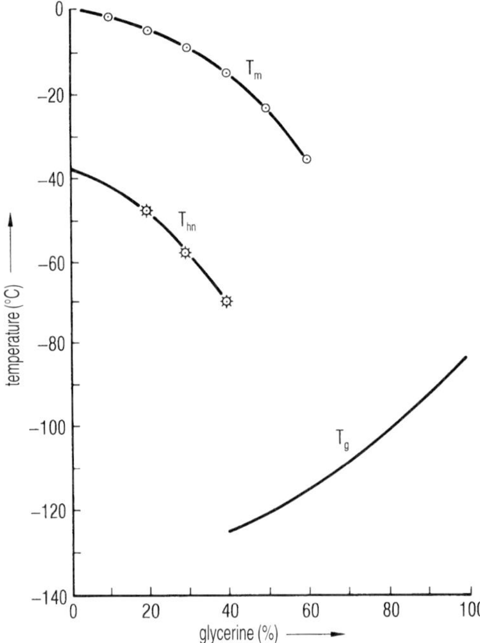

Fig. 1.17.1. Phase diagram for glycerine – water. T_m, melting temperature; T_{hn}, temperature of homogeneous crystallization; T_g, devitrification temperature (Fig. 3 from [1.14]).

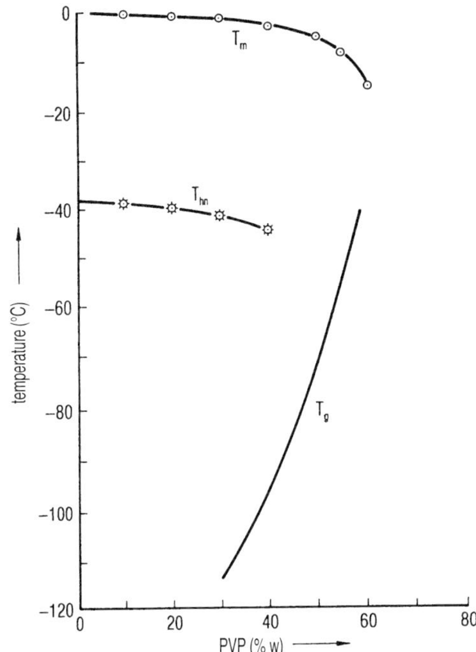

Fig. 1.17.2. Phase diagram for polyvinylpyrrolidone (PVP) – water. Explanations, see Fig. 1.17.1 (Fig. 4 from [1.14]).

Sutton [1.15] studied the question of how quickly solutions with certain CPAs {GL, dimethylsulfoxide (DMSO) and others] have to be cooled in order to avoid crystallization. At 100 °C/min concentration of 42.1 % DMSO and 48.5 % for GL are necessary to achieve the glass phase. With a 32.5 % solution of (2R.3R)-(–)butan-2,3-dio, the same effect can be accomplished at ≈ 50 °C/min. In Fig. 1.18 Sutton (Fig. 11 from [1.114]) showed, that polyethylene glycol with a molecular weight of 400 (PEG 400) reduced the critical cooling rate down to approx. 25 °C/min. The addition of PEG 8000 [1.115] improved the protection of lactate dehydrogenase (LDH) by maltodextrins, if maltodextrins with low dextrose equivalents are used.

Levine and Slade [1.16] investigated the mechanics of cryostability by carbohydrates. Figure 1.19.1 shows an idealized phase diagram developed from differential scanning calorimetry (DSC) measurements for hydrolyzed starch (MW > 100) and for polyhydroxy combinations having a small molecular mass. With slow cooling (quasi in equilibrium conditions), no water crystallizes below the T_g curve.

The terms 'antemelting' and 'incipient melting' describe phenomena, which occur at temperatures near T_g. Also the 'eruptive' crystallization, during the main drying of the freeze drying process is the consequence of a collapse of the matrix, allowing the water molecules to diffuse to the ice crystals. This may also free volatile substances, enclosed in the amorphous phase. These phenomena can occur also only in a part of the product, especially if the temperature gradient in the product is substantial. Measurements of T_r and T_c by other authors agree well with $T_{g'}$ – measurements made Levin [1.16], e. g. for sucrose –

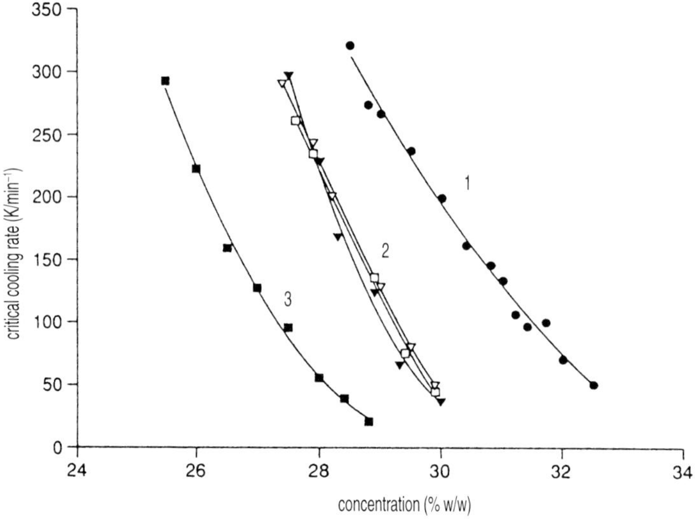

Fig. 1.18. Critical cooling rates for butane-2,3-diol and dextran as a function of butane-2,3-diol concentration.
1, butane-2,3-diol; 2, butane-2,3-diol and dextran 20; 3, butane-2,3-diol and polyethyleneglycol (PEG) 400 (Fig. 11 from [1.114]).

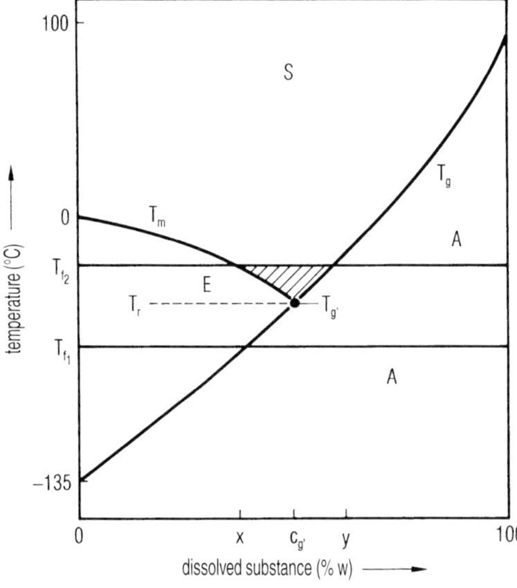

Fig. 1.19.1. Idealized diagram to show generally the dependence of phase on temperature and concentration. The dissolved, hypothetical substance consists of small carbohydrates, as found in food. The figure illustrates the meaning of $T_{g'}$. If a temperature range between T_{f1} and T_{f2} is applied, the product can above $T_{g'}$ recrystallize, start melting, or remain in the amorphous phase, depending on the concentration of the dissolved substance. Below $T_{g'}$ and at concentrations smaller than $c_{g'}$, crystallization is possible.
A, amorphous solid; E, ice; S, solution area (Fig. 1 from [1.16]).

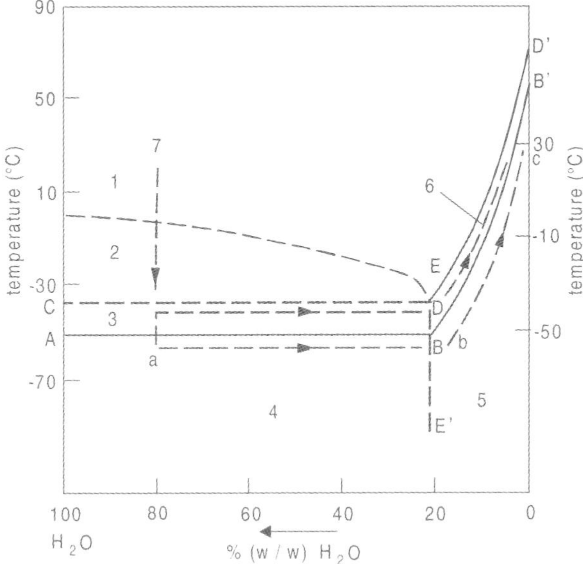

Fig. 1.19.2. Isoplethal section of the solid-liquid state diagram for $R = 0.1$. Section fields: 1, solution (viscous-flow state); 2, ice; 3, ice and amorphous phase (mechanical properties of a solid); 4, ice + glass, 5, glass; 6, amorphous phase (mechanical properties of a solid); 7, freeze drying path way (Fig. 11. from [1.116]).

32 °C/–32 °C, for maltose –32.2 °C/–29.5 °C, for lactose –32 °C/–28 °C (Tables 5 and 6 [1.16]). $T_{g'}$, $c_{g'}$ and g UFW/g carbohydrate are characteristic data for such solutions. Shalaev et al. [1.116] investigated the solid-liquid state diagram of the ternary system water-glycine-sucrose with DTA and X-ray diffraction. Figure 1.19.2 shows the isoplethal section of the solid-liquid state diagram for the ratio of glycine/sucrose = $R = 0.1$. The line EE' divides the two-phase (ice + amorphous phase) and single-phase (amorphous phase) fields. The lines ABB' and CDD' subdivide the fields with different states of the amorphous phase. Below $T_{g'}$ (ABB-line) the amorphous phase is in a structural-solid state, and the translational motion of the molecules is retarded. Above $T_{g'}$ the product transforms into a structural-liquid state, but the sample keeps its form because the viscosity is so high. At $R = 1$ a devitrification is found at –30 °C and ≥ 50 % water.

Jang et al. [1.117] investigated the glass transition temperature, $T_{g'}$, for FK 906, (a synthetic peptide) during rewarming from the frozen state and T_g of the dry product in the presence of sucrose, maltose, trehalose and lactose as well as polymers with different molecular weight, and three salts. For the first group of disaccharide the Gordon-Taylor equation [1.118] is describing the glass transition temperatures as a function of the FK906 content if the T_g of each component is known. The three salts have an eutectic temperature of approximately: NaCl –21 °C, NaBr –31 °C, and KCl –11 °C, while FK 906 is approximately –19 °C. NaCl (0.1–0.3 %) lowers the $T_{g'}$ of a 10 % FK 906 solution from –19 °C to –27 °C. NaBr (0.1–1 %) solutions have approximately the same effect, while KCl induces a small decrease from –19 °C to –23 °C at concentrations of 0,1–0,2 % in a 10 % FK 906 solution. However, by increasing the concentration to 1.5 %, T_g increases by approximately 2 °C. The glass transition temperature of the freeze dried product is not changed by NaCl content up to 0.6 %. Nicolajsen et al. [1.119] described a similar effect of NaCl in the trehalose-NaCl-water system. At a trehalose concentration above 3.5 % frozen to –70 °C and heated at 2 °C/min, no eutectic transition was found, indicating that all NaCl is trapped in the glass phase. However the glass transition temperature was lower the larger the ratio % NaCl/% trehalose became. NaCl appears to be a destabilizing factor in the glass phase.

Since the definition of the glass phase is 'fuzzy' as Reid et al. [1.17] note, and the methods of measurement can be chosen, the results for $T_{g'}$ will have a range of variations. For freeze drying one can summarize:

- If the product temperature approaches $T_{g'}$ from lower temperatures, the viscosity changes rapidly (within a few degrees) by several decimal powers. Since also the product temperature can be measured only with a certain accuracy (see Section 1.2.3), one has to account for both uncertainties. It is recommended that the maximum product temperature during main drying 3 to 5 °C below $T_{g'}$ be chosen.
- The addition of certain carbohydrates increases, by varying degrees, the $T_{g'}$ values and decreases the amount of UFW. These stabilize the glass phase to higher temperatures and permit higher drying temperatures. They can also bind volatile components.

Carpenter et al. [1.18] showed, that the stabilization of proteins, using the enzymes phosphofructokinase (PFK) and (LDH) as models, during freezing and thawing and freeze drying is based on two different mechanisms. In Table 1.7 nine substances provide a relative good protection against denaturation during freezing and thawing (40 to 85 % of the activity remain). This does not apply for 3 mol L^{-1} NaCl. (The 10 substances listed are an selection of 28 substances studied). Timasheff et al. [1.19] explained the stabilizing or destabilizing effect of the additives by the combination of the additive molecule with the protein (destabilizing) or its rejection by the protein (stabilizing). The predominant effect of the additive depends on its chemical qualities and the surface structure of the protein. For example NaCl bound predominantly to the LDH and destabilized it, while urea had the same effect in an LDH solution.

A different effect stabilized PFK during freeze drying and subsequent storage. As shown in Table 1.8, only disaccharide can protect PFK. Since only a special group of CPAs is effective, one can assume that these CPA combine with the protein. If water molecules of the hydrated envelope of the protein are removed during freeze drying, the molecules of the effective CPA can replace these water molecules, keeping the protein stable. Pretrelski et al. [1.123] demonstrated by infrared spectroscopy, that the addition of 10 mM mannitol, lactose, trehalose or 1 % PEG to lactate dehydrogenase and phospho-fructokinase attenuated the unfolding, but spectral differences in the dried state are still observed. However a combination of 1 % PEG with either 10 mM mannitol, lactose or trehalose preserved full enzymic activity upon reconstitution of the freeze dried product.

Carpenter et al. [1.120] found that certain polymers (e. g. PVP) could stabilize multimeric enzymes during freezing and freeze drying by a different mechanism: They cannot replace water molecules in the dried state; therefore it is assumed that they inhibited the dissociation of the enzymes molecules induced by freezing and freeze drying.

Table 1.7: Activity of lactate dehydrogenase after freezing and thawing in the presence of dissolved substances, which are mostly repelled by proteins (Table III from [1.18]).

Dissolved substance	Highest tested concentration (mol/L)	Remaining activity (%)
None	0.0	21.5
Sucrose	1.0	85.4
Lactose	0.5	74.2
Glucose	1.0	60.2
Glycerine	1.0	71.4
Sorbitol	1.0	75.3
Mannitol	1.0	67.3
Glycine	2.0	39.1
Sodiumacetate	2.0	81.2
$MgSO_4$	2.0	61.7
NaCl	3.0	20.7

Table 1.8: Activity of phosphofructokinase after freeze- and air-drying in the presence of different CPAs (Table IV from [1.18]).

CPA (0.5 M)	Remaining activity	
	Freeze-dried	Air-dried
None	0.0	
Trehalose	56.0	68.4
Maltose	69.2	51.4
Sucrose	56.3	67.7
Glucose	3.3	2.1
Glycerine	0.0	0.0
Glycine	0.0	2.8

1.1.4 Freezing of Cells and Bacteria

In 1968, Meryman [1.20] presented his idea about the 'minimum cell volume' and hypothesized, that during feezing cell are damaged in two steps. Initially water diffuses from the cell to the surrounding, the freezing solution concentrating the solution in the cell. However only a certain amount of water can be withdrawn from the cell until it has shrunken so much, (minimum cell volume), that any further withdrawal takes water from the membrane molecules, that are an essential part of the cell's structure. The removal of this water leads to an irreversible change in the membrane structure.

As shown in Section 1.1.3, these structural changes can be avoided if certain sugars or other CPAs replace the water molecules. Pushkar et al. [1.21] showed with a cryomicroscope and X-ray analysis of structures that 15 % glycerol in suspensions of marrow cells lowered the beginning of ice nucleation to –15 °C. Down to –10 °C, no nucleation in the cells was observed. With polyethylenoxide (10–15 %) a few crystals of 10^{-3} cm to 10^{-4} cm haven been detected in an amorphous surrounding. Under these conditions the function of the cells after thawing remained mostly normal. Nei [1.22] studied the nucleation of ice crystals during rapid freezing (100 to 1000 °C/min), using electron microscopy. Ice crystals can be observed in the cells of yeast, while in most bacteria (e. g. coli forms) almost no ice crystals could be detected. In Nei's opinion, ice crystals were more likely to have been produced in animal cells than in those of micro organisms. Cosman et al. [1.28] showed using photographs taken with a cryo microscope, that the volume of mouse islet cells shrank to 40 % during cooling from 0 to –10 °C. Figure 1.20 indicates how many cells in rat liver contained ice as a function of cooling rate, the cells have been cooled from –1 °C to – 21 °C: Up to 50 °C/min, no ice wasformed in the cells, while at approx. 160 °C/min all cells contained ice crystals.

De Antoni et al. [1.23] demonstrated, that the addition of trehalose during freezing and thawing of two strains of *Lactobacillus bulgaricus* improved the survival rate differentially, but in both cases considerably. The samples (1 mL) were frozen at 18 °C/min to – 60 °C and thawed to 37 °C at 15 °C/min. The solution consisted of distilled water, culture medium and 10 % milk with or without trehalose. It was shown, that after three freezing-thawing cycles, milk alone resulted in a survival rate of 24 % or 65 %, while with trehalose this was can be improved to 32 % and 100 % respectively. The efficacy in the case of both strains was clearly different. De Antoni et al. suggested, that the efficiency of milk was related to its Ca^{2+} content, while the trehalose could replace water molecules in the phospholipids of the membranes. However no mention was made wether other sugar molecules in milk showed any effect.

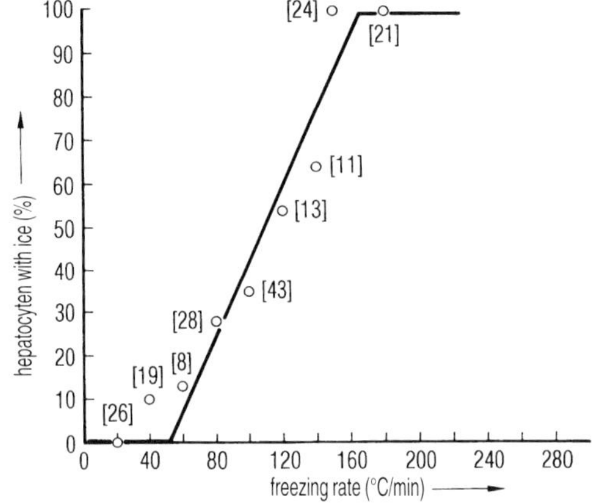

Fig. 1.20. Percentage of rat hepatocytes which show intracellular ice as a function of freezing rate in the range of –1 °C to –21 °C. The values in [ ] are the numbers of hepatocytes participating in each test (Fig. 8 from [1.28]).

1.1.5 Methods of Structure Analysis

The knowledge that successful freeze drying depends largely on the structure of the frozen product has inspired the development of methods to analyze and understand these structures more quantitatively. Rey [1.24] has shown that as well the electrical resistance (ER) of a freezing substance, thermodynamic behavior can be used to study the freezing process and the frozen product. Measurement of ER is done in an apparatus (Fig. 1.21) in which the sample is cooled by LN_2 and electrically heated. Two electrodes are immersed in the sample, which is filled into a vial. The resistance between the two electrodes is measured with an alternating current of 50 Hz. For complete information, high resistance up to 10^{12} ohm should be measurable. This requires high resistance, a factor of ten higher than 10^{12} Ohm, not only between the electrodes itself but also to the temperature sensor and surroundings. As shown in Fig. 1.22 a unit capable of measuring only 10^8 Ohm would supply incomplete information. The product is not completely solidified at either 10^8 ohm (–43 °C) or at $2 \cdot 10^9$ Ohm (–60 °C). The derivation curve during rewarming between –60 °C and –50 °C shows a smaller resistance change than between –50 °C and –37 °C. Above –37/35 °C the resistance changes starts to increase rapidly, the mobility of the ions increases, and the beginning of a collapse can be expected during main drying . Such an instrument is illustrated in Fig. 1.23.

The idea by Rey to use ER during the main drying for control of the process has not been introduced largely, because as the measurement is not done in a solid block, the product becomes increasingly porous with unknown configurations. The ER measurement of physiological NaCl solution (Fig. 1.24) can be used to check the system. The ER measurement

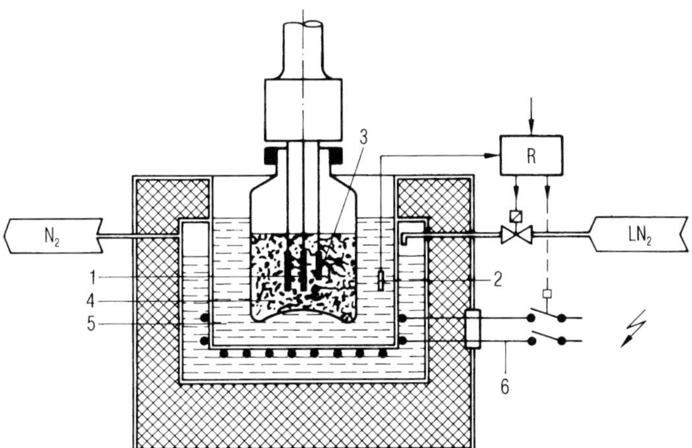

Fig. 1.21. Schematic drawing of an instrument to measure the electrical resistance (ER) of a sample during cooling and rewarming (Fig. 2 from [1.27]).
1, Platinum electrodes; 2, temperature sensor in the heat transfer medium; 3, resistance thermometer; 4, product sample; 5, heat transfer medium; 6, resistance heating.

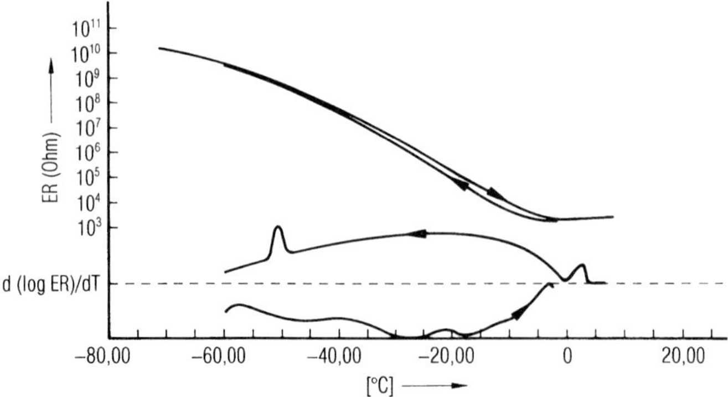

Fig. 1.22. Resistance as function of the temperature during cooling at 1 °C/min and rewarming at 3 °C/min.

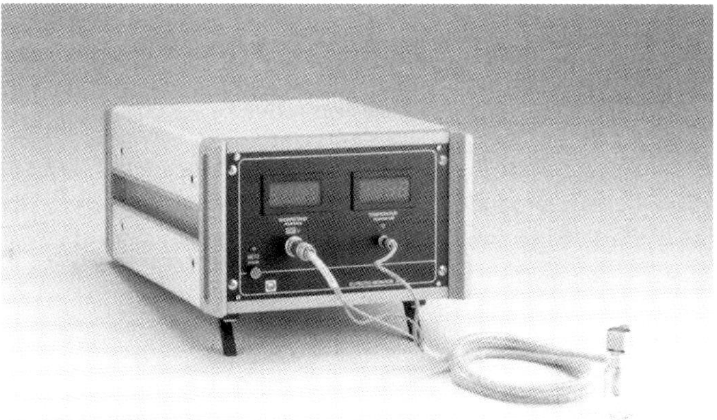

Fig. 1.23. Monitor AW 2. In the foreground right: Sample vial with measuring electrodes and resistance thermometer, behind to the left the control- and analysis unit. The storage of LN_2 and its control valve are not shown. The resistance in the measuring head has to be large compared with the resistance to measure e. g. 10^{11} Ω (photograph: AMSCO Finn-Aqua, D-50354 Hürth).

analyzes the mobility of ions or dipoles in the solidifying solution. The reasons for the change in ER can only be deduced by experience, or from other methods. Such comparisons, e. g. with DTA, have been made by Mac Kenzie [1.25].

DTA (Fig. 1.25) measures the different course of temperature between the sample and a probe, which changes its thermal behavior constantly but does not have a phase transition in the measured temperature range. Such an instrument is illustrated in Fig. 1.26.

Using DTA and ER measurements of quickly (200 °C/min) frozen sucrose-NaCl solutions, Mac Kenzie presented the different events occurring during slow rewarming. Among others two sucrose-NaCl solutions were studied: a 24 % sucrose solution with 6 % NaCl

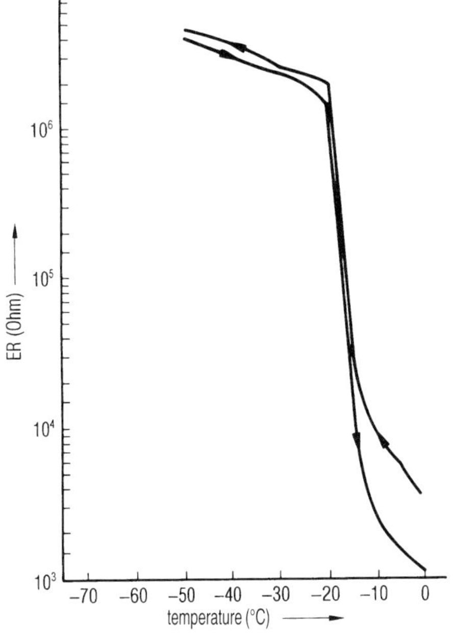

Fig. 1.24. Resistance of a physiological NaCl solution as function of the temperature during cooling and warming (Fig. 3 from [1.27]).

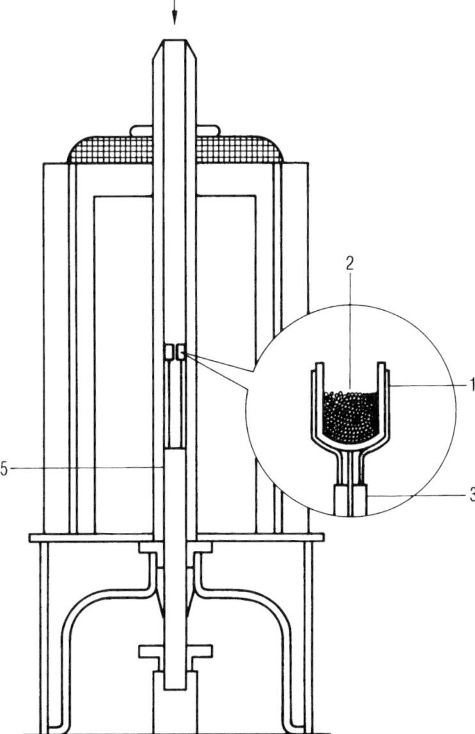

Fig. 1.25. Schema of a DTA-measuring cell.
1, Crucible with sample; 2, sample; 3, thermocouple (reference crucible not enlarged); 4, gas inlet; 5, ceramic support.

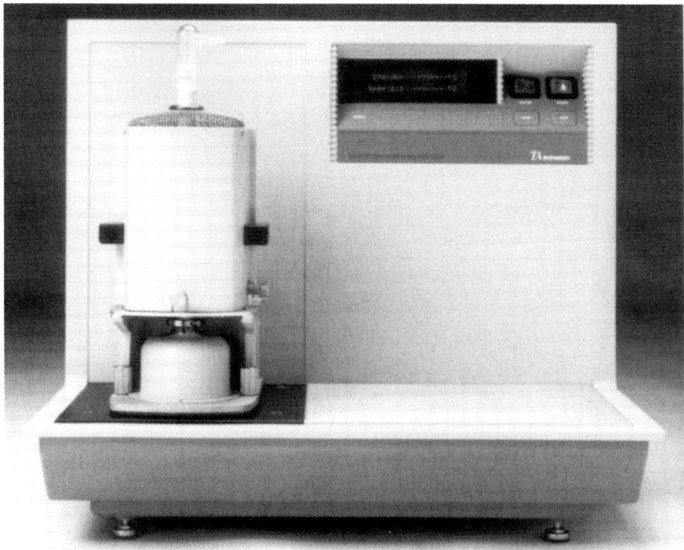

Fig. 1.26. Instrument DSC 2920 with DTA-measuring cell. This installation is not designed for low temperatures, but can be modified for this purpose. (Commercial standard installations for low temperatures have not been found by the author.) (Schema and photograph: TA Instruments, Inc. New Castle DE 19720, USA.)

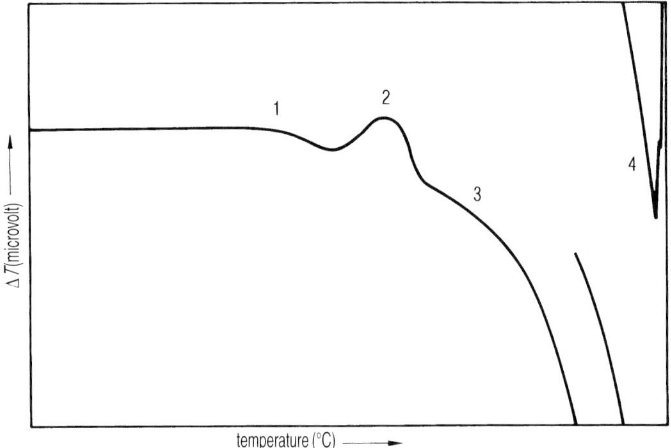

Fig. 1.27.1. DTA measurement of a 24 % sucrose – 6 % NaCl solution during slow rewarming after quick (200 °C/min) freezing.
1, glass transition at approx. –78 °C; 2, growth of crystals (exothermic) at approx. –52 °C; 3, increase of c_p, water is formed between the crystals; 4, ice melts at approx. –7 °C.

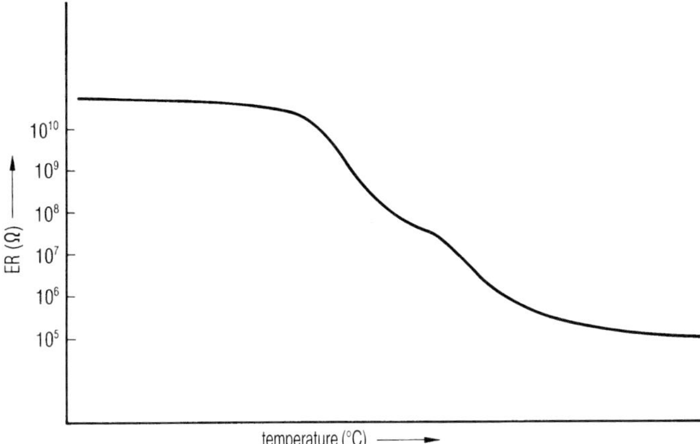

Fig. 1.27.2. Electrical resistance (ER) measurement of the same solution as in Fig. 1.27.1.; identical temperature scale (Fig. 1and 2 from [1.25]).

(sucrose : NaCl ratio = 80 : 20) Figs. 1.27 and a 10 % sucrose solution with 10 % NaCl (ratio 1:1) Figs. 1.28. In Fig 1.27.1, event 1 at ≈ –78 °C can be explained as glass transition. In Fig. 1.27.2 event 1 reduces the ER significantly. In event 2 at ≈ –50 °C, the mobility of the molecules has increased so much as to allow the growth of crystals (exothermic), and the resistance drops more slowly. At event 3, some water is formed between the crystals, and $c_{p.}$ rises. c_p of water is about twice that of the c_p of ice. At event 4 (at ≈ –7 °C) the ice melts.

In Fig. 1.28.1 event 1 is at ≈ –93 °C, event 2 at ≈ –66 °C, and the exothermic event 3 at ≈ –44 °C results from the crystallization of NaCl. Event 4 at ≈ –22 °C represents the eutectic melting and event 5 corresponds to event 4 in Fig. 1.27.1.

In Fig. 1.28.2, the softening of the glass phase can be seen in the change of the ER, while at event 2 the resistance changes more slowly, corresponding to Fig. 1.28.1. The crystallization of NaCl can be seen from the increase in ER at event 3, which does not exists in Fig. 1.27.2. Events 4 and 5 can be observed in the ER curves shown. The interpretation of ER measurements is substantially improved by using the derivate of the ER curve, as shown in Fig. 1.22. In Fig. 1.29.1 and 1.29.2, the possibilities of this method are demonstrated: in Fig. 1.29.1 the bath (5 in Fig. 1.21) and the product are cooled at about the same rate (1 °C/min). In Fig. 1.29.2 the wall of the vial with the probe is isolated by plastic up to the height of the probe, to simulate the cooling of a vial on the shelf of a freeze drying plant. The heat removal is mainly through the bottom of the vial. The data in Fig. 1.29.1 for log $R = f(T)$ between –5 °C and –15 °C can barely been differentiated. The derived function d (log ER)/dT shows that during the rewarming at event 1, the resistance drops, then rises at event 2 and then remains almost constant for a period. At event 1, (approx. –12 °C) part of the product softens and permits a crystallization during event 2 (–11 °C to –8 °C). In Fig. 1.29.2 these events do not appear, since the product had time to crystallize fully during the

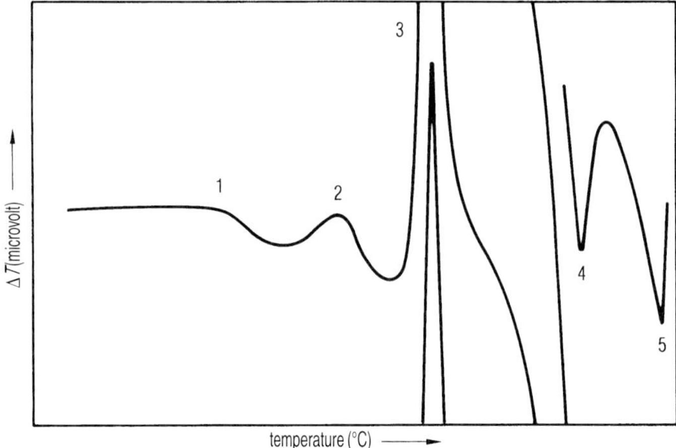

Fig. 1.28.1. DTA measurement of a 10 % sucrose – 10 % NaCl solution during slow rewarming after quick (200 °C/min) freezing.
1, glass transition at approx. –93 °C; 2, crystal growth (exothermic) at –65 °C; 3, significant exothermic event, crystallization of NaCl at approx. –44 °C; 4, eutectic melting at approx. –22 °C; 5, melting of ice at approx. –7 °C.

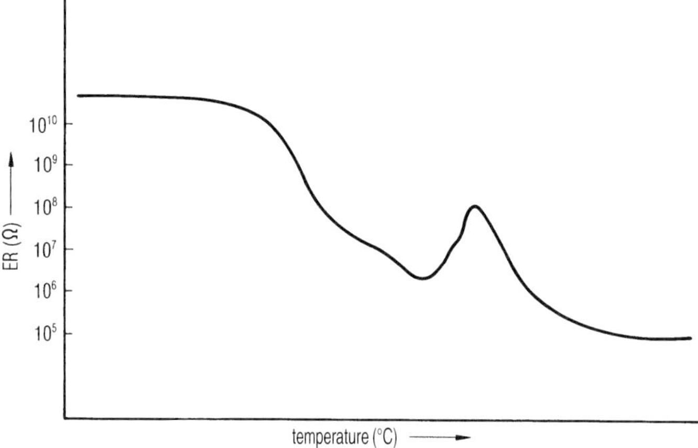

Fig. 1.28.2. Electrical resistance (ER) measurement of the same solution as in Fig 1.28.1.; identical temperature scale (Fig. 5 and 6 from [1.25]).

slower heat removal from the bottom. This influence of the isolation can be clearly seen in Fig. 1.30. The suspension subcools twice at events 1 and 2. Also, event 3 is a mixture of subcooling and crystallization. During rewarming, no event can be detected, since the product has been frozen to the maximum possible. The product then starts to melt at approx. –12 °C:

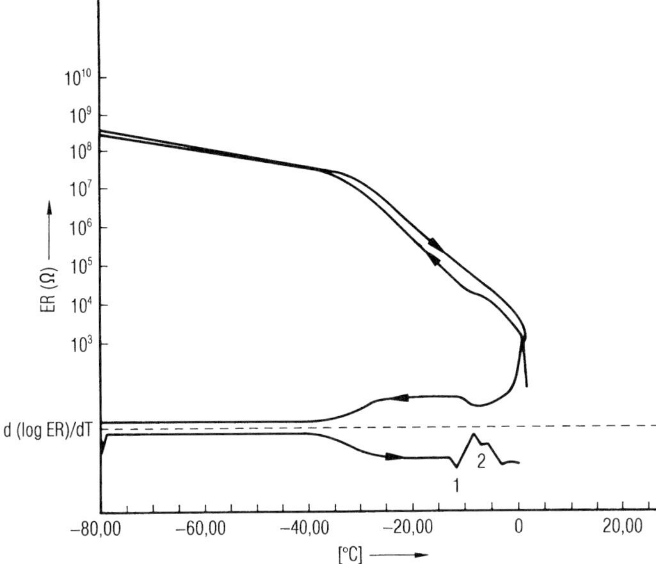

Fig. 1.29.1. Electrical resistance of a pharmaceutical product as the function of temperature during cooling at 1 °C/min and rewarming at 3 °C/min. The heat transfer medium and product are approx. uniformly heated.

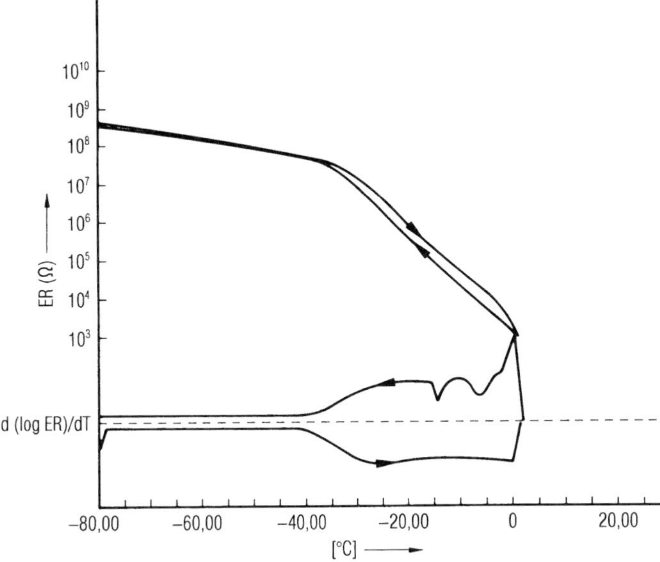

Fig. 1.29.2. Measurement of the electrical resistance as in Fig. 1.29.1. However the wall of the vial is insulated by a plastic tape up to the filling height of the product. Therefore the heat is mostly removed through the bottom of the vial (from [1.102]).

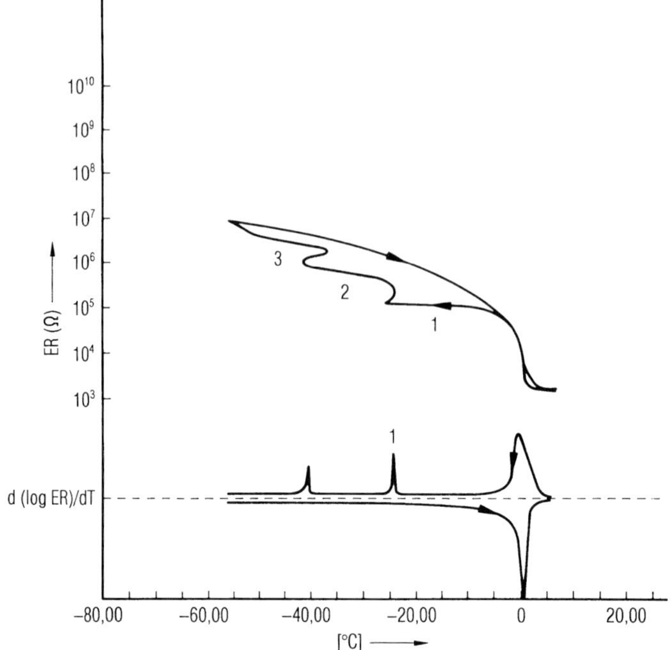

Fig. 1.30. Electrical resistance as a function of temperature of a suspension cooled at 0.8 °C/min and heated at 3 °C/min. The vial is insulated as described in Fig. 1.29.2 (from [1.102]).

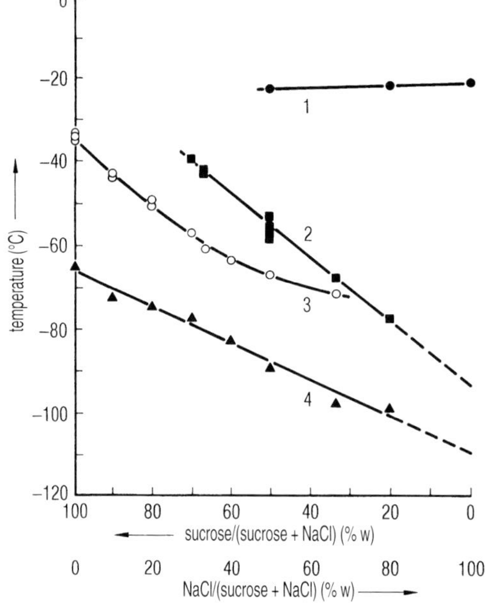

Fig. 1.31. Behavior of a sucrose – NaCl solution at different sucrose – NaCl concentrations and temperatures after quick freezing (200 °C/min), during slow rewarming (Fig. 8 from [1.25]).

1, Eutectic melting temperature of NaCl; 2, crystallization temperature of NaCl: 3, temperature at which the glass phase starts to soften; 4, glass transition temperature.

The results of several measurements by Mac Kenzie [1.25] are shown in Fig. 1.31. The glass transition line 1 exists over the whole concentration range studied, while the lines 2 and 3 are absent in the area of high sucrose and respectively high NaCl concentrations. Later measurements prove, that the mobility in the solid matrix is too reduced, to observe the events during the observation time used. Only the rotation of the water molecules is still possible. With an increase of temperature (line 3), the energy increases sufficiently to allow some movements of the molecules, which can also be seen as a decrease of ER. Mac Kenzie denotes this temperature as 'antemelting', though Mac Kenzie (Note in [1.38]) subsequently suggested, that term should not be used, but should be replaced by 'collapse temperature' (T_c). (For alternative opinions on the subject see Section 1.1.3.).

An other event called 'incipient melting' at temperature T_{im} is the melting of ice crystals between crystallized eutectic mixtures, or the dissolution of crystals surrounded by highly concentrated inclusions, known as interstitial melting of ice. Luyet and Rasmussen [1.26] have studied the phase transitions by DTA of quickly (75 °C/min or 200 °C/min) frozen water solutions of glycerol, ethyleneglycol, sucrose and glucose during rewarming (5 °C/min). Figure 1.32 shows a typical DTA curve, if measurable amounts of amorphous product have been formed during freezing, which starts to crystallize after $T_{g'}$ is exceeded at a temperature T_d (exothermic event). At the temperature T_d one can expect a viscosity of approx. 10^9 Poise. At $T_{g'}$ the viscosity, in agreement with other authors, is in the order of 10^{13} Poise. This concept is shown in Fig. 1.33: If the solution freezes relatively slowly

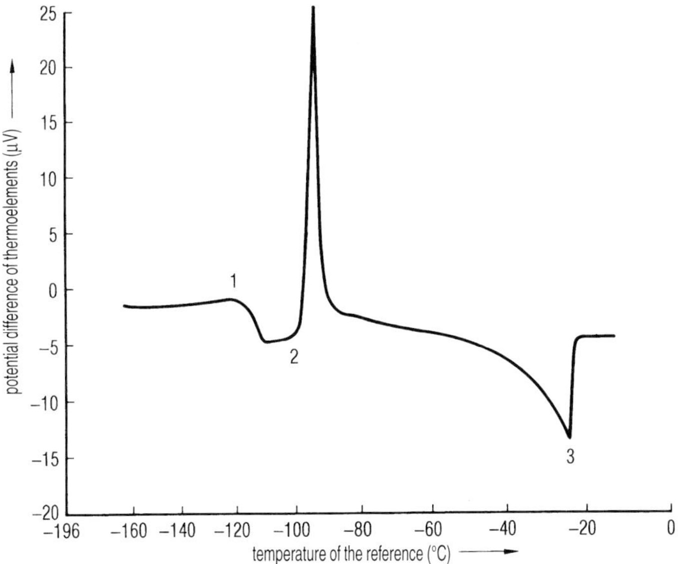

Fig. 1.32. Plot of the DTA measurement of a 50 % glycerine-solution during slow rewarming after quick freezing at 75 to 200 °C/min.
1, T_g; 2, T_d; 3, T_m (Fig. 1 from [1.26]).

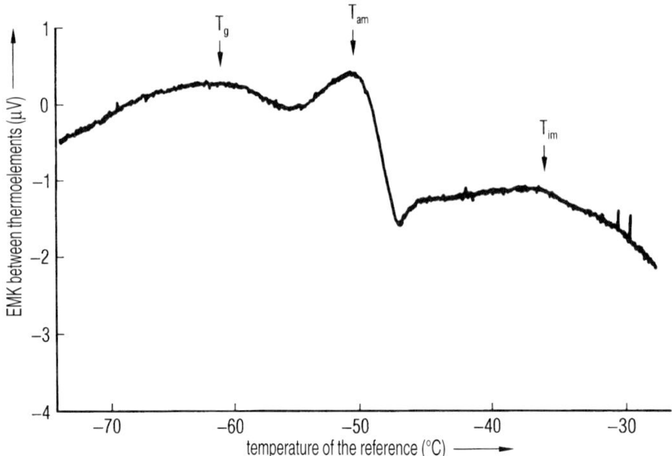

Fig. 1.33. DTA plot of a 50 % glucose solution, frozen at 3 °C/min, during rewarming. Arrow T_g, Start of devitrification; arrow T_{am}, start of ante-melting; arrow T_{im}, start of incipient melting (Fig. 1 from [1.103]).

(3 °C/min) all freezable water is crystallized, or if the rewarming is interrupted before the melting starts and the product cooled down again to e. g. –150 °C, the rewarming curves resemble that in Fig. 1.33. There is no water left, which can crystallize at T_d. There are only two events, which are denoted (as by Mac Kenzie) with antemelting and incipient melting.

From DTA measurements phase diagrams can be constructed as shown for ethyleneglycol in Fig. 1.34. A solution of 40 % ethyleneglycol is only stabile in the glass phase below ≈ –135 °C, at ≈ –120 °C unfrozen water starts to crystallize, at ≈ –65 °C a recrystallization is found, and at ≈ –45 °C melting will start. As recrystallization is the growing of existing crystals, and not the nucleation of new ones, this event cannot be detected by DTA, but can be observed in a microscope when a transparent area becomes opaque.

Hsu et al. [1.121] observed recrystallization on the recombinant CD4-IgG with a cryomicroscope cooled to –60 °C by a cascade of four Peltier modules. The observation cell can also be evacuated for freeze drying studies.

Willemer [1.27] compared ER measurements with photographs made by a cryomicroscope, the schema of which is shown in Fig. 1.35. ER measurements of complex products are some times difficult to interpret. Figure 1.36.1 shows the ER curve of a cryoprotectant solution for a virus. The solution freezes partially by cooling to –10 °C, subcools thereafter down to ≈ –46 °C, and crystallizes until ≈ –65 °C. Upon rewarming the resistance changes rapidly at ≈ –32.5 °C. The photographs taken using a cryomicroscope show at –40 °C a uniform structure as well in the already dried as in the frozen part (Fig. 1.36.2). At –30 °C, both parts show a mix of dark and gray zones, indicating that some ice is molten and is also diffused into the dried part. In such cases ER measurements can be used as a relatively quick method to study the influences of different CPAs, varying their

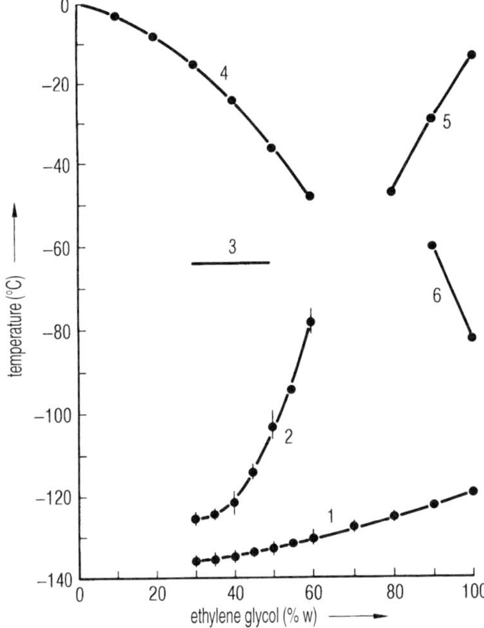

Fig. 1.34. Phase diagram of ethylene glycol, in which the following events are shown:
1, glass transition; 2, devitrification; 3, recrystallization; 4, melting; 5 and 6, represent the devitrification and melting of ethylene glycol (Fig. 4 from [1.26]).

concentrations and select an optimal freezing rate. The finally selected combination can be tested in the cryomicroscope.

Nunner [1.104] photographed with a special cryomicroscope the change of the planar front of a 0.9 % NaCl solution during directional freezing in 360 s to a stable dendritic ice structure (Fig. 1.37). The concentrated NaCl (dark border) can be seen on the surface of the ice crystals.

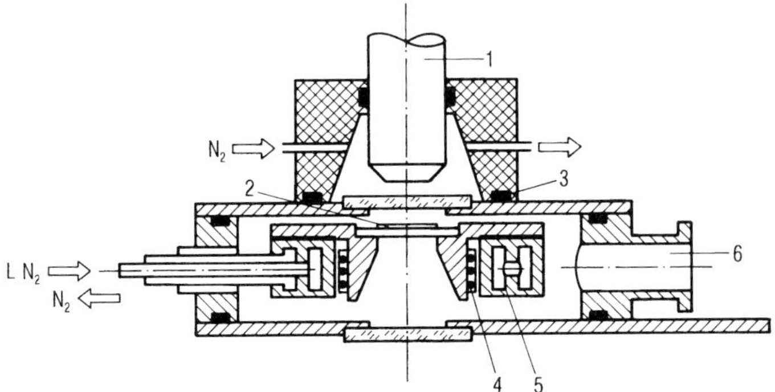

Fig. 1.35. Schema of a cryomicroscope.
1, Objective of the microscope; 2, sample; 3, sample support; 4, electrical heating; 5, cooling chamber with LN_2 connection; 6, vacuum connection (Fig. 1 from [1.27]).

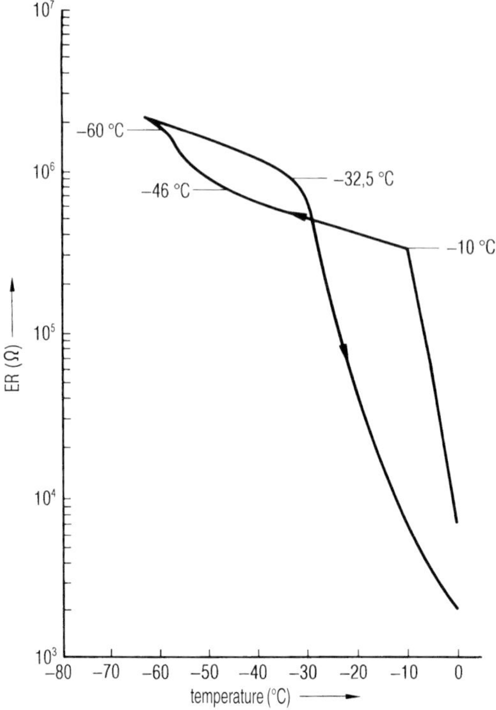

Fig. 1.36.1. Electrical resistance (ER) as function of temperature during cooling and rewarming of a virus suspension. The suspension subcools from –10 °C to approx. –46 °C and freezes at –60 °C to –65 °C. During rewarming the resistance drops clearly at approx. –33 °C. This product should be freeze dried at T_{ice} = –40 °C or a little higher (Fig. 7 from [1.27]).

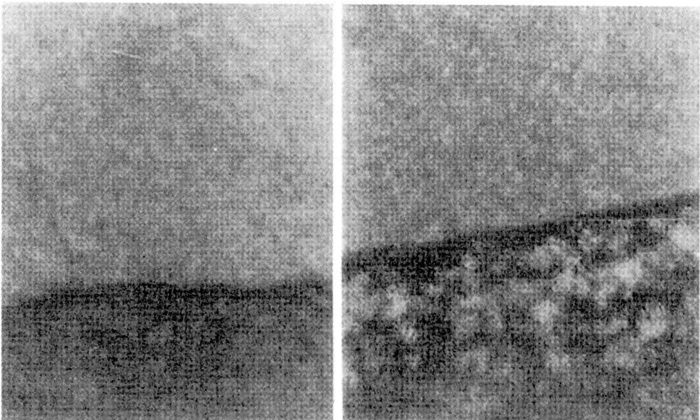

Fig. 1.36.2. Photographs of the virus suspension (Fig. 1.36.1.) by a cryomicroscope during slowly increased temperature: left: –40 °C, right: –30 °C. In the right photo, bright inclusions in the dark lower zone can be seen. These represent partially molten product between the remaining ice. Water is also diffused into the already dried product, partially dissolving it (Fig. 8 from [1.27]).

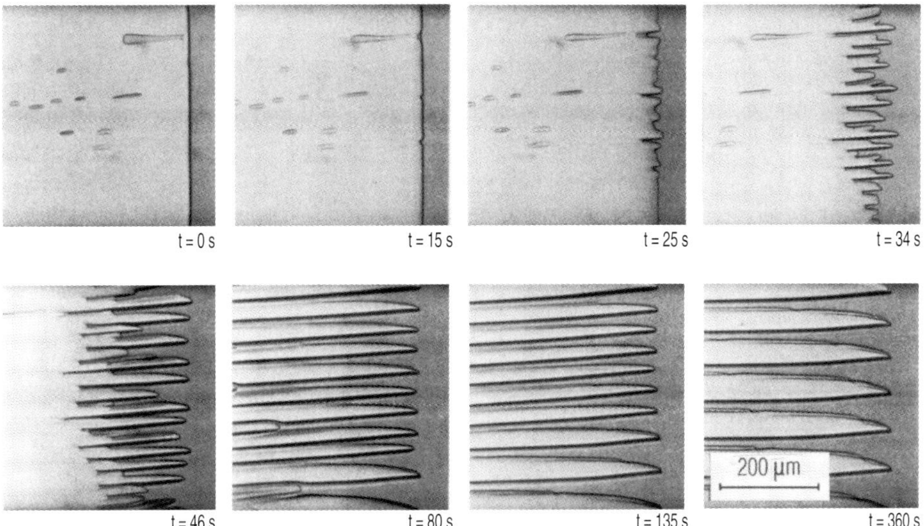

Fig. 1.37. Change of the planar front of ice (t = 0 s) through an unstable phase (t = 34 s) into a dendritic structure. 0.9 % NaCl solution is directionally frozen in a temperature field having a gradient of 67 K/cm. The sample is moved with a speed of 15 µm/s through the temperature field (from [1.104]).

A cryomicroscope, which permits quantitative evaluation of the pictures is described by Cosman et al. [1.28]. The unit has four distinctive features:

- temperature generation, measurement and control are programmable;
- the picture of the microscope is documented for later use;
- the documentation can be partially used for automatic picture recognition;
- the amount of data can be reduced in such a way, that a freezing process can be described mathematically and the behavior of cells predicted.

Figure 1.38 shows the layout of the system. By the use of a very good heat-conducting sapphire window and a cooling system with LN_2, the authors reach cooling rates of several hundred degrees per minute down to –60 °C, and temperature gradients in the sample of 0.1 °C at a temperature of approx. 0 °C.

Three examples will show how freezing processes can be studied quantitatively and documented using this microscope system. Figure 1.39 shows the change in volume of an isolated islet cell of a mouse as a function of temperature. The different permeability of cell membranes for H_2O and CPAs are important for freezing of cell, as Fig. 1.40 shows.

The volume of oocytes of a rhesus monkey placed in 10 % (v/v) dimethylsulfoxide (DMSO) is reduced to almost one third, since the water can diffuse out of the cell into the surrounding, but the DMSO cannot enter the cell during the same time (measured at +23 °C).

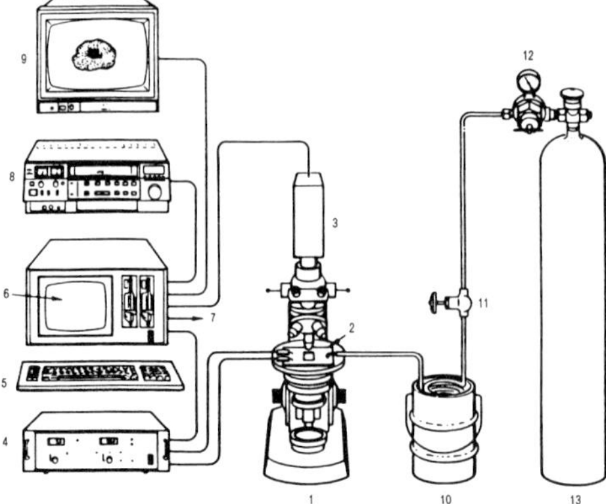

Fig. 1.38. Schema of a cryomicroscope research system.
1, Microscope; 2, cryostat; 3, video camera; 4, temperate control; 5, keyboard; 6, menu display; 7, printer connection; 8, video recorder; 9, video monitor; 10, Dewar flask with LN_2; 11, metering valve; 12, pressure reducer; 13, N_2 cylinder (Fig 1 from [1.28]).

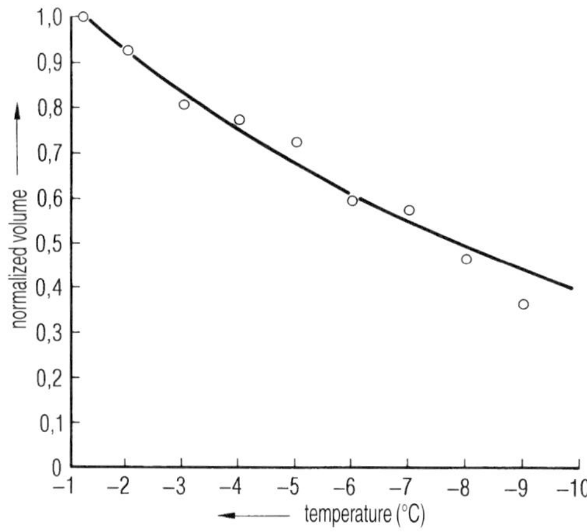

Fig. 1.39. Volume change as a function of temperature of an isolated islet cell of a mouse (Fig. 4 from [1.28]).

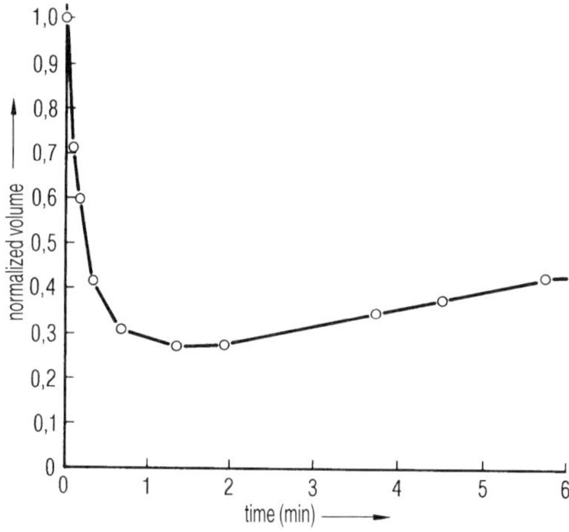

Fig. 1.40. Volume changes of oocytes of rhesus monkeys as a function of the time elapsed after their exposure to 10 % dimethylsulfoxide solution (Fig. 6 from [1.28]).

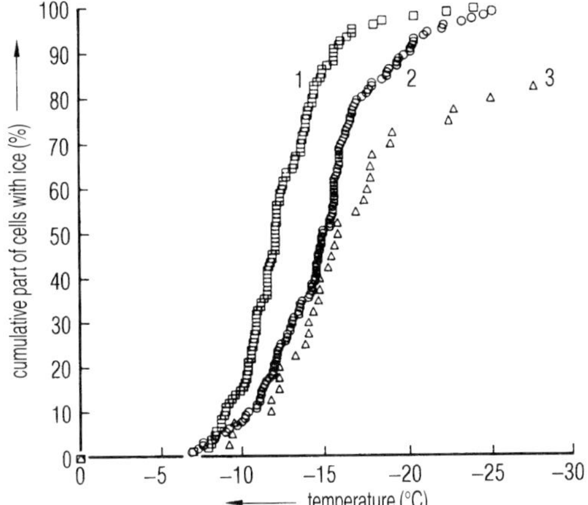

Fig. 1.41. Cumulative abundance of intracellular ice forming in mice oocytes as a function of temperature with three different cooling rates:
1, 111 oocytes at 120 °C/min; 2, 132 oocytes at 5 °C/min; 3, 34 oocytes at 3.5 °C/min (Fig. 9 from [1.28]).

The nucleation of ice in the cell is looked at as the cause of cell damage. Figure 1.41 indicates in how many mouse oocytes intracellular ice is found as a function of temperature at different cooling speeds. In hepatocytes of rats no ice could be detected during cooling to –21 °C up to a cooling rate of ≈ 40 °C/min, while at a rate of 140 °C/min practically all cells contained ice. The water did not have sufficient time to diffuse into the surrounding, and froze in the cells. Figure 1.41 also demonstrates how the intracellular nucleation of ice depends on the absolute temperature and cooling rate: at ≈ –25 °C and a rate of 5 °C/min practically all cells contain ice, while at 3.5 °C/min, approx. 20 % of the cells were without ice.

Dawson and Hockley [1.29] have used scanning electron microscopy (SEM) to show the morphological differences between quick (150 °C) and slow (1 °C/min) freezing of trehalose and mannitol solutions. Figure 1.42 shows the surface of slowly (a) and (b) a quickly frozen center part of a 1 % trehalose solution. On the slowly frozen sample a cracked surface (c) can develop by concentrated solids, while the structure in the quickly frozen sample is amorphous and fibrous. Figure 1.43 shows the coarse (a) and fine (b) structure in the center part of slowly (a) respectively quickly frozen (b) 1 % lactose. A collapsed part of a trehalose solution can be found in Fig. 1.44 (a), while (b) shows the dried product stored with too high a moisture content for 6 months. The pictures prove, that different freezing rates will result in different structures and may concentrate solids on the surfaces, which reduce the drying speed or prohibit low residual moisture content during drying.

Differential scanning calorimetry (DSC) compares the two different heat flows: one to or from the sample to be studied, the other to or from a substance with no phase transitions in the range to be measured e. g. glassmaking sand. Figure 1.45 is the scheme of a DSC system; Fig. 1.46 is a commercial apparatus for DSC measurements.

Gatlin [1.30] measured not only T_g' for mannitol and Na-cefazolin by DSC, but also the dependence of the exothermic crystallization energy from the rewarming rate (Fig. 1.47).

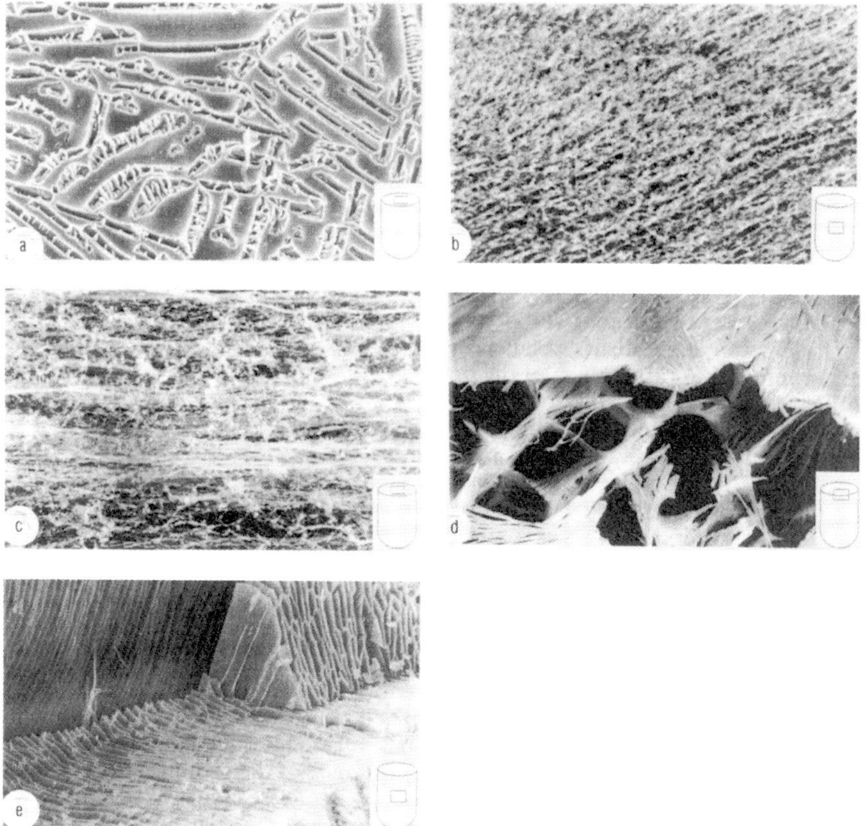

Fig. 1.42. Scanning electron-microscopic photographs of different freeze dried products. (a) 1 % trehalose solution, 1 °C/min, cutout of the surface; (b) 1 % trahalose solution, 150 °C/min, cutout of the center; (c) 1 % trehalose solution, 1 °C/min, cutout of the most upper surface; (d) 1 % mannitol solution, 1 °C/min, shows sugar crystallization; (e) serum, 150 °C/min, morphology similar to plasma (part of Fig. 1 from [1.29]).

The crystallization energy, extrapolated for the warming rate zero is calculated for mannitol (13.5 kJ/Mol) and for Na-cefazolin (39,1 kJ/Mol). These data agree with measurements by other methods. The activation energies are generated with certain assumptions to be 335 kJ/Mol for mannitol and 260 kJ/Mol for Na-cefazolin. DeLuca [1.31] derived at with slightly different data: at a warming speed of 0.625 °C/min, he found 16.3 kJ/Mol for mannitol and 41.8 kJ/Mol for Na-cefazolin.

Na-cefazolin is instable in its amorphous state. Takeda [1.32] described a method to ensure complete crystallization in which micro crystalline Na-cefazolin were added to at 0 °C supersaturated Na-cefazolin solution, frozen and freeze dried. The product did not contain amorphous or quasi-crystalline components.

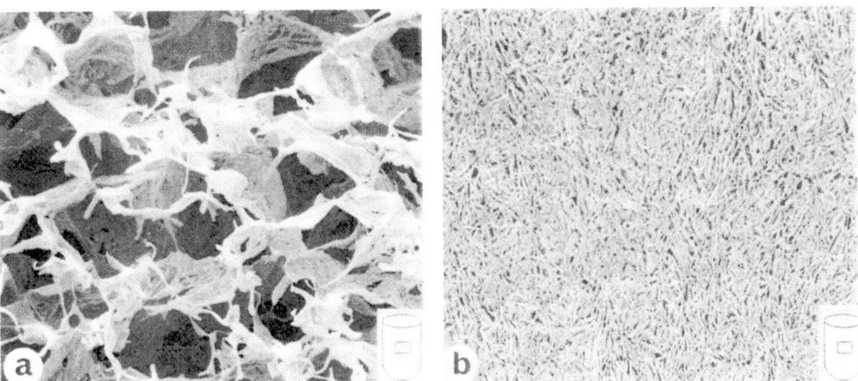

Fig. 1.43. Scanning electron-microscopic photographs of a 1 % lactose solution. (a) Frozen at 1 °C/min; (b), frozen at 150 °C/min (Fig. 3 from [1.29]).

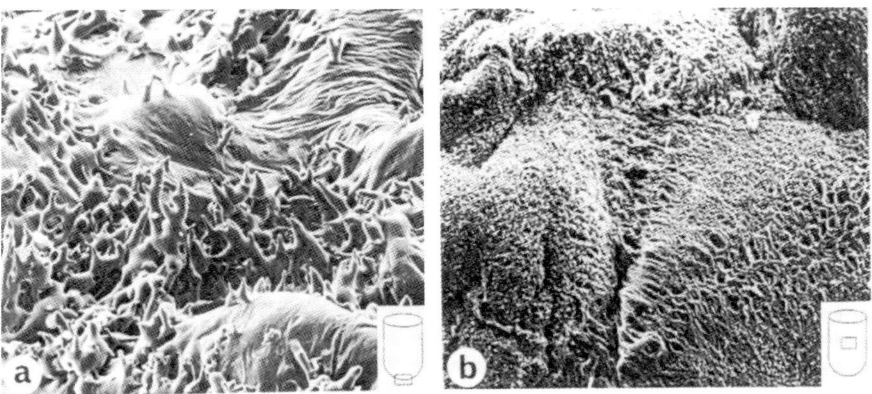

Fig. 1.44. Scanning electron-microscopic photographs of a vial containing freeze dried trehalose solution. (a), collapsed product from the bottom of the product; (b), shrunk product after 6 months of storage at +20 °C with a RM too high and stored at a too high a temperature (Fig. 6 from [1.29]).

Roos and Karel [1.33] measured $T_{g'}$ of fructose and glucose by DSC and showed the influence of annealing/heat treatment. In Fig. 1.48, DSC curves are shown for 60 % solutions, cooled at 30 °C/min down to –100 °C and rewarmed at a rate of 10 °C/min to –48 °C and cooled again to –100 °C at a rate of 10 °C/min. T_g of the non annealed products was ≈ –85 °C respectively ≈ –88 °C. In the range of –50 °C the crystallization of unfrozen water was seen as an exothermic event in both solutions (curves A). If rewarming is interrupted at ≈ –48 °C, the product remained at that temperature for ≈ 15 min and cooled again to –100 °C. When curves B were measured, $T_{g'}$ was increased to ≈ –57 °C, the exothermic of crystallization had disappeared, and all freazable water was frozen to ice. The temperature

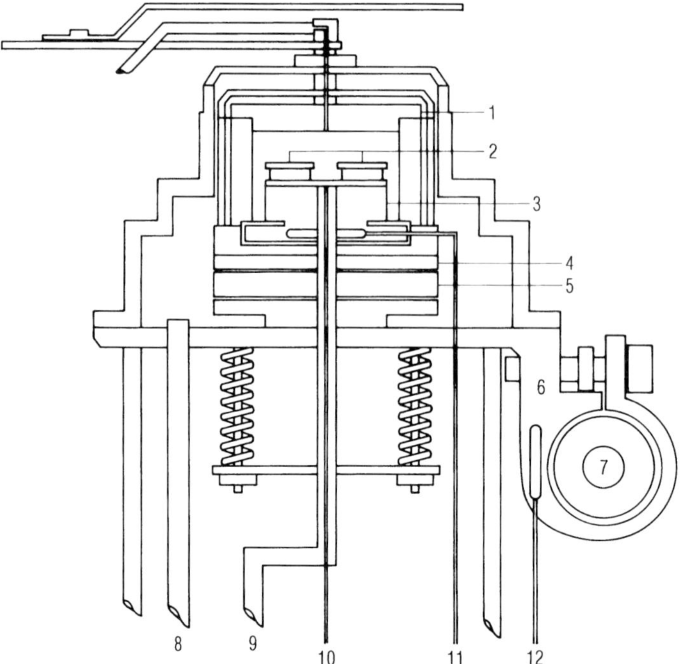

Fig. 1.45. Schema of a differential scanning calorimetric (DSC) apparatus. 'Geräte DSC 821®'. Temperature range with LN_2 cooling –150 °C to 500 °C; accuracy of temperature ±0.2 °C; resolution 0.7 μW in the measuring range of +/–350 mW; cooling rate from +100 °C to –100 °C approx. 13 °C/min; size of sample, several mg to 200 mg.
1, Automatic closure; 2, crucible on top of the DSC-sensor; 3, furnace made of silver; 4, heating element; 5, thermal resistance after the cooling flange; 6, cooling flange; 7, cooling element; 8, inlet of dry gas (to avoid condensation); 9, inlet of purging gas; 10, DSC signal to the amplifier; 11, Pt 100 of the oven; 12, Pt 100 of the cooling flange (Meßmodul DSC 821®, Mettler-Toledo AG, CH-8603 Schwerzenbach, Switzerland).

T_m is the on-set temperature of the softening process in the product and close to the collapse temperature, T_c.

Talsma et al. [1.34] described the freezing behavior of certain liposomes by DSC measurements. Besides the expected influences of freezing and rewarming speeds, and of the CPAs (mannitol and mannitol in Tris-buffer solutions) it was shown, that the heterogeneous and homogeneous crystallization in mannitol solutions exists and the nucleation of ice depends also on the liposome size: In small liposomes (e. g. 0.14 μm) mannitol suppressed the heterogeneous crystallization more effectively than in large (0.87 μm) liposomes. If in certain substances no crystallization or eutectic mixtures can be found by DSC (cephalosporin, Williams [1.35]) with the used experimental conditions, one has to seek different conditions [1.32].

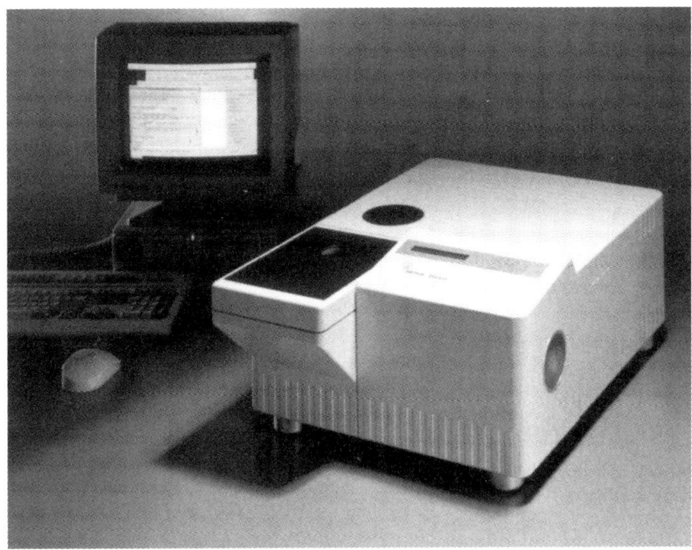

Fig. 1.46. DSC 821® with interpretation and display unit (photograph: Mettler-Toledo AG, CH-8603 Schwerzenbach, Switzerland).

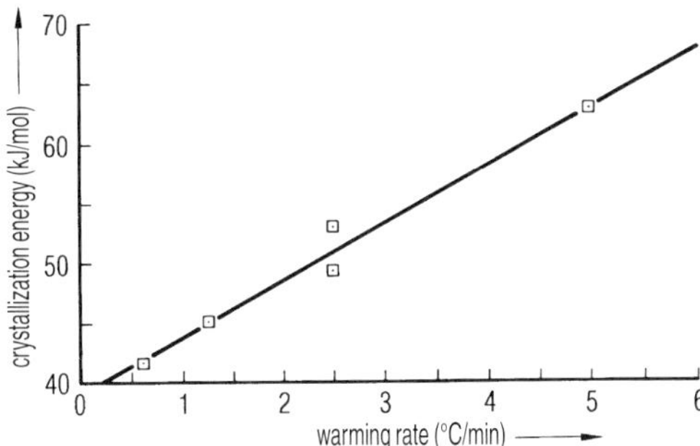

Fig. 1.47. Crystallization energy of sodium cefazolin as a function of the warming rate, measured by DSC (Fig. 2 from [1.30]).

Nuclear magnetic resonance (NMR) is a highly sensitive analytical method. It can be used, to study the way that water behaves during freezing in aqueous saccharide and protein solutions as well as in coffee extracts. Using NMR it is possible to determine whether water is bound to other molecules (e. g. proteins) and cannot crystallize, how the collapse temperature T_c is influenced by unfrozen water, and the changes in a glass of highly concentrated solutions during warming from low temperatures below and above T_g.

NMR spectroscopy (a commercial unit shown in Fig. 1.49) uses the fact that some atomic nuclei have a magnetic moment, e. g. very distinct in a proton, the nucleus of hydrogen, but also in ^{13}C, ^{31}P, ^{14}N, and ^{33}S. In an external magnetic field the energy levels split, as described in quantum mechanics. The size and extend of the split is given by Eq. (9)

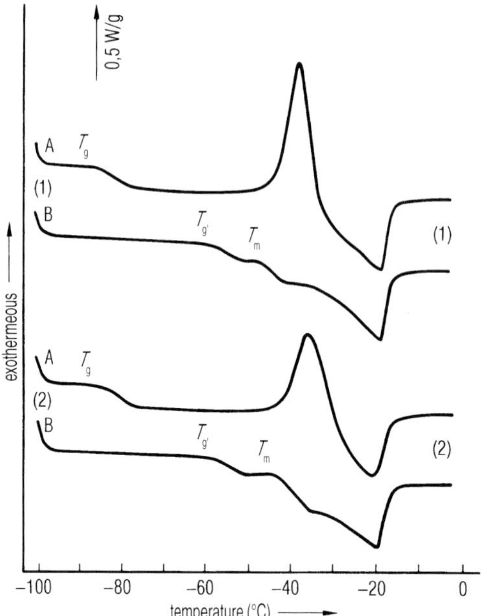

Fig. 1.48. Results of annealing (thermal treatment) on the formation of ice in a 60 % fructose solution (1) and in a 60 % glucose solution (2).
Curve A: After cooling at 30 °C/min down to –100 °C, the DSC plots have been recorded during rewarming with 5 °C/min. T_g approx. –85 °C, respectively –88 °C. At approx. –48 °C, respectively –44 °C, ice crystallization starts clearly, followed by the beginning of the melting of ice. (During freezing only a part of the water has been crystallized.)
Curve B: After cooling down to –100 °C, the product has been warmed up at 10 °C/min to –48 °C, kept for 15 min at this temperature (thermal treatment), cooled down again at 10 °C/min to –100 °C, and the DSC plot (B) measured during rewarming. During thermal treatment all freezable water is crystallized, and *Tg'* is increased to –58 °C, respectively –57 °C. During rewarming, no crystallization can be detected (Fig. 2 from [1.33]).

$$\Delta E = \mu B \text{ g } H_{\text{eff}} \tag{9}$$

where μB is nuclear magneton; g is a constant (characteristic for the magnetic quality of a given nucleus), and H_{eff} is the effective strength of the magnetic field at the location of the nucleus.

The transition energy can also be described as a frequency of electromagnetic radiation

$$\Delta E = h^* f \tag{10}$$

where
h^* is Planck's quantum of action, and
f is the frequency of radiation, or

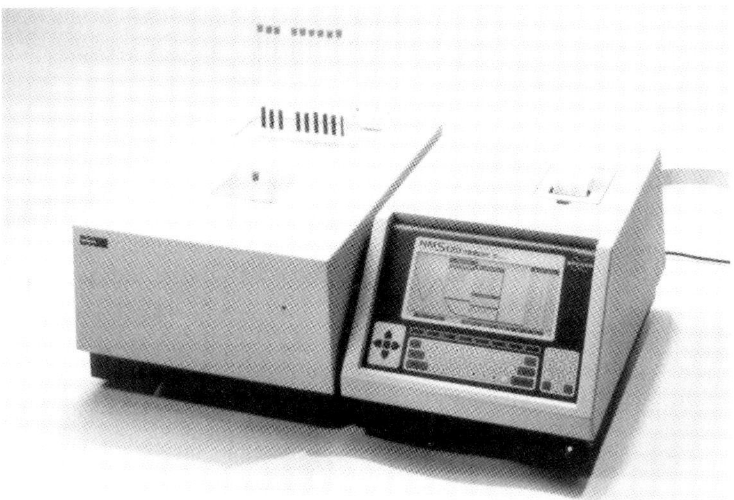

Fig. 1.49. NMR analyzer 'NMS 120 minispec' from the NMS 100 minispec serie. Measurements below room temperature require a special measuring head, which is cooled by a cryostat (photograph: Bruker, Analytische Meßtechnik GmbH, D-76287 Rheinstetten, Germany).

$\Delta E = h^* c/\Lambda$

where c is the speed of light, and Λ wavelength.

The energy difference between the levels depends on the field strength of the external magnetic field. To use 60, 100 or 270 MHz for NMR measurements with protons, the magnetic field strengths must be 14.1, 23.5 or 63.4 · 10^3 Gauss (G). The latter value is only possible with superconducting magnets. Since all other nuclei have a magnetic momentum which is small compared with that of the proton, still higher magnetic field strengths are necessary. The magnetic momentum of a nucleus is, by the electrodymic laws, the consequence of a rotating electric charge. This rotation is described in quantum mechanics as spin (S^*) of the nucleus. Spin can only have discrete, defined energy levels parallel or vertical with the direction of the magnetic field ($S^* = \pm 1/2$). Transitions, e. g. to higher levels (absorption) are only possible (they happen with a certain degree of probability) if S is not changed with the transition ($\Delta S^* = 0$) and the projection of the spin in the direction of the magnetic field changes by ±1 ($\Delta S^*_z = +/-1$). If a sample with a magnetic momentum is irradiated by ultrashort waves in an external magnetic field, only radiation of a defined wavelength, defined energy can be absorbed. This wavelength at a given external magnetic field is characteristic for the isolated nucleus.

Is the nuclear (e. g. proton) part of a molecule the external field is changed by factors which are characteristic for that molecule. The resonance frequency of isolated protons is shifted in a way typical for the chemical compound in which the proton is located. This shift is called chemical shift of the resonance frequency (at a given external magnetic field). The chemical shifts are small, e. g. at a proton up to 30 ppm of the used frequency, if 100 MHz (10^8 Hz) are used 10 ppm correspond to 10^3 Hz. The shift is normally not absolutely measured, but compared with known frequency of a reference substance, e. g. for protons tetramethylsilane (TMS). The area of the resonance is proportional to the number of nuclei which give rise to it.

Besides the chemical shift of the resonance line, under certain conditions the lines split in two or more lines. This reflects the influence of the spin orientation of two or more neighboring nuclei on the magnetic field in the surrounding nuclei. The size of the splitting is called the coupling constant *J*. *J* represents the quantity of influences between the nuclei, while the number of split lines and their intensity represent the number of the influencing nuclei. The lines in a NMR spectrum are not infinitely small, but show a certain linewidths, since the magnetic field at the location of a nucleus changes slightly, albeit constantly. After the high frequency impulse is terminated, the earlier equilibrium is reinstalled by magnetic noise, and the system relaxes. Bloch has connected the two possible relaxation processes with two characteristic times: t_{SGR}, the spin-lattice-relaxation time and t_{SSR} the spin-spin-relaxation time. Half the width of the resonance line measured at half the height of the peak equals $1/t_{SSR}$. As shown in Fig. 1.50, for very small molecular correlation times t_c, t_{SGR} and t_{SSR} are identical. The correlation time is the time that one molecule requires to travel the distance of its own diameter; it is a measure for the mobility of the molecules.

In aqueous solutions with small molecules the relaxation is slow (0.1 to 0.5 s), while t_{SSR} of ice is very small (some 10^{-3} s) [1.36]. Close to the glass temperature of a substance the relaxation time does not decrease exponentially, and thus a different means of description must be used [3.9].

Hanafusa [1.36] showed with this method, how the amount of unfrozen water in a 0.57 % solution of ovalbumin reaches practically zero at –20 °C, if 0.01 M sucrose is added (Fig. 1.51). For globular proteins Hanafusa described the freezing process as follows: between 0 °C and –20 °C, water molecules from the multilayer hydrate shell are decomposed. Be-

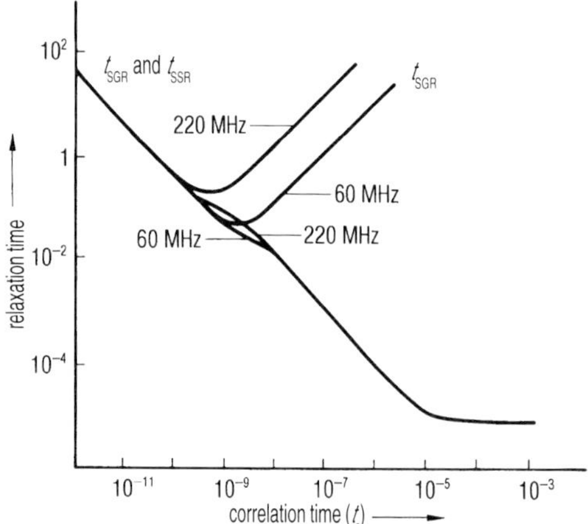

Fig. 1.50. Relaxation time as a function of the molecular correlation time for two spectrometer frequencies: 60 MHz and 220 MHz. t_{SGR}, spin-lattice relaxation time; t_{SSR}, spin-spin relaxation time (Fig. 2.24 from [1.105]).

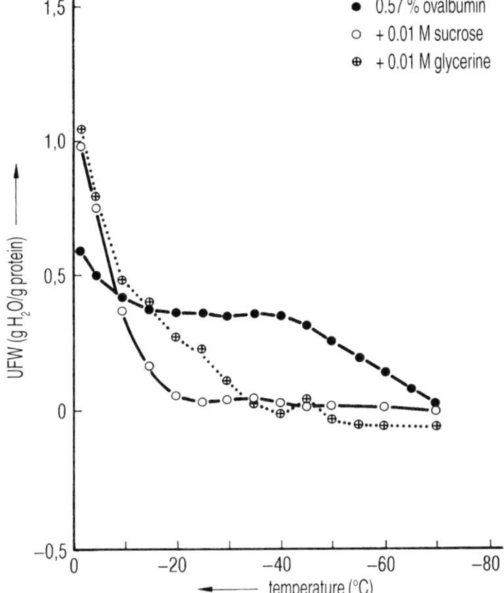

Fig. 1.51. Unfreezable water (UFW) in a 0.57 % ovalbumin solution as a function of the freezing temperature with different CPAs (Fig. 4. from [1.36]).

low –45 °C, molecules from the monohydrate shell are removed, thereby destroying the shell; between –20 °C and –45 °C, an equilibrium exists between the hydrate bond to the protein molecule and the forces to insert additional water molecules into the ice crystals. By adding CPAs, the amount of bound water is much reduced. Water molecules are replaced by CPA molecules and form a 'quasi-hydrate' shell, which protects the protein during freezing and freeze drying against denaturation. Hanafusa showed (Fig. 1.52) a simplified, graphic picture how, with rising concentration (1) to (3), the CPA molecules form a new shell for the protein. Some water molecules are so strongly incorporated, that they can no longer diffuse to the ice crystals.

Nagashima and Suzuki [1.37] used NMR to show the interdependence of UFW, $T_{c,}$, the cooling rate and the concentration before freezing. The amount of UFW in g H_2O/g dry substance is measured e. g. of coffee extract with 25 % solids (Fig. 1.53), which at –20 °C has approx. 30 % UFW (0.3 g/g) but is reduced at –50 °C to 0.1 g/g. Above –20 °C the UFW rises rapidly. During freeze drying above –20 °C, the structure is going to collapse. Nagashima demonstrated that, after quick freezing (3 to 5 °C/min) of mannitol solution, crystallization of mannitol can be seen during rewarming. UFW rose to ≈ 50 %, water than crystallized and UFW is reduced to a few percent. The crystallization temperature measured agreed well with other reports (e. g. Hatley [1.38]) using DSC. During slow freezing mannitol crystallizes, and there is no hysteresis (Fig. 1.54). Figure 1.55.1 shows the strong dependence of UFW with the Japanese miso-sauce. At –50 °C and a concentration of 52.7 % UFW is ≈ 5 units, while at 26.4 % solids in the original product, UFW remains ≈ 2 units, and only at 13.2 % solids ≈ 0.6 units UFW do not freeze.

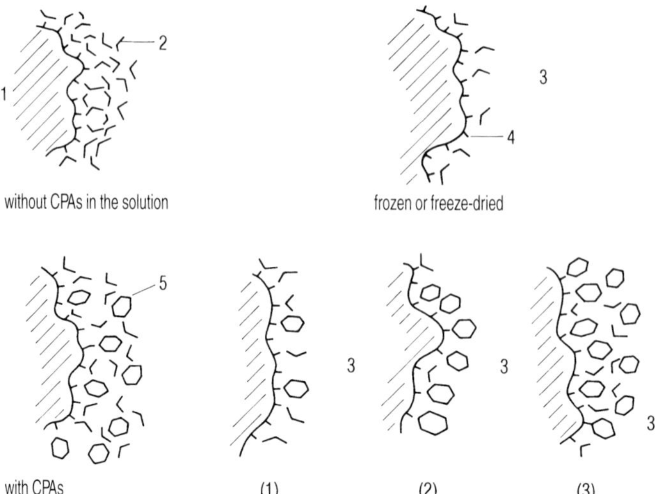

Fig. 1.52. Schematic model of CPA action in protein solutions during freezing and freeze drying. (Fig. 10 from [1.36]). Top row: Without CPA; the hydrate water of the ovalbumin has migrated into the ice and the freed valences are exposed to the influence of the environment. Second row: With CPA; a part of the hydrate water of the proteins becomes replaced by CPA molecules. These, together with the remaining water molecules and the protein molecule, form a 'quasi' (replacement) hydrate layer.

1, Protein; 2, water molecule; 3, ice or air; 4, exposed valence; 5, CPA molecule.

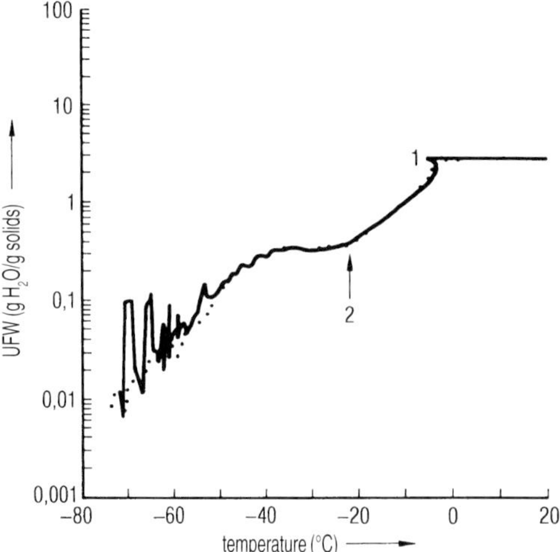

Fig. 1.53. Freezing and thawing plot of coffee extract with 25 % solids. UFW (g H_2O/g solids) as a function of temperature (Fig. 2 from [1.37]).

1, Subcooling; 2, collapse temperature.

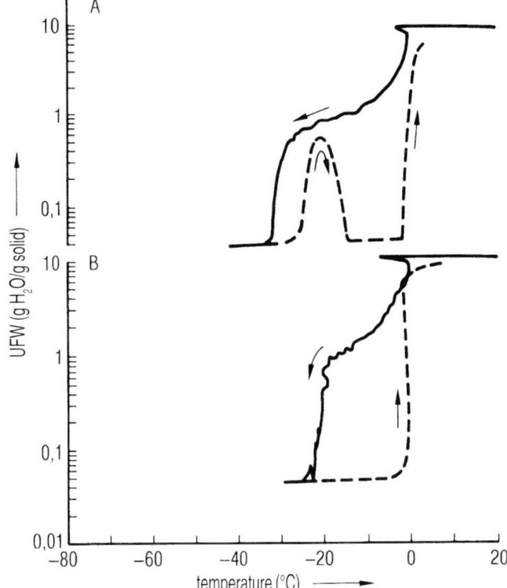

Fig. 1.54. Freezing and thawing plot of a 9.1 % D-mannitol solution.
A, freezing rate 5 °C/min; B, freezing rate 0.5 °C/min (Fig. 4 from [1.37]).

Harz et al. [1.39] demonstrated by NMR spectroscopy that freezing of food (e. g. grapefruit juice) almost never followed the ideal expectation. The crystallization of carbohydrates is much hindered and further reduced by the high viscosity of the solutions. Water crystallizes much below the eutectic temperature, producing further increases in viscosity and leading to a glass phase during further cooling. Depending on the carbohydrates, this metastable phase can at –18 °C last for weeks or, on occasions up to one year.

Girlich [1.40] studied by NMR the molecular dynamics of aqueous saccharins solutions. At a concentrations down to 30 % solids, the S molecules do not influence each other, while with decreasing temperature the existing H-bond-bridges prevent a reorientation of the H_2O molecules. Dissolved S molecules can destroy the H bonds, such that subcooling becomes possible. At > 40 % solids, associations of S molecules are formed. H_2O is increasingly bound by H bridges and loses translation- and rotations-mobility. With increasing concentration of the solution, the S molecules cross-link, hydrate-water becomes freed and can lead locally to low concentrations. $T_{g'}$ of water becomes different from that of S hydrates. Below 70 % solids, the cross-linked system of S molecules develops into a gel. During the observation time no crystallization takes place, and a metastable glass exists with a viscosity > 10^{12} Poise. The mechanical behavior is like that of solids.

Kanaori et al. [1.122] studied the mechanism of formation, and association of human calcitonin (hCT) fibrils using NMR. hCT associates and precipitates during storage in aqueous solution. The freeze dried hCT and its behavior is described.

Carrington et al. [1.124] usd thermomechanical analysis (TMA) to study the ice-crystallization temperature of 30 % (w/w) fructose, sucrose and glucose with and without sodium carboxy methyl cellulose (CMC). TMA has been used to measure the expansion of

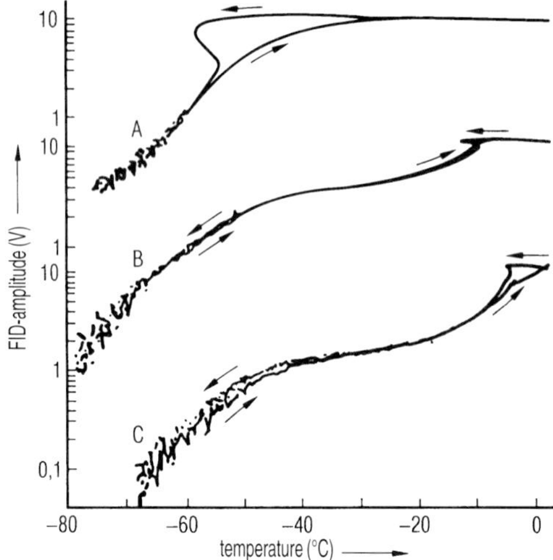

Fig. 1.55.1. UFW content of miso sauce (A) and two dilutions (B and C) as a function of temperature. The solid content is: A, 52.7 %; B, 26.4 %; C, 13.2 % (Fig. 5 from [1.37]).

the sample during freezing and rewarming. Parallel studies have been done using DSC. A typical result of TMA measurements during freezing is shown in Fig. 1.55.2 for fructose with and without CMC during freezing with a rate of 5 °C/min. Figure 1.55.3 shows the plot of the warming profile of slowly frozen and annealed 30 % sucrose solution, as determined by TMA. Figure 1.55.4 shows the warming DSC curve of 30 % sucrose solution slowly frozen and annealed. In comparing the two temperatures T_{r1} and T_{r2} (as shown in

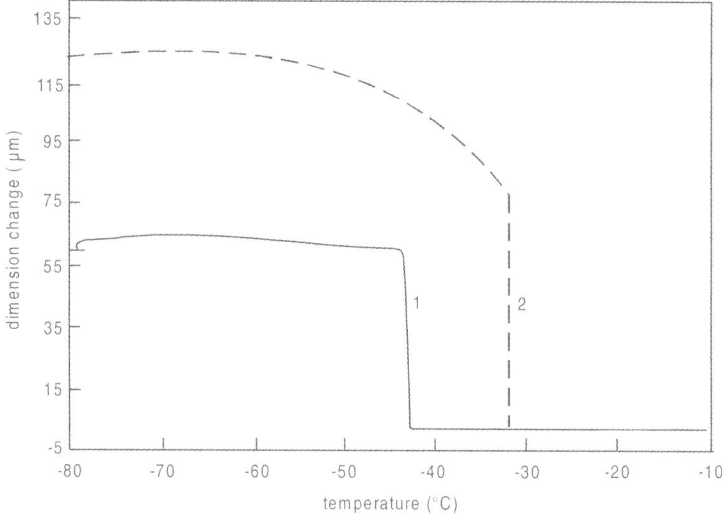

Fig. 1.55.2. Dimension change as a function of temperature for 30 % fructose solution during freezing at 5 °C/min down to –80 °C.
1, Fructose alone; 2, fructose plus 0.25 % sodium carboxymethyl cellulose (CMC).

Figures 1.55.3 and 1.55.4) by both methods for sucrose: $T_{r1} \approx -60$ °C (TMA) and –41.2 °C (DSC), $T_{r2} \approx -35$ °C (TMA) and –32.6 °C (DSC) it is obvious, that several factors influences the resulting data as discussed by the authors (onset data for DSC from a table in the publication, TMA estimated from the plot).

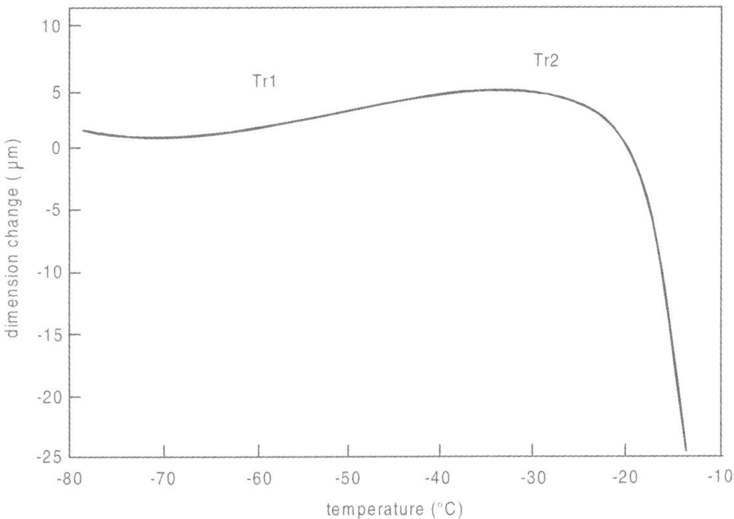

Fig. 1.55.3. Dimension change as a function of temperature for a 30 % sucrose solution during warming at 2 °C/min after slow freezing to –80 °C and annealing up to –35 °C.

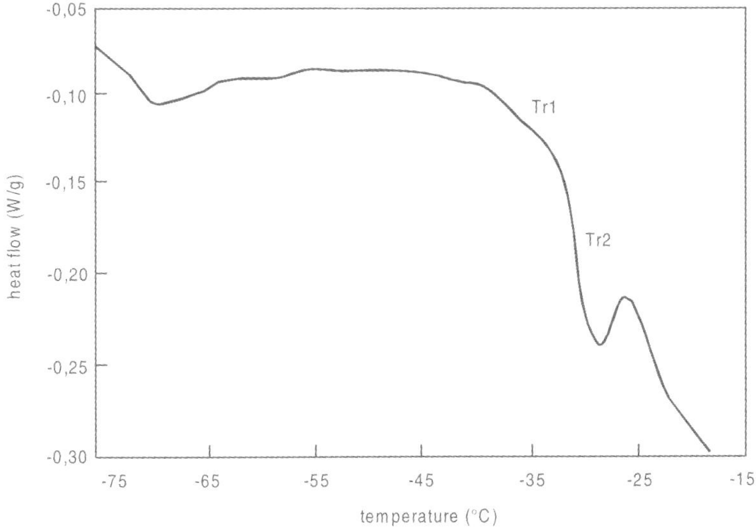

Fig. 1.55.4. Heat flow as function of temperature in the DSC thermogram of 30 % sucrose solution frozen at 5 °C/min to –80 °C during warming (5 °C/min) after annealing up to –35 °C (Fig. 1.55.2–1.55.4 from [1.124]).

TMA measurements have been helpful in explaining the breakage of vials during the warming of frozen solutions of mannitol and other stereoisomers [1.125]. For example, above –25 °C mannitol expands 30 times more than standard type 1 flint glass. Depending on the filling volume and the concentration, 10–40 % of the vials break when filled with 3 % mannitol solution.

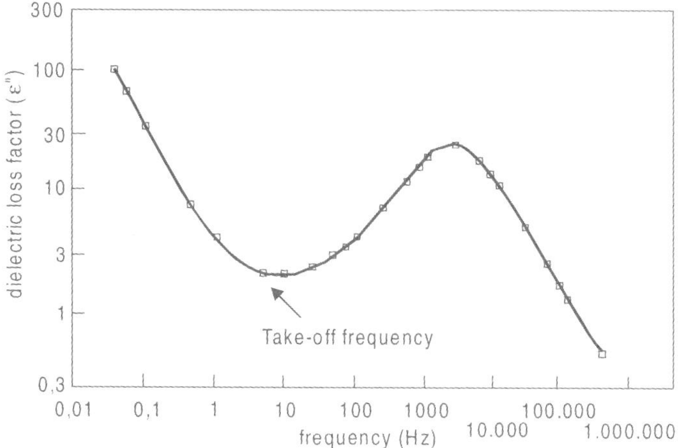

Fig. 1.55.5. The take-of frequency, at a given temperfature, occurs at the first minimunm in the dielectric loss factor (ε") versus frequency curve as frequency increases (Fig. 9 from [1.126]).

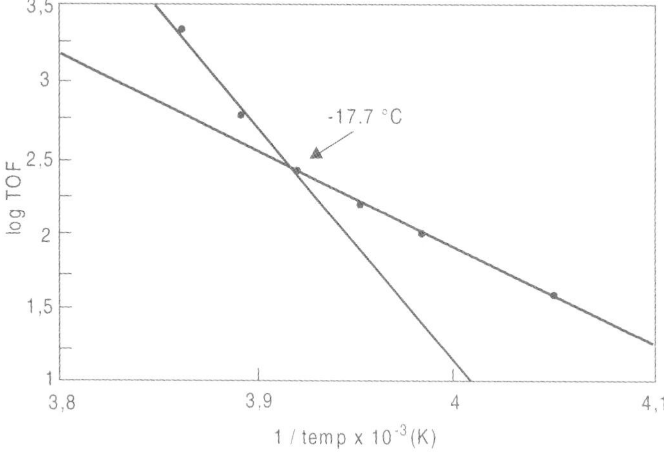

Fig. 1.55.6. Collapse plot in log (TOF) versus 1/*K*. The extrapolated intersection of the two linear portions identifies the collapse temperature of the system. (Fig. 10 from [1.126]).

Morris et al. [1.126] proposed to use dielectric analysis (DEA) to predict the collapse temperature of two component systems. The background of DEA is explained and the 'take off frequency' (TOF) is chosen as the best analytical method to identify the collapse temperature. Figure 1.55.5 shows the dielectric loss factor as a function of the frequency.

The frequency at the minimum of this curve is called TOF by the authors. TOF varies with the temperature as shown in Fig. 1.55.6. The extrapolated intersection of the two linear portions identifies the collapse temperature. The predicted T_c by TOF for 10 % sucrose, 10 % trehalose, 10 % sorbitol and 11 % Azactam™ solution deviates from observations by a freeze-drying microscope (Table 1, from [1.126]) to slightly lower temperatures, the differences are: –3 °C, –1.4 °C, 2.2 °C and 0.7 °C.

Smith et al. [1.127] reviewed the dielectric relaxation spectroscopy (DRS) as a method for structural characterization of polymers and proteins providing, among others, information about the water content and states of water.

1.1.6 Changes of Structure in Freezing or Frozen Products

Independent of the growth of ice crystals (Section 1.1.2), which can be observed down to approx. –100 °C, and a possible recrystallization (Section 1.1.3), this chapter describes only such developments or changes of structures that can be influenced by additives. The addition of CPAs to albumins, cells or bacteria influences the nucleation of ice – or at least its growth – in such a way that their natural structures are retained as much as possible. On the other hand, additives are introduced to crystallize dissolve substances. If this method does not help, e. g. with antibiotics, the solution concentrates increasingly until a highly viscous, amorphous substance is included between ice crystals. This condition has disadvantages:

- The water is not crystallized to its maximum and can be removed during freeze drying only with difficulty, or not at all. The residual moisture content remains undesirably high.
- Drugs are often less stable in the amorphous state than as crystals [1.41 to 1.44].

The phase transition from amorphous to crystalline can sometimes be promoted by thermal treatment (annealing) [1.45]. In a laboratory scale, this can be done relatively simple. In a production scale the process must be proven as reproducible and reliable by a validation process, which is time consuming. It is therefore recommended, that a search for CPAs and process conditions, which would lead to crystallization be carried out, using methods such as DTA, DSC, ER and DRS (see Section 1.1.5); also see Yarwood [1.46]. If this is not successful, time and temperature for TT should be chosen in such away, that the tolerances for time and temperature are not to narrow, e. g. –24.0 °C ± 0.5 °C and 18 min ± 1 min are difficult to operate, while –30 °C ± 1.5 °C and 40 min ± 2 min might be easier to control.

A suitable freezing rate, start-up concentration and an amount of product per vial (for example for Na-ethacrynate) can be selected that results in a stable, crystalline phase. However, the addition of CPAs may provide another means of achieving crystallization as seen for several pharmaceutical products [1.47].

De Luca et al. [1.48] showed, that the addition of 5 % tertiary butyl alcohol (tBA) to aqueous sucrose and lactose solutions (up to 40 %) resulted in a frozen matrix, which could be easily freeze dried. De Luca demonstrated by DSC that the melting point rose distinctly (with 60 % solution to –10 °C), but the endothermic of melting returned to 25 %, indicating that not much water had frozen. In solutions with 5 % tBA the exothermic of crystallization became more visible and the melting of tBA could be recognized.

Kasraian and De Luca [1.128] developed a phase diagram by DSC for tBA. Two eutectics were observed at 20 % and at 90 % tBA concentrations. Using a freeze drying microscope, the change of ice crystals by tBA became visible, though 3 % tBA was required to form large needle-shaped ice crystals. A solution with 10 % tBA grew finer, needle-shaped ice crystals, and a 70 % tBA solution formed very large, hydrate crystals. The rate of sublimation of water and tBA depended on the concentration The crystallization behavior of the water tBA mixtures could explain the influence of tBA on the freeze drying of sucrose and lactose, when used in certain concentrations. Oesterle et al. [1.144] showed, that not only tBA can speed up the sublimation of ice from amorphous freeze-concentrated mixtures, but similar effects can be achieved by volatile ammonium salts such as ammonium acetate, bicarbonate and formate. 0.1 M ammonium salt solutions and 5 % tBA were studied in an 8.5 % excipient solution. The onset temperatures of $T_{g'}$ are determined by DSC: sucrose –33.6 °C, PVP –21.1 °C and lactose –29.7 °C. The onset temperatures for the tBA/ammonium mixtures are between 3 and 14 °C lower than without additives. MD has been carried out 5 °C lower than the respective $T_{g'}$. The percent of weight losses during the first approx. 7 h of drying are the largest with 5 % tBA in PVP and lactose solutions. In the sucrose solution tBA and ammonium salts show approx. equal effects. The authors conclude, that the sublimation rates can be enhanced by tBA and other additives, but the influence of this additives on the stability and activity of proteins is not clear and the problem of possible residuals of the additives in pharmaceutical preparations needs further studies.

1.2 Drying

Drying is basically well understood, and is governed by two transport mechanisms: (i) the energy transport to transform the ice into water vapor (between –21 °C and –30 °C approx. 2805 kJ/kg); and (ii) the transport of the water vapor from the sublimation surface through the already dried product into the drying chamber to the condensation – or absorbing system for the vapor. Figure 1.56.1 shows the process of the main drying (MD) observed with a cryomicroscope. A 10 % aqueous solution of hydroxyethyl starch (HES) has been directionally frozen (see Fig. 1.37). The ice dendrites are surrounded by the concentrated

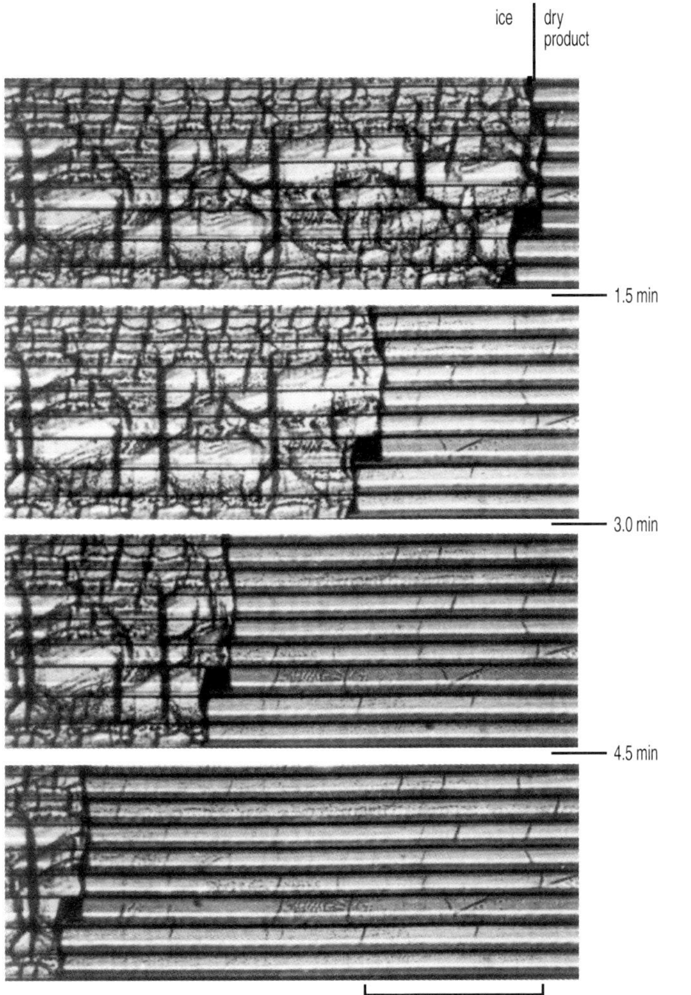

Fig. 1.56.1. Course of main drying observed by a cryomicroscope, in which the freeze drying is carried out. The hydroxy ethyl starch solution is optimally frozen (see Fig. 1.37). The dark lines show the form of the sublimated ice crystals (Fig. 9 from [1.106]).

solid, which can be seen as a dark line after the ice is sublimated. In spite of optimal freezing the sublimation speed is not uniform for all dendrites.

After the ice has been sublimated, the adsorbed water is desorbed from the solid. This process is governed by laws different from those of the main drying. This step is called secondary- or desorption drying. During the secondary drying (SD) the energy transport does not play an important role, since the amount of water is normally less than 10 % of the solids. Nevertheless, SD time-wise can be an important part of the total process and consume half or the same time as MD. Figure 1.56.2 shows the typical run of a freeze drying process which can be divided into the two parts: In MD, large amounts of water vapor (e. g. 900 % (w/w) of the solids) are sublimated and transported at an almost constant tempera-

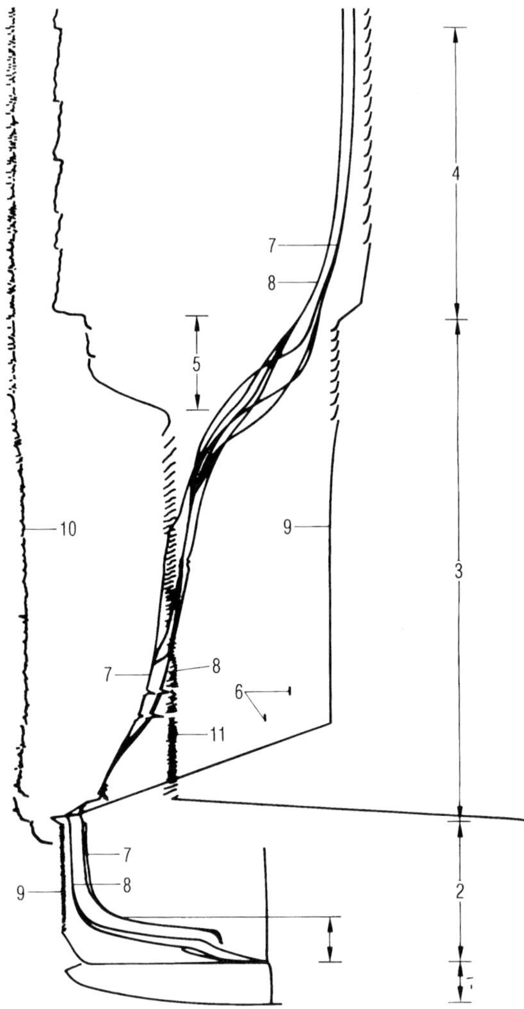

Fig. 1.56.2. Course of a freeze drying process.
1, Precooling of the shelves; 2, freezing of the product; 3, evacuation and main drying (MD); 4, secondary drying (SD); 5, change over from MD to SD; T_{sh} raised to maximum tolerable product temperature; 6, T_{ice} measurements by barometrifc temperature measurement (BTM); 7, temperature sensors RTD in the product; 8, temperature sensors Th in the product; 9, temperature of the shelves (T_{sh}); 10, ice condenser temperature; 11, pressure in the drying chamber (p_{ch}).

ture at the sublimation front, (T_{ice}). In SD, the product temperature rises to the maximal tolerable temperature of the product, and the water content is lowered e. g. by 9 % (w/w) of the solids.

1.2.1 Main Drying (Sublimation Drying)

The amount of energy necessary for the sublimation depends on the sublimation temperature, but between –10 °C and –40 °C the energy varies by less than 2 %. Furthermore energy is consumed to heat the vapor during the transport through the already dried product or in contact with the warmer shelves. The specific heat of water vapor is 1.67 kJ/kg and the maximum increase in temperature is up to +20 or +40 °C. The energy consumption by this process can almost be neglected compared with the sublimation energy; heating the vapor from –30 °C to +30 °C results in ≈ 100 kJ/kg or ≈ 3.5 % of the sublimation energy.

The necessary energy can be transduced to the ice in four different forms:

(a) by radiation of heated surfaces;
(b) by conduction from heated plates or gases;
(c) by gas convection: or
(d) by dielectric losses in the ice in a high-frequency field. The method is not discussed, since high-frequency fields with the necessary field strength in the pressure range of MD freeze drying (1 to 0.01 mbar) tend to start gas discharges.

(a) An infinite plate with the temperature K_{str1} and the radiant efficiency ε_1 will transmit, independent of the distance, to a similar plate of frozen product with the temperature K_{str2} and a radiant efficiency ε_2 an amount of radiation energy. The surface heat flux q is:

$$q = \delta\,(K^4_{Str1} - K^4_{Str2}) \cdot 1 / [(1/\varepsilon_1) + (1/\varepsilon_2) - 1] \qquad (11)$$

In this equation $\delta = 2.05 \cdot 10^{-7}$ kJ/m^2 h K^4.

This presentation is simplified. In practice a part of the energy, depending on the distance of the plates, will hit the walls of the chamber. This effect is small as long as the dimensions of the plates are large compared with their distance. For two plates of 1 m · 1 m and a distance of 0.1 m the effective radiation is ≈ 0.8 q and with two plates 0.5 m · 0.5 m ≈ 0.7 q [1.49].

The values of ε for products will normally be close to 1, which is also true for anodized aluminum and varnished steel. For polished steel, $\varepsilon = 0.12$. Figure 1.57 shows the energy transmission by radiation if both ε are 1.

At shelf temperatures of 100 °C, approx. 2000–4000 kJ/h m^2 are transmitted, depending on the product temperature. At lower shelf temperatures, as is usual in freeze-drying plants for pharmaceutical products, q values between 500 and 1500 kJ/m^2 can be expected. However for $\varepsilon = 0.12$, these data are reduced by a factor of 0.12.

At a shelf temperature of 100 °C and both $\varepsilon = 1$, at the beginning of a freeze drying cycle (surface temperature –20 °C) approx. 1.4 kg ice/h m^2 can be evaporated, at +30 °C and –30 °C remain approx. 0.4 kg/h m^2. If $\varepsilon_1 = 0.12$, the sublimation rate is reduced to approx. 200 g/h m^2, respectively 50 g/h m^2.

(b and c) An important part of the energy transfer is by conductivity, as well by direct contact of the product container with the shelf, as by the gas. Furthermore, the gas transports energy by convection, which becomes an essential factor, if the distance between the shelf and tray or vial becomes small. Figure 1.58 [1.50] shows, that for distances larger than 10 mm the energy transfer is independent of the pressure and becomes very small.

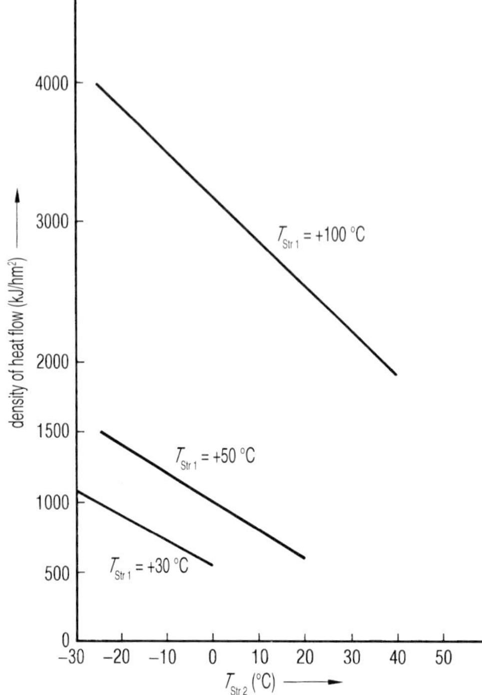

Fig. 1.57. Heat transfer by radiation only. The figure shows only the order of magnitude.
Density of heat flow (q) as a function of T_{Str2} (ice- or product surface temperature) with three temperatures T_{Str1} of the radiation surface (+100 °C, +50 °C, +30 °C) as parameter.

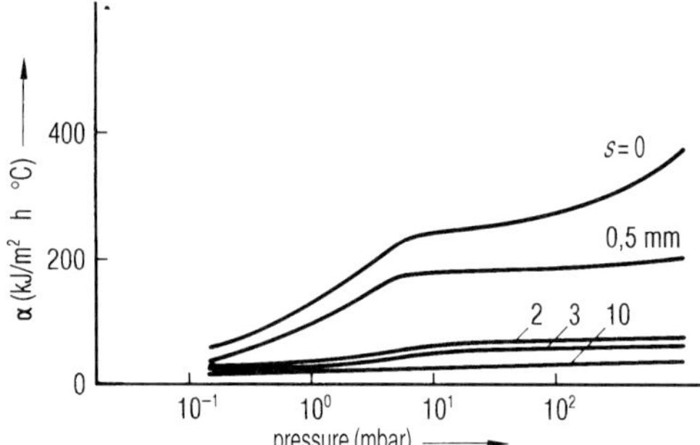

Fig. 1.58. Heat transfer coefficient α as a function of pressure. Parameter s = distance between shelf and bottom of the product container (mm). Data measured in air (Fig. 4a from [1.50]).

However for small distances, e. g. 0.5 mm, the heat transfer coefficient rises between 0.13 mbar and 1 mbar by a factor of 4.

If the shelf and the tray are as planar as technically possible, the plot marked s = 0 applies. At 0.2 mbar, a heat transfer coefficient of approx. 85 kJ/m^2 h °C can be achieved, rising by a factor of two at 1 mbar. In a well designed freeze drying plant with planar trays or vials a heat transfer coefficient of 160 kJ/h m^2 °C at 0.9 mbar is possible (Fig. 1.59), while at a pressure of 0.45 mbar, approx. 120 kJ/h m^2 °C (Table 1.9) is measured for the heat transfer coefficient K_{tot}. To sublimate 1 kg of ice per hour and m^2 with a coefficient of

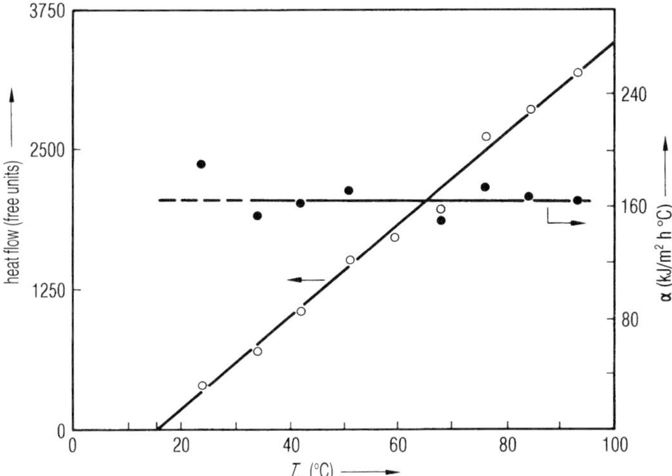

Fig. 1.59. Heat transfer from a shelf with the temperature T_{sh} to the bottom of the product container with the temperature T_{bo} at pressure $p_{H2O,ch}$ = 0.9 mbar (modified from Fig. 4b in 1.50]).

120 kJ/h m^2 °C the temperature difference between T_{ice} and T_{sh} (temperature of the shelf) must be an average of 23 °C.

Until now, only the heat transfer from the shelf to the tray or vial has been looked at. The heat transfer to the sublimation front and the transport of the water vapor from the sublimation front into the chamber will now be included. Wolff et al. [1.129] described a model for a uniformly-retreating ice front, and experiments with milk and water to confirm the usefulness of the model. Three parameters were studied: the water diffusion in the dried layer; the external mass transfer; and the heat transfer from the shelf to the product. The last parameter was found to control the dehydration kinetics. Ybeme et al. [1.130] used a conductive paste on the shelves to reduce the heat transfer resistance between the shelf and the vials. The resistance towards mass transport has been varied by using different restrictive capillaries. The conclusion of the experiments confirms, that the heat transfer to the vials limits the rate of sublimation. Chang and Fuscher [1.131] showed how and under which circumstances it was possible to use the T_{sh}, applied during secondary drying, already during the main drying. Recombinant human interleukin-1 receptor antagonist (rhIL-1ra) in various concentrations has been studied in a solution of 2 % (w/v) neutral glycine, 1 % (w/v) sucrose and 10 mM sodium citrate buffer at pH 6.5 (25 °C). At a 100 mg/mL rhIL-1ra concentration no devitrification was seen to start at –37 °C and no recrystalization began at –27 °C as measured at lower concentrations (10 to 50 mg/mL). For this product a temperature of –22 °C was considered low enough to avoid collapse. This temperature was controlled by pressure control as described in Table 1.10.1, its related text and Figs. 2.37 1. and 2. The shortest drying time, keeping the temperature below –22 °C, was found to be at a shelf temperature of +40 °C, which was also used during SD. As the authors note, this method cannot be applied to all formulations and is also dependent on the whole system, on the type of vial, heat transfer from the shelf to the sublimation front of the ice, water vapor transport, etc. In Fig. 2.37.2, an example is given in which the ice temperature range can be adjusted by pressure, valid for one product in one type of vial in one plant and with one shelf temperature.

The author has used a model and an equation developed by Steinbach [1.51] for many years and for many experiments in a wide field of applications. The model, shown in Fig. 1.60, uses an infinitely expanded plate of the product with the thickness d. Equation (14) describes the time of the main drying part of the freeze drying cycle:

$$t_{md} = (\rho_g \, \xi_w \, LS \, \Delta m \, d) \, / \, T_{tot} \, \{(1 \, K_{tot}) + (d/2 \, \lambda_g) + (d/2 \, LS \, b/\mu)\} \qquad (12)$$

where

ρ_g = density of the frozen product (kg/m^3)
ξ_w = part of water (kg/kg)
LS = sublimation energy (2.805 kJ/kg)
T_{tot} = temperature difference ($T_{sh} - T_{ice}$)
K_{tot} = total heat transmission coefficient from the shelf to the sublimation front of the ice
λ_g = thermal conductivity of the frozen product
d = thickness of the layer (m)
Δm = content of frozen water = 0,9 [1.2]
b/μ = permeability (kg/m h mbar) for water vapor through the dried product

In this equation the following simplifications are made

- The layer is endless, energy is only transmitted from the shelf to one side of the layer.
- The vapor is only transported from the ice front through the porous dried layer.
- The frozen layer is not porous.
- The heat transport in the already dried layer is neglected.

The error resulting from the last assumption at T_{tot} = 100 °C is approx. 4 % and at T_{tot} = 50 °C approx. 2 % [1.51].

For the evaluation of the equation, four data are necessary in addition of those already known:

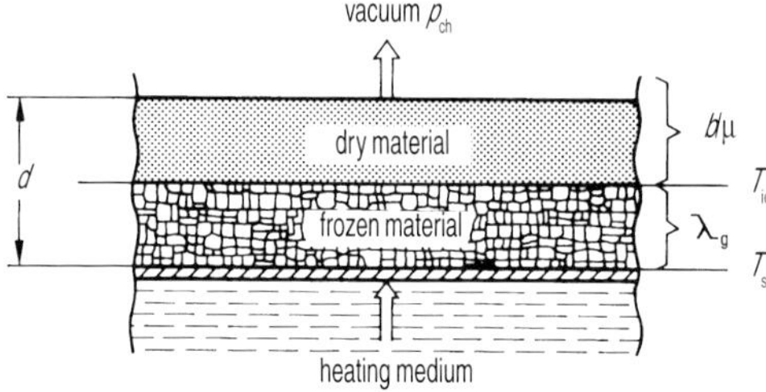

Fig. 1.60. Schema for the calculation of the drying time t_{MD} for the main drying. The product is frozen in plates (Fig. 6 from [1.51]).

T_{tot} $T_{sh} - T_{ice}$

K_{tot} heat transfer coefficient by conduction and by convection from the shelf to the sublimation front
λ_g heat conductivity of the frozen product (ice)
b/μ mass transport coefficient

The equilibrium vapor pressure (p_s) can be measured by the barometric temperature measurement (BTM) and be converted into temperature by the water vapor pressure diagram (see Section 1.2.3).

To develop an idea how the various terms of Eq. (14) influence the drying time, some experimental date are used, described in this Section.

m_{ice}	= mass of the frozen water	1,243 kg
F	= used surface area of the shelves	0.2193 m^2
d	= thickness of the product layer	$7 \cdot 10^{-3}$ m
T_{sh}	= temperature of the shelf during MD maximal	+29 °C
T_{ice}	= temperature of the ice at the sublimation front	–22 °C
T_{tot}	= $T_{sh} - T_{ice}$, average temperature difference during main drying	43.88 °C
$p_{H2O,ch}$	= partial vapor pressure in the chamber during MD	0.245 mbar
p_s	= equilibrium vapor pressure at T_{ice}	0.85 mbar
Δp	= $p_S - p_{H20,\ ch}$	0.605 mbar
t_{MD}	= time of MD (frozen water is sublimated) (see Fig. 1.74)	2.5 h
LS	= sublimation energy of ice	2.805 kJ/kg
ξ_w	= part of water in the initial product	0.931
ρ_g	= density of the frozen product, assumed as	900 kg/m^3
Δm	= part of freezable water, assumed as	0,9
λ_g	= heat conductivity in the frozen product	6.28 kJ/m h °C

With these data

$$K_{tot} = (m_{ice}\ \mathrm{LS}) / (t_{MD}\ F)\ 1 / (T_{tot}) = 144{,}9\ \mathrm{kJ/m^2\ h\ °C} \tag{12a}$$

$$b/\mu = (m_{ice} / (t_{MD}\ F)\ (d/2 / \Delta p) = 1.3 \cdot 10^{-2}\ \mathrm{kg/h\ m\ mbar} \tag{12b}$$

Using Eq. 12 t_{MD} is calculated

$$t_{MD} = \rho\ (\xi_w\ \Delta m\ \mathrm{LS}\ d) / T_{tot}\ \{(1/K_{tot)} + (d/2\ \lambda_{g)} + (d/2\ \mathrm{LS}\ b/\mu)\} \tag{12c}$$

term A	term B	term C	term D
374.5	$6.9 \cdot 10^{-3}$	$0.56 \cdot 10^{-3}$	$0.096 \cdot 10^{-3}$

$t_{HT} = 374.5\ (6.9 \cdot 10^{-3} + 0.56 \cdot 10^{-3} + 0.096 \cdot 10^{-3})$
$t_{HT} = 2.8$ h (calculated)

Equation (12) describes the main drying reasonable well, if some experimental data are used.

The influence of different parameter on the MD is discussed:

Term A: a variation of d changes this term proportionally, T_{tot} influence is the inverse ratio. Term B: this is approximately 12 times larger than C, and 70 times larger than term D. In both terms d enters a second time, but as the absolute numbers are much smaller than A, the influence of d in these terms is reduced.

If term A is constant, term B, the influence of the heat transfer term on t_{MD} is in the example 91 %. The term C, influence of the heat conductivity, ten fold decade smaller than term B. The heat conductivity depends on the characteristics of the product. In [1.50] an average value 6.28 kJ/m h °C is used at –30 °C, were in [1.2] a value of 5.9 kJ/m h °C is reported. However, even if λ_g varies by ±50 %, term C would vary approx. between $0.37 \cdot 10^{-3}$ and $1.1 \cdot 10^{-3}$ and the influence on drying time is little noticeable (±5 %).

Term D, also linear with d, shows the influence of the water vapor transport from the sublimation surface through the dried product into the vial or tray on t_{MD}. b/μ (kg/m h mbar) has often been measured: Steinbach [1.52] measured $1.3 \cdot 10^{-2}$, Gehrke and Deckwer [1.53] found for different groups of bacteria an order of magnitude of $4 \cdot 10^{-2}$, Sharon and Berk 1.54] demonstrated, how b/μ decreases for tomato pulp from $3 \cdot 10^{-2}$ to $0.8 \cdot 10^{-2}$ if the solid concentration rose by a factor of 4, while Oetjen and Eilenberg [1.50] used $1.3 \cdot 10^{-2}$ as an average value. Kasraian and De Luca [1.132] measured the resistance of the vapor transport through the dried cake of a 5 % (w/w) sucrose solution with and without 3–5 % tertiary butyl alcohol (TBA), and obtained the following results: In the absence of TBA, and with a skin on the surface $b/\mu = 0.13 \cdot 10^{-2}$ kg/m h mbar After the skin had cracked $b/\mu = 0.77 \cdot 10^{-2}$, and with BTA b/μ is in the range of $15.4 \cdot 10^{-2}$ to $2.5 \cdot 10^{-2}$ kg/m h mbar. By using the two extreme data for b/μ $0.13 \cdot 10^{-2}$ and $15.4 \cdot 10^{-2}$ kg/m h mbar t_{MD} rises to ≈ 2.8 h or remains ≈ 2.5 h. The influence of b/μ becomes measurable in the example given only at very small values of b/μ resulting from a skin on the surface. With the normal variation of b/μ its influence remains in the region of a few percent.

As long as the sublimation energy has not to be transported through the already dried layer of product (see Fig. 1.60) the heat transfer (term B) is the decisive factor. For a layer thickness of 25 mm (for freeze drying a large thickness), term A = $1.205 \cdot 10^3$, term B remains (heat transfer is not modified) = $6.9 \cdot 10^{-3}$, term C = $1.99 \cdot 10^{-3}$ and term D = $0.34 \cdot 10^{-3}$ t_{MD} than becomes 11.1 h. The drying time is not extended by a factor of 25/7 = 3.6, but by a factor of 4.4, due to the increase mainly in term C and slightly in term D.

If the material is granulated, e. g. frozen and granulated coffee extract, having a solid content perhaps of 40 % and a density of 0.6 g/cm^3 (Fig. 1.61), Eq. 12 is still applicable, but the product data are different. The heat transfer is through the dried product with a $\lambda_{tr} = 8.37 \cdot 10^{-2}$ kJ/m h °C. To make the results better comparable, d = 0.7 cm and T_{tot} have been retained from the earlier example, even if T_{tot} were normally higher, e. g. 100 °C.

$$t_{MD} = (\rho_g\ \zeta_w\ LS\ d)\ /\ T_{tot}\ (1/K_{tot} + d/2\ \lambda_{tr} + d/2\ LS\ b/\mu) \qquad (13)$$

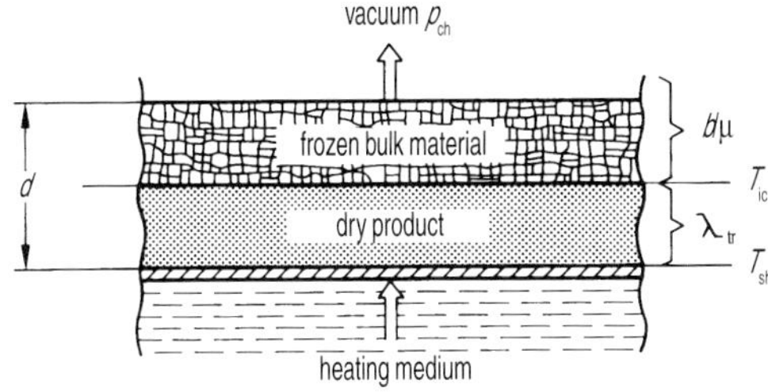

Fig. 1.61. Schema for the calculation of the drying time t_{MD} for the main drying of granulated product (Fig. 4 from [1.51]).

$t_{MD} = (0.6 \cdot 10^3\ 0.6\ 2805\ 7 \cdot 10^{-3}) / 51\ (6.9 \cdot 10^{-3} + 41.8 \cdot 10^{-3} + 0.096 \cdot 10^{-3})$

term A	term B	term C	term D
138.6	$6.9 \cdot 10^{-3}$	$41.8 \cdot 10^{-3}$	$0.096 \cdot 10^{-3}$

t_{MD} = approx. 6.8 h

In this example of the granulate, the main drying time (term A is assumed constant) depends largely on term C, while the vapor transport has virtually no influence.

λ_{tr} is given in [1.50] as 8.4 to $16.8 \cdot 10^{-2}$ kJ/m h °C, while Magnussen [1.55] uses for freeze dried beef at 0.4 mbar a value of $15.5 \cdot 10^{-2}$ and at 1.1 mbar a value of $17.2 \cdot 10^{-2}$. Sharon and Berk [1.54] have found for concentrated tomato pulp with 28 % solids, at 0.5 mbar $28.5 \cdot 10^{-2}$ kJ/m h °C and $31.8 \cdot 10^{-2}$ at 1 mbar. If the concentration of solids is only 6 %, the values were $12.6 \cdot 10^{-2}$ respectively $15.9 \cdot 10^{-2}$ kJ/m h °C. Steinbach uses $16.7 \cdot 10^{-2}$ and Gunn [1.56] has found 5.9 and $9.2 \cdot 10^{-2}$ kJ/m h °C for turkey meat at 0.5 mbar respectively 1 mbar.

With both the extreme values $5.9 \cdot 10^{-2}$ respectively $31.8 \cdot 10^{-2}$ kJ/m h °C the term C becomes $59.3 \cdot 10^{-3}$ or $11 \cdot 10^{-3}$ and $t_{MD\,(5.9)} \approx 9.1$ h, respectively $t_{MD\,(31.8)} \approx 2.5$ h. The heat conductivity in the product becomes the decisive value. It is a function of the chamber pressure, but changes in the interesting pressure range of 0.5 mbar to 1 mbar by only 15 %. However it varies with the solid content by a factor of 2 and is dependent on the structure. The λ_{tr} of turkey meat parallel to the fiber structure is three times larger than given above.

In Fig. 1.62–1.64, three runs of freeze drying in two different plants are shown. Figure 1.65 gives the scheme of the plant for the run in Fig. 1.62, and Fig. 1.66 the scheme of the runs plotted in Fig. 1.63 and 1.64. Table 1.9 summarizes the plant, the experimental data, and the relevant results. From these data, the values to calculate t_{MD} in Eq. (12) can be deducted as far as they are unknown.

Data for the three test runs:

ρ_g	density of frozen product	$0.9 \cdot 10^3$ kg/m^3
ζ_w	part of water	0.931 kg/kg
Δm	part of frozen water	0.9 kg/kg
LS	sublimation energy	2805 kJ/kg
λ_g	heat conductivity in the frozen product	6.3 kJ/m h °C

Table 1.9: Summary of conditions during test runs and their results derived from Figs. 1.62–1.64. $p_{H2O,\ ch}$ has been measured by a hygrometer during MD.

	Test run, Fig. 1.62	Test run, Fig. 1.63	Test run, Fig. 1.64
Used shelf area (m^2)	0.166	0.343	0.172
Heating of shelf	electrical resistance	circulated brine	circulated brine
T_{co} (°C)	down to –45	down to – 53	down to –53
Product	skimmed mild, 9.67 % solids		
Freezing method	freezer down to –30 °C	precooled shelves, –45 °C	
T_{ice} (°C)	–22.3	–27	–21
p_s of T_{ice} (mbar)	0.83	0.51	0.94
average p_{ch} during MD (mbar)	0.45	0.15	0.36
K_{tot} (kJ/m² h °C)	119.3	65.7	90.4
b/μ (kg/h m² mbar)	0.010	0.011	0.011
t_{MD} taken from run, approx. (h)*	3.5	5.0	3.0
t_{MD} calculated Eq. (14) (h)	3.0	4.3	3.4
pressure control (mbar)	No	No	0.36

* MD assumed terminated if T_{ice} is approx. 1–1.5 °C smaller as the average of all T_{ice} measured during MD. The amount of sublimated ice during MD is assumed to be 90 % of the freezable water.

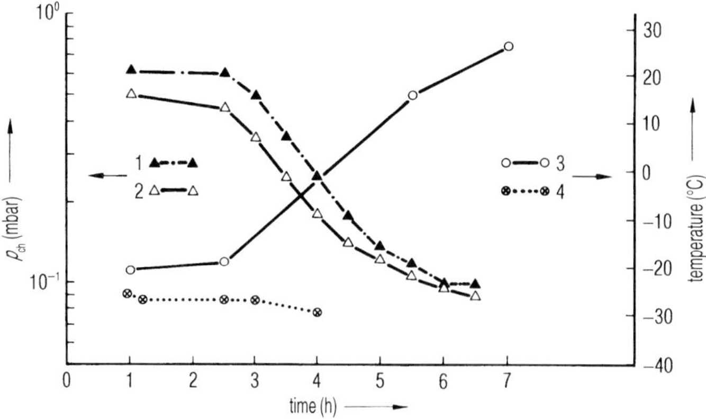

Fig. 1.62. Course of a freeze drying process in a plant, as shown in Fig. 1.65. Pressure control during MD is not activated. 1.0 kg product in four aluminum trays with machined bottoms, T_{sh} after evacuation controlled at +29 °C, d = 0,6 cm.

1, p_{ch} heat conductivity gauge (TM); 2, p_{ch} capacitive gauge (CA); 3, T_{Pr} resistance thermometer (RTD); 4, T_{ice} by barometric temperature measurement (BTM). End of MD 3.5 h.

Data for test run Fig. 1.62:

d	thickness	$6 \cdot 10^{-3}$ m
T_{tot}		38.6 °C
t_{MD}	time of main drying	3.5 h

Data for test run Fig. 1.63:

d	$7 \cdot 10^{-3}$ m
T_{ot}	56 °C
t_{MD}	5.0 h

Data for test run Fig. 1.64:

d	$7 \cdot 10^{-3}$ m
T_{tot}	51 °C
t_{MD}	3.0 h

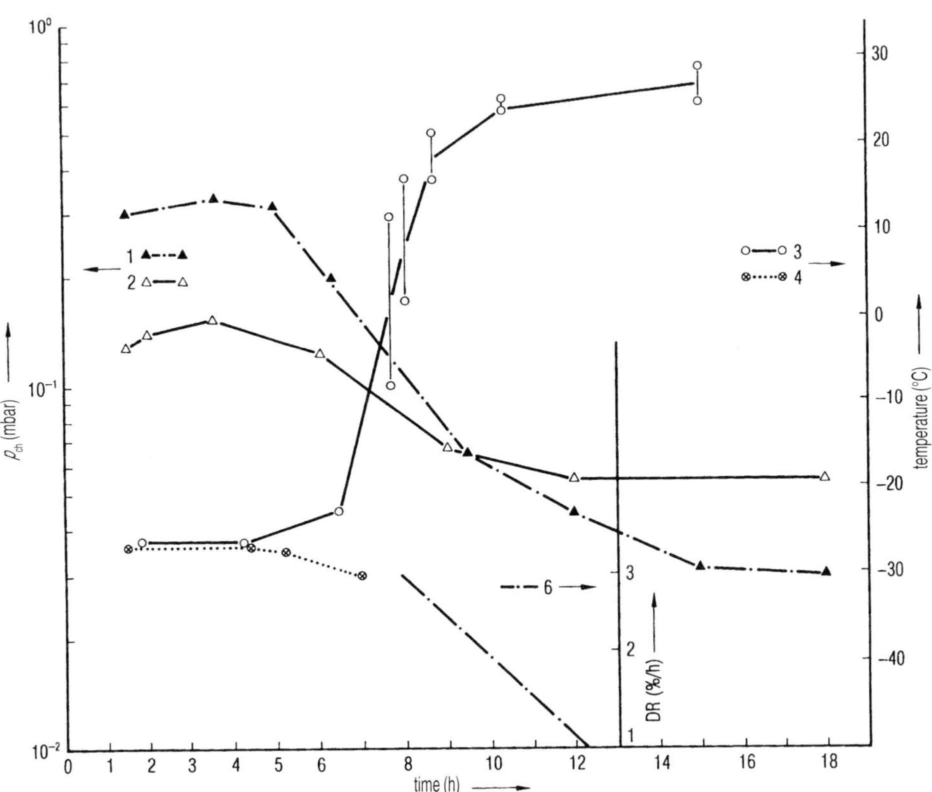

Fig. 1.63. Course of a freeze drying process in a plant, as shown in Fig. 1.66. Pressure control is not activated. 2.4 kg product in three welded, stainless steel trays with flattened bottom. T_{sh} after evacuation controlled at +29 °C; $d = 0{,}6$ cm.

1, 2, 3, 4, as in Fig. 1.62. 6, Desorption rate (DR), desorbebable water in % of solids (see Section 1.2.2). End of MD 5 h.

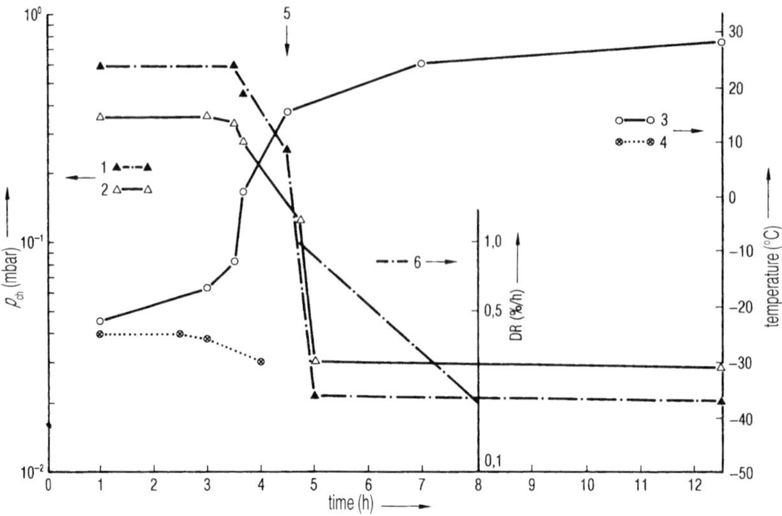

Fig. 1.64. Course of a freeze drying process in a plant, as shown in Fig. 1.66. Pressure control during MD, 0.36 mbar (CA). 0.678 kg of product in 404 vials. T_{sh} controlled up to +29 °C in such a way, that p_{ch} = 0.36 mbar have not been exceeded during MD. d = 0.64 cm.
1, 2, 3, 4, as in Fig. 1.62. 5, End of pressure control; 6, DR (see Section 1.2.2). End of MD 3.0 h.

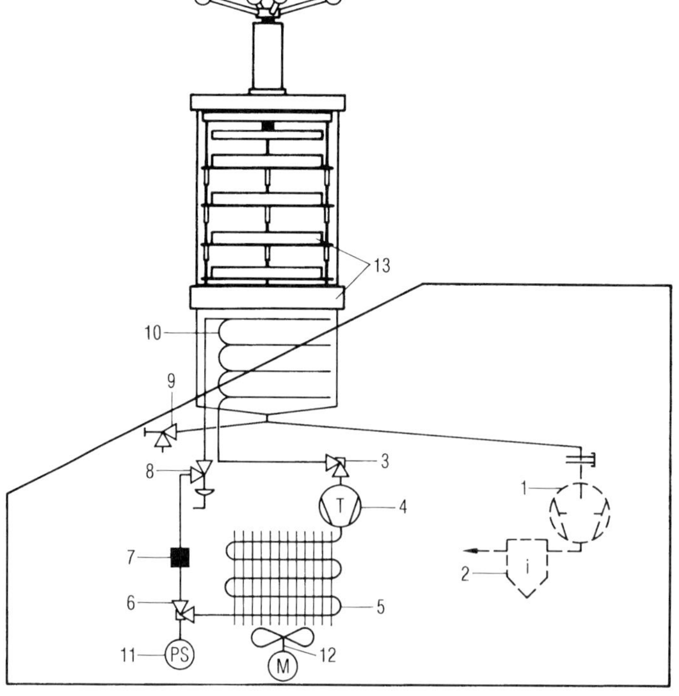

1, Vacuum pump; 2, exhaust filter; 3, valve; 4, refrigeration machine; 5, liquifier; 6, valve; 7, filter; 8, expansion valve; 9, drain valve; 10, ice condenser; 11, pressure switch; 12, ventilator; 13, drying chamber with heated shelves (Lyovac® GT 2, AMSCO Finn-Aqua, D-50354 Hürth, Germany).

Fig. 1.65. Schema of the freeze drying plant in which the tests shown in Fig. 1.62 have been carried out.

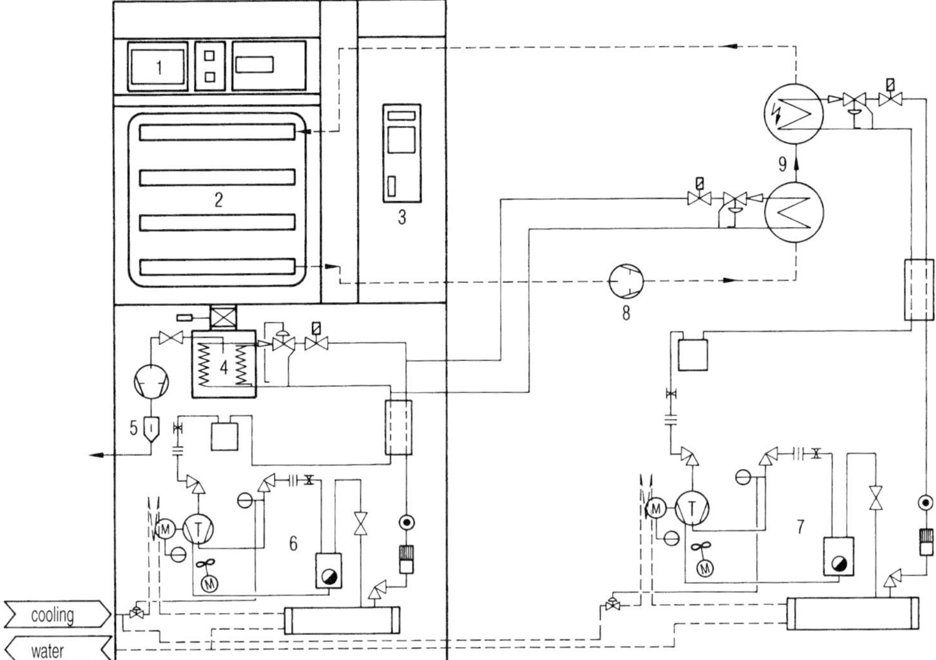

Fig. 1.66. Schema of the freeze drying plant, in which the runs of Fig. 1.63 and 1.64 have been carried out.
1, Process documentation; 2, drying chamber with shelves; 3, operation control; 4, ice condenser; 5, vacuum pump with exhaust filter; 6, refrigeration machine for the ice condenser; 7, refrigeration machine for the shelves; 8, circulation pump for the brine; 9, heat exchanger (Lyovac® GT 6, AMSCO Finn-Aqua, D-50354 Hürth, Germany).

With the help of these data K_{tot} and b/μ can be calculated, if the water pressure at the sublimation front (p_s) and the partial vapor pressure in the chamber, measured by a hygrometer, is taken from the respective curves.

The results are summarized in Table 1.9: K_{tot} values show the dependence of this data (see results in Table 1.10.1), it rises from ≈ 62.4 kJ/m h °C at 0.15 mbar to ≈ 119.3 kJ/m h °C at 0.45 mbar. The accuracy of the measurements to determine K_{tot} can be estimated in the two runs 1.62 and 1.62/W (Table 1.10.2) as approx. ±5 %.

The permeability (kg/m h mbar) for water vapor through the dried product fluctuates by a larger margin, which can be estimated from all six test runs (Table 1.9 and 1.10.1) as $1.1 \cdot 10^{-2} \pm 25$ %. However, the measurement of this data is of interest to judge whether the b/μ will influence the process time, as can be the case with products which have a high solid content and are dried with a large thickness. Small b/μ can also result from a skin on the surface of the product (see [1.132]).

Table 1.10.1: Comparison of four test runs in the same plant (type as in Fig. 1.65) with the same product and the same product thickness as in the run of Fig. 1.62 in Table 1.9.

	Test run Fig. 1.62	Test run Fig. 1.62/W	Test run Fig. 1.62/ 0.36 mbar	Test run Fig. 1.69/ 0.20 mbar
T_{ice} (°C)	–22.3	–22.5	–26.8	–30.5
p_s (mbar)	0.83	0.81	0.53	0.36
p_{ch} (mbar)	0.45	0.50	0.36	0.21
p_{H2O} (mbar)	0.31	0.31	0.23	0.19
K_{tot} (kJ/m² h °C)	119.3	114.5	79.1	62.4
b/μ(Kg/m² h mbar)	0.01	0.009	0.011	0.014
$t_{MD,}$ taken from run (h)*	3.5	3.5	4.5	6.0
t_{MD} calculated Eq. (14)	3.0	3.1	4.4	5.8
$T_{Sh} - T_{ice}$ (°C)	38.6	39.1	38.1	36.3

Test run Fig. 1.62/W: best possible repeat of test run Fig. 1.62, both runs not pressure controlled.
Test run Fig. 1.62/0,36 mbar and 1.62/0.20 mbar differ from 1.62 and 1.62/W by pressure control 0.36 mbar respectively 0.21 mbar

The results show:

1. The deviation between the two repeated runs is < 5 %. (The higher p_{ch} could indicate a larger content of permanent gases in the product).
2. T_{ice} can be controlled by pressure in a temperature range of 8–10 °C (with otherwise equal conditions).
3. K_{tot} decreases with decreasing p_{ch} (from 0.5 mbar to 0.21 mbar) by approx. 50 %.
4. t_{MD} decreases with increasing pressure to approx. 50 %, since the decisive data in Eq. (14) is K_{tot}. $T_{sh} - T_{ice}$ changes only a little, the control avoids exceeding the controlled pressure. T_{sh} increased a little more slowly; this means that during MD $T_{sh} - T_{ice}$ is smaller than without presser control.
5. A pressure control does not always shorten the MD, as can be read frequently. The pressure control adjusts a desired T_{ice}. If T_{ice} without pressure control is larger than with pressure control (as in this example) MD, would only decrease when the run would have been operated at 0.7 mbar (if the product tolerates the increased T_{ice}.

Table 1.10.2: Freezing and subcooling

	5 % mannitol (6R – design)		10 % sucrose (10 R – design)	
	Freezing rate °C/min (+10 °C/–30 °C)	subcooling °C	Freezing rate (+10 °C/–30 °C)	subcooling °C/min °C
s-vial	0.92	–10	1.06	–7
qc-vial	1.07	–7	1.11	–8
p-vial*	0.79	–10	0.67	–12

s-vial: tubing glass vials, qc-vial: quartz-coated tubing glass vials,
p-vial *: resin vials, 6 R-design

Schellenz et al. [1.133] confirmed that the assumption of an infinite plate in Eq. (12) is a reasonable approximation, even for drying of products in vials. They show by the measurement of temperature profiles and by X-ray photos during drying of a 5 % mannitol solution, 23 mm filling height, that the sublimation front retreats mostly from the top parallel to the bottom. The heat transfer from glass vials deforms the flat surface only to some extent close to the wall.

Drummond and Day [1.134] have studied the influence of different vials: molded glass, glass tubing and molded resin on the freeze drying of 5 % solutions of maltose and mannitol with arginine added. The freeze drying performances, the inter- and intravial uniformity and the morphology have been compared. The lyophilization performance was best for glass tubing vials, with molded vials only marginal lower but better than resin vials. Intravial uniformity was found to be best for the resin vials. However during drying this depends on the cycle parameters used. Intervial uniformity differences were measured both by the time at which nucleation of ice occurred, and by temperatures during the drying process, but the differences for the three type of vial were in same range. The morphology of mannitol in resin vials was found to be similar to the morphology in both types of glass vials. The authors concluded, that the temperature distribution in a vial indicates a greater degree of uniformity in the cake, the temperatures from vial to vial are more consistent with resin vials, and the morphology in resin, molded and tubing vials was not significantly different.

Willemer et al. [1.135] have studied the influence of tubing glass, coated tubing glass and resin as vial material on the freezing and freeze drying of 1 % and 5 % mannitol and 10 % sucrose solution. During freezing, the different forces between the walls and liquid influence the structure of the freezing product and its subcooling, as shown in Table 1.10.2. The freezing speed in the coated vials was up to 16 % greater than in standard vials, but in the resin vials the freezing speed was 14 % lower. As shown in Table 1.10.3, the main drying is ≈ 20 % longer in resin vials than in glass vials. The weight loss in the quartz-coated vials during sublimation is faster than in the other vials; also the standard deviation of weight loss in the quartz-coated vials is by far the lowest, indicating high inter-container homogeneity of the product. The secondary drying time is almost the same for glass- and polymer-vials. and partially reduces the difference for the total drying time.

The discussions so far about main drying have assumed that trays or vials are exposed to uniform temperatures on the shelves. Kobayshi [1.57] has shown that this condition does

Table 1.10.3: Weight loss during main drying of a 5 % mannitol solution in R6- vials.

Drying time (h)	3.5	4.5	6.0	6.5	3.5	4.5	6.0	6.5
	Weight loss (% of initial weight)				Standard deviation of 12 vials			
s-vial	91.85	94.40	94.81		2.89	0.8	0.7	
qc-vials	90.24	95.05	94.47		1.59	0.35	0.63	
p-vials	71.87	87.68		95.22	2.53	3.0		0.41

not exist in some freeze drying plant, as the walls and doors of chambers can have different temperatures from the shelves. During freezing, this could lead to a slower freezing at the edges of the shelves.

During main drying, the influence of the wall temperatures can be small, as long as the shelves are only heated up to +15 °C or +25 °C. If the shelf temperature during main drying is, e. g. –10 °C the vials at the edge will receive more energy than planned, which could lead to collapse or melting of the product in these vials. In contrast during secondary drying the vials at the edge will be warmed up more slowly than those in the center if the shelf temperature is e. g. +35 °C. Kobayashi proposes that all walls and the door are operated at the same temperatures as the shelves, though this is not always necessary if the shelves are shielded from the walls and the door, as shown schematically in Fig. 1.67. The curve in Fig. 1.68 indicates that this shielding is sufficient in most cases: during freezing the vials at the edges of the shelves are exposed practically to the shelf temperature, while during secondary drying, the influences of the walls and the door are reduced.

Wall and door influences can exist mainly by radiation or by a small heat conductivity of the gas. It can be seen from Fig. 1.68 that the shielding in the temperature range between –40 °C and 0 °C is effective. However the shielding becomes more important with an increasing temperature difference between the shelves and surrounding and It is especially necessary if the vials contain only a small amount of product.

In Table 1.9 two different plants and three different types of container have been used, in Table 1.10.1 it is always the same plant. During main drying (MD) different pressures have been applied.

The control of a desired constant total pressure is called pressure control (PC), and can be operated by two methods:

- A dry inert gas, e. g. nitrogen is fed into the chamber by a needle-valve in such a way that the desired total pressure is built up.

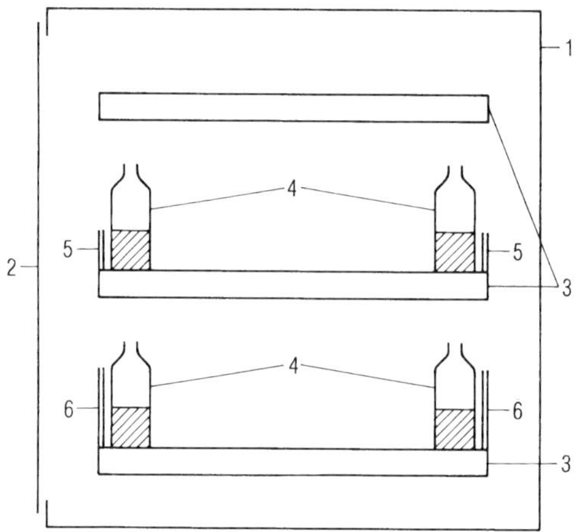

Fig. 1.67. Schema related to the plots of Fig. 1.68.
1, Chamber wall; 2, chamber door; 3, shelves; 4, vials with product; 5, radiation shield, height ≥ filling level of the vials; or 6, radiation shield, height ≈ cylinder length of the vials.

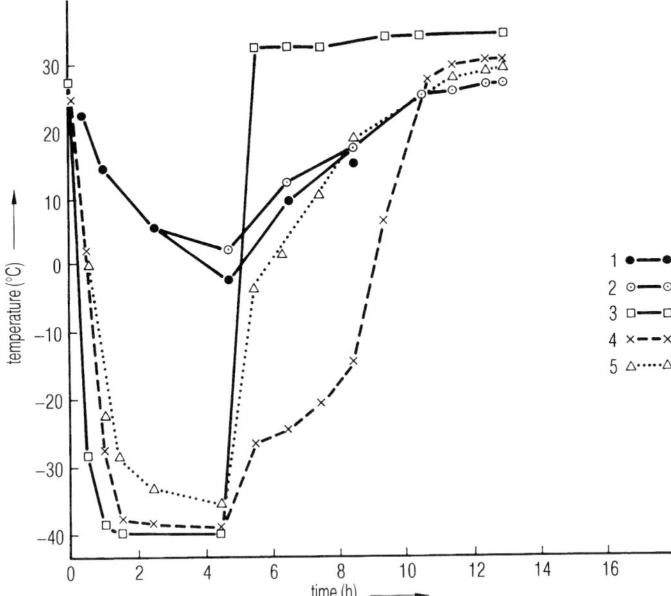

Fig. 1.68. Temperature distribution during freezing (4.5 h) and freeze drying (8 h) Each shelf carries a frame for radiation shielding. The design can be different, depending on the kind of vials and the loading and unloading system.
1, Internal temperature of chamber wall; 2, internal temperature of door; 3, $T_{sh;}$ 4, T_{pr}; 5, temperature on inner side of the radiation shield.

- The valve between condenser and vacuum pump set (Fig. 1.65 and Fig. 1.66) is closed until the desired total pressure is built up. If the controlled pressure is exceeded, the valve opens.

In the second case, the gas included in the product is used and is released during sublimation of the ice. In every liquid product, some gas is dissolved. Indeed, liquids [1.58] may contain gas contents from $5 \cdot 10^{-5}$ kg/kg up to $1 \cdot 10^{-3}$ kg/kg, often very close to the upper value, though the actual content depends very much on the history of the product.

The test run in Fig. 1.64 contained 1.535 kg product having a water content of 1.39 kg and containing 0.7 g air. The total pressure of 0.37 mbar, including 0.25 mbar water vapor pressure, was to be controlled. The air had to be pumped off at a partial pressure of air of 0.12 mbar; 0.7 g air at 0 °C and 0.12 mbar represent a volume of approx. 7.3 m^3. During the main drying time of 3.0 h, 2.4 m^3/h must be removed. If the vacuum pump has an effective pumping speed for air of e. g. 4.8 m^3/h the pump would operate in average of 50 % of MD. If the dissolved amount of gas is 10 fold larger this must to be considered in the lay out of the vacuum pump.

The second method has two advantages:

- only gas of the product is used, and no additional inert gas supply is needed;
- at the end of MD, when less and less dissolved gas is freed, the chamber pressure drops automatically, as it is necessary for secondary drying.

The advantage of pressure control is the improved heat transfer leading to shorter drying times, or possibly lower shelf temperatures. On the other hand, this is equally important, the ice temperature can be accurately controlled by the controlled pressure: in Table 1.10.1, the ice temperature at the sublimation front is –26,8 °C at a pressure of 0.36 mbar and 0.21 mbar, T_{ice} = –30,5 °C (see column 3 or 4). In the test run in Fig. 1.64, 0.36 mbar results in T_{ice} –21 °C, since K_{tot} is larger between vials and shelves at the same shelf temperature of +29 °C. Should the total pressure exceed the desired value, the shelf temperature must be reduced accordingly. Fig. 1.69 illustrates how I_{ce} is reduced in 4 hours from –26,8 °C to –28 °C. The shelf temperature should have been raised slightly, or the controlled pressure of 0.36 mbar increased. The change in shelf temperature is usually much too slow for such small changes, however the change in pressure is quick and can easily be performed in small steps. The method of pressure control can only be applied as long as the partial vapor pressure is large compared with the pressure of permanent gases. If the pressure of permanent gases is of the same magnitude, or larger as the vapor pressure, the water vapor transport is hindered and the ice temperature is no longer a well-defined function of the control pressure.

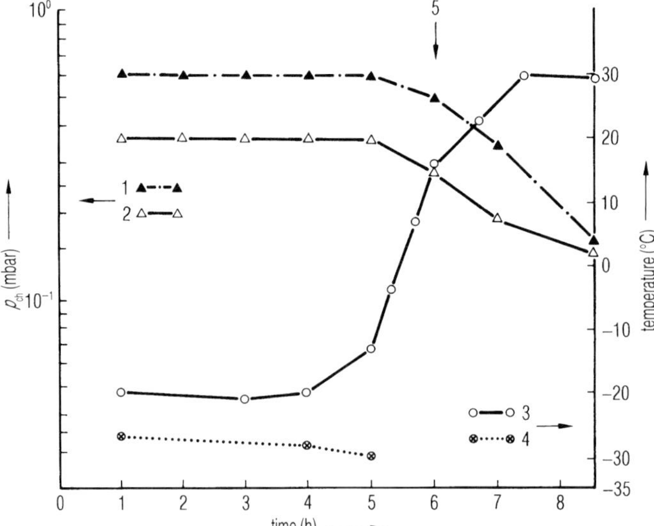

Fig. 1.69. Course of a freeze drying process during which T_{ice} has not been kept constant at –26.8 °C. To avoid the declining temperature, either T_{sh} could have been increased after 2 h (difficult to control, the inertia of the heating system is substantial) or p_c increased until T_{ice} is constant at –26,8 °C. 1, p_{ch} (TM); 2, p_{ch} (CA); 3, T_{pr} (RTD); 4, T_{ice} (BTM); 5, end of pressure control.

1.2.2 Secondary Drying (Desorption Drying)

During the secondary drying, the water will be removed which interacts with the solids such that the water cannot crystallize. The water can be bound to the surface of crystals in a crystallized product, or can be included in amorphous product. Pikal et al. [1.59] have listed four statements concerning secondary drying:

1. At the beginning of the secondary drying, the decrease in water content is rapid, but thereafter remains on a plateau-level.
2. This plateau moves with increasing temperature in the direction of smaller water contents.
3. The drying speed increases with increasing specific surface of the product.
4. A change of the chamber during secondary drying to less than 0.26 mbar, e. g. to 0.065 mbar, does not change the drying speed.*

* Point 4 above may be correct, but only if the partial water vapor pressure of the product at the product temperature is larger than the existing partial water vapor pressure in the chamber, otherwise the chamber pressure has to be lowered. Since the energy input during secondary drying is not decisive, it would be safer to use a chamber pressure as small as the condenser temperature allows.

Pikal [1.60] described two possibilities to define the change from main to secondary drying:

- the increase in product temperature
- a decrease of the partial water vapor pressure.

As shown in Fig. 1.62 and 1.63, the product temperature increases at the end of main drying, the measurements by temperature sensors vary widely and are therefore a relatively uncertain indicator for the end of the main drying. The partial water vapor pressure changes during the transition from main drying to secondary drying over a period of several hours depending on the process conditions (e. g. 2–3 hours in Fig. 1.62 and 1.63). In practice one may have to several hours before the higher temperature for the secondary drying can be applied in order to avoid partial collapse. It is well known that pressure rise measurements for a given time can be used to determine the change-over. The method can be applied more generally if the amount of water desorbed per time is measured and related to the solid content. This can be defined as desorption rate (DR):

$$\mathrm{DR} = \frac{\text{amount of water desorbed} \cdot 100}{\text{time} \cdot \text{mass of solid}} \ (\%/\mathrm{h}) \tag{14}$$

The amount of water desorbed can be calculated by the pressure rise after the valve between the chamber and condenser has been closed, divided by the time of closure and the chamber volume (see Section 1.2.3).

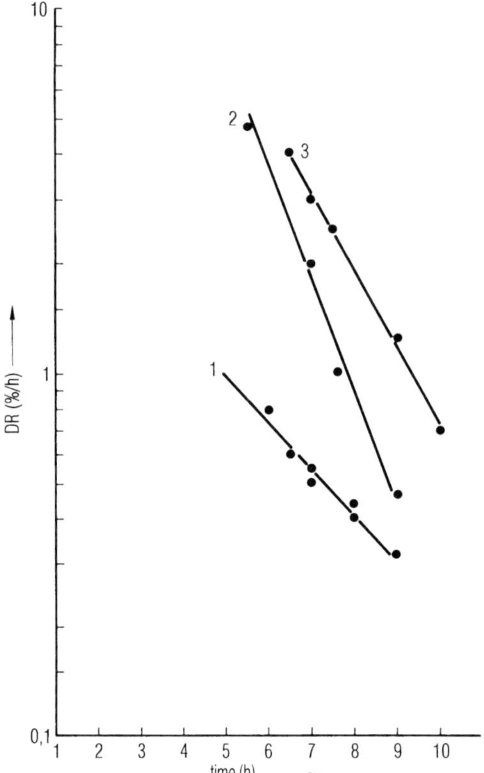

Fig. 1.70. Desorption rate (DR) data as a function of drying time.
1, Three repeated tests, pressure control not activated, process data as in the two left columns in Table 1.10. 2, process data as in Table 1.10, pressure control activated at 0.36 mbar; 3, process data as in Table 1.10, pressure control activated at 0.21 mbar.

Figure 1.70 shows the three times repeated measurement of desorption rates, without pressure control to demonstrate the reproducibility, and two measurements where the main drying has been pressure-controlled at 0.36 and 0.21 mbar. The process conditions for these five measurements correspond with those in Table 1.10.1.

By barometric temperature measurements (BTM) and the measurements of the desorption rate (DR) the influence of varied drying conditions can be seen and analyzed. Figure 1.71 compares four different test runs:

1. Test run (see Fig. 1.63 and Table 1.9): Without pressure control, in this installation – with the given shelf area, condenser temperature, the dimensions of the connection between chamber and condenser – a total pressure of 0.15 mbar exists for approx. 5 h. The gas in the chamber is always pure water vapor. The ice temperature is during this time almost constant at approx. –27 °C. The heat transfer coefficient at this pressure is small at approx. 65.7 kJ/m^2 h °C. The product temperature (resistance thermometer) increases only after these 5 h above the ice temperature. After12 h the desorption rate (DR) is only approx. 0.7 %/h. The total drying time, depending on the desired residual moisture content is between 13 and 15 h.

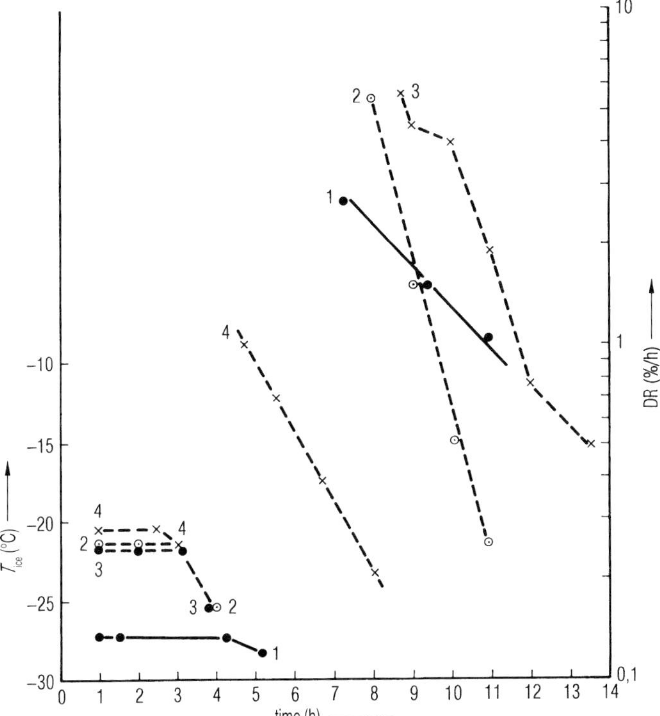

Fig. 1.71. Synopsis of T_{ice} and desorption rates (DR) of the two tests in Fig. 1.63 (1) and Fig. 1.64 (4) and comparison with two other tests: (2) carried out as (1) but with activated pressure control at 0.36 mbar and (3): only one tray used (instead of three trays in Fig 1.63), which has been placed at such a slope that the thickness of the product has been 0.5 cm at one side and 0.9 cm at the other. The course of the pressure (see Figs. 1.63 and 1.64) permits quantitative judgment of the SD. The DR data measure, independent of the chosen process data, the amount of desorbed water per hour in % of the solid content. It is visible, that a DR value of 5 %/h in test (4) is reached in 6.2 h, in test (2) in 10.2 h, in test (1) in 13.5 h, but in test (3) the time cannot be estimated. Because of the unequal product thickness, the DR values can change (9.5 h), the desorption process is not uniform for such a product.

2. Test run: the curves 2 in Fig. 1.71 are taken from the test run, as shown in Fig. 1.63, but with pressure control 0.36 mbar (total pressure measured with Capacitron). The ice temperature has been –22 °C (constant) for 3 h and DR reached 0.05 %/h after 10 h. Secondary drying could have been started much more early, thus shortening the drying process.

3. Test run: The results of this run are only shown in Fig. 1.71 in curves 3. In this run the shelf with the tray was inclined in such a way that uniform thickness of 7 mm was changed from 5–9 mm. Otherwise, the conditions equaled those of the second test (Fig.

1.64). The ice temperature during main drying was very similar but the DR-value of 5.5 %/h at 9.5 h shows the variation of thickness of the layer. DR of 0.5 %/h has been reached not in 10 h, but in 13 h. The test also showed (not in the figure) that the product temperatures (T_{pr}) varied at 9 h from 0 °C to +22 °C.

4. Test run (see Fig. 1.64 and Table 1.9): The analysis of the run shows that the relatively high heat transfer coefficient of 90.4 kJ/m^2 h °C at a controlled pressure of 0.36 mbar resulted in a constant ice temperature –22 °C for 2.5 h. Secondary drying has been started after 3.5 h and DR of 0.5 %/h is reached at approx. 6 h.

The measuring of desorption rates can be used, as the above examples show, to determine the amount of desorbable water if the following prerequisites are fulfilled:

- the product shows a reproducible desorption isotherm, meaning that it is not measurably changed at the applied temperature.
- the final temperature must be applied for some time, depending on the layer thickness, in order to reduce the temperature gradient in the product.

Pikal et al. [1.59] showed that a small water content below a certain value changes little at a given time. The DR-values in a semi-logarithmic scale plotted as a function of time (Fig. 1.72) can be straight lines as long as the temperature of the product is approx. constant and the DR-values not less than as 0.1 %/h (sometimes 0.05 %/h). In Fig. 1.72 the change in

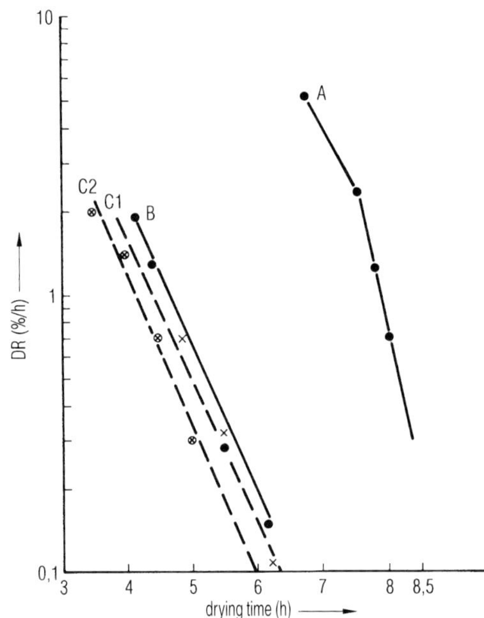

Fig. 1.72. Desorption rates (DR) as a function of time for three different foodstuffs, A, B and C. C1 and C2 differ by the product temperature +42 °C respectively +47 °C. Product A contains approx. 85 % water, while B and C contain only approx. 30–40 % water. The slope change in A at 7.5 h indicates that the final product temperature has been reached around that time. The plots are based on measurements by Dr. Otto Suwelack, D-48723 Billerbeck.

inclination in plot A at 7.5 h indicates that the final temperature was only reached at approx. 7.5 h. After a drying of approx. 8 h at a temperature of approx. 50 °C, product A showed a desorption rate of 0.7 %/h, and after 8.5 h, a DR of approx. 0.3 %/h.

Product B has, at 42.5 °C shortly before 6 h, a desorption rate of 0.25 %/h and at 6.3 h a DR of 0.14 %/h. The desorption rate of product C1 at 42 °C has reached at 6.3 h, while at 47 °C (C2) already after 5.8 h only 0.1 %/h can be desorbed.

By integrating the DR-values it is possible to calculate the residual moisture content (RM). The integral is calculated from the last measurement of DR over time up to any other measured DR. The integral is RM at the time up to which the integral has been calculated. The RM calculated in this way is too small by the amount of water which would have been desorbed after the measured DR. Thus a method of calculation can be deduced: The straight line of the DR values is extrapolated until the still desorbable RM is small compared with the RM to be measured.

Example: By integrating the DR value for product C1 from 0.1 %/h up to 2 %/h, the line C1 in Fig. 1.73 is obtained; at 3,6 h the RM was 2.5 %. The RM at 3.6 h is too small by the amount of water which would have been desorbed after 6.2 h. Between 6.2 and 7 h, 0.08 %/h would have been desorbed. The RM 3.6 h would not have been 2.5 % but 2.56 %. This example shows that it is always possible to extrapolate the desorption rates, as long as the error by the integration can be made small compared with RM to be calculated (see also discussion of Eqs. (18 a), (18 b) and (20)).

It should be clear that the RM measured in this way, e. g. 0,1 %, must not be identical with residual moisture contents measured with other methods (see Section 1.3.1) because there will be always some water which cannot be desorbed at the end temperature of the drying. This content of bound water for one product and one temperature is a stable value which can be taken from the measurements of absorption isotherms.

The residual moisture content measured by desorption is therefore called desorbable water (dW), and it indicates how much water can still be desorbed at that temperature – or put other way – how much water could be desorbed by further drying, e. g. a product having dW = 0.5 % can only be further dried by this 0.5 %. This is of interest for such products in which the water content should not be lower than predetermined value. Pikal [1.60] missed the exact proof, that overdrying (removal of a certain amount of bound water) is detrimental to the product. Hsu et al. [1.61] have shown that freeze drying of tissue-type plasminogen activator (tPA) below 7.6 % RM denatures the product, as 7.6 % RM corresponds to a mono-layer of water molecules on the tPA molecule. However, the dried product having 7.6 % RM at a temperature of 50 °C during 50-day storage loses more of its activity, than a product with a lower water content. Hsu recommended examining to look at an optimum water content which cannot be reached on the basis 'the lowest residual moisture is the best'.

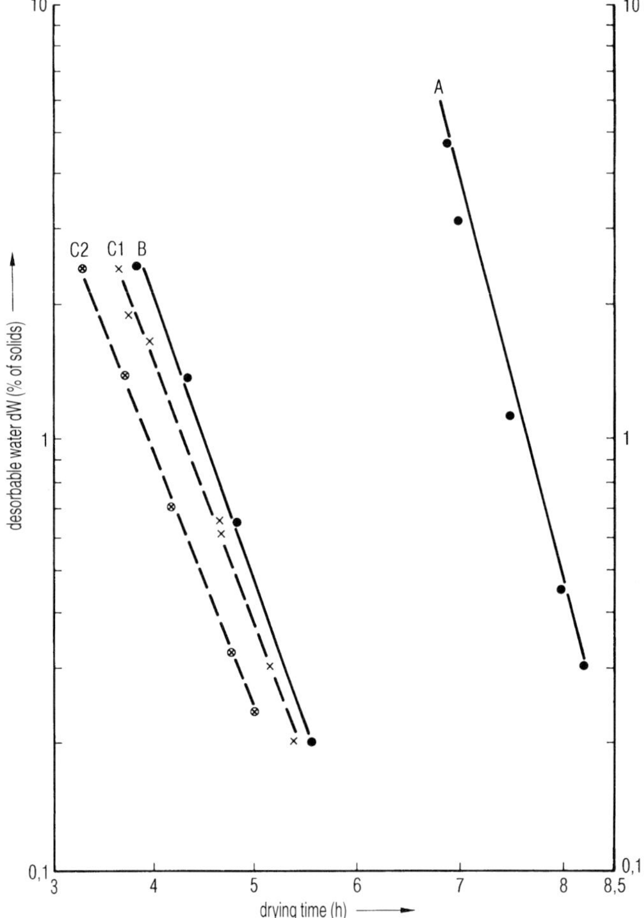

Fig. 1.73. Desorbable water in % of solids (dW) as function of the drying time. The dW-values have been calculated from the data of Fig. 1.72. In plot A, after 7.5 h only 1 % (of solids) water can be removed by further drying at this temperature. If e. g. 0.3 % are required, the drying can be terminated at 8.3 h.

1.2.3 Temperature and Pressure Measurement

Temperature and pressure measurements during freeze drying are recognized as difficult tasks. Thermal elements (Th) and temperature-depended electrical resistance (RTD) systems measure only their own temperature and that of its surroundings only if they are in very close contact with it. Furthermore they are heating themselves and their surroundings by current flow through the sensors. Also they influence the crystallization of the product in their surrounding:

- by the energy they produce;
- by inducing heterogeneous crystallization which can be different in the product without sensors (Nail and Johnson [1.62]) and

- by different subcooling, which can be smaller around the sensor and result in a more coarse structure.

These structure changes and the heat input by the sensors also influences the main drying of vials with sensors. In addition to these problems of principle, there are also practical ones: Ths and RTDs have to be inserted into the product and connected with vacuum-tight leadthroughs to the measuring system. During freezing, the type of sensors used can have the influence shown in Fig. 1.74 [1.63]. The position of the sensors during freezing has a limited influence [1.64] as shown in Fig. 1.75. During freezing, temperature sensors provide a reasonably accurate temperature picture, even if the product with sensors reacts differently from the way it does without sensors.

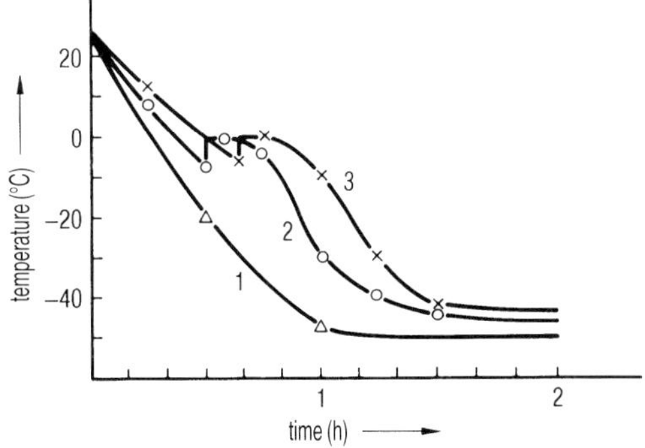

Fig. 1.74. Temperature in the product as a function of freezing time, measured by one RTD and one Th each in three vials. The vials had been distributed diagonally on one shelf. (The differences between the three vials are within the accuracy of the drawing).
1, T_{sh}; 2, T_{pr} measured by RTD; 3, T_{pr} measured by Th (Fig. 3 from [1.63]).

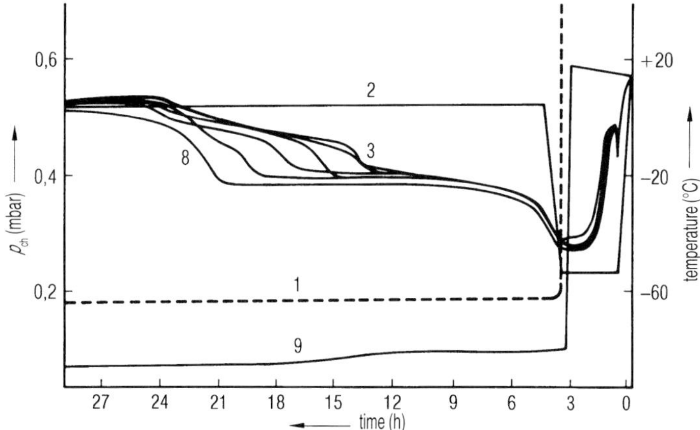

Fig. 1.75. Temperature and pressure as a function of the process time. During freezing, the data measured by six temperature sensors are reasonably close together. The split up after 12 h shows that the progress of MD has reached a level at which the different locations of the sensors becomes relevant.
1, p_{ch}; 2, T_{sh}; 3–8 temperature sensors; 9, T_{co} (Fig. 3 from [1.64]).

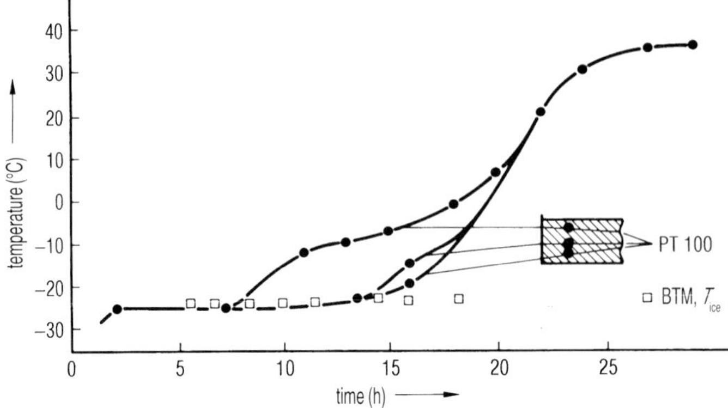

Fig. 1.76. Product temperature measured in three different locations in the product as function of the drying time. T_{ice} is measured simultaneously by BTM (Fig. 3 from [1.107]).

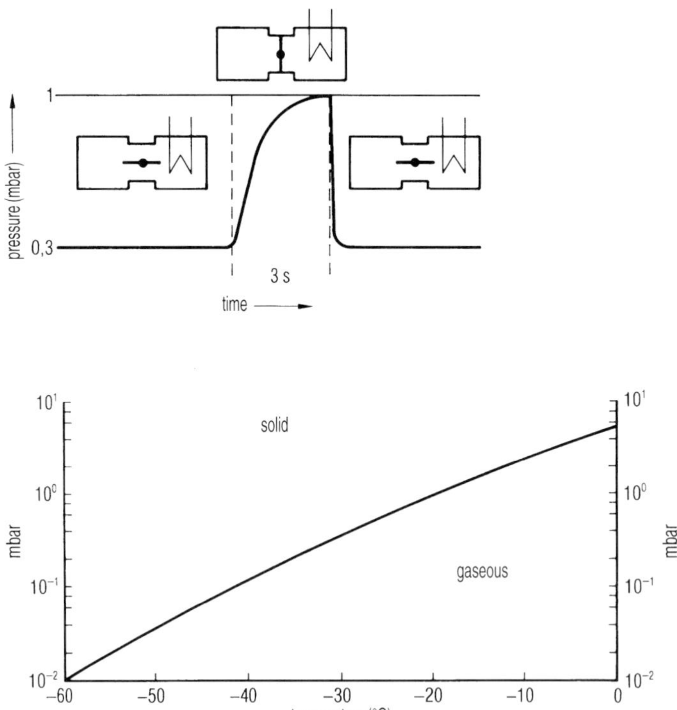

Fig. 1.77. Schema of the 'barometric temperature measurement' (BTM) and plot of the water vapor pressure of ice (Fig. 4 from [1.108]).

During main drying, the situation is very different: the condition of a close contact with the product is only true on the beginning of MD; thereafter the measured temperature depends on circumstances which are difficult to analyze. The position of the sensor on top, in the center, or near the bottom contact with the vial wall decide the measured data, as shown in Fig. 1.76. If the filling volume of vials is small (a few millimeter layer thickness), or if the product is granular, it is a specially difficult to attain useful data. Also, in homogeneous layers with a thickness of 6–10 mm, temperature differences during main drying can be 10–20 °C when measured with three RTDs, as shown in Fig. 1.63. Such differences can also be found between two vials in the same charge during main drying.

The most important data during main drying is the temperature at the moving sublimation front which cannot be measured by Ths or RTDs. In 1958, Neumann and Oetjen [1.65] showed that the barometric temperature measurement (BTM) measures exactly this data. In Fig. 1.77 this is schematically shown: if the drying chamber is separated from the condenser by a valve for a short time the pressure in the chamber rises to the saturation vapor pressure (p_s) corresponding to the temperature of the sublimation front. p_s can be converted into the ice temperature by the water vapor- temperature diagram (e. g. 0.3 mbar = –30 °C). Data for accurate conversion are given in Table 1.11 the temperatures between –100 and –1 °C.

Milton et al. [1.136] used this methods and refer to it as manometric temperature measurement. They used times of pressure rises of up to 30 s. During this time, the ice temperature will increase, mainly due to continued heat flow. Therefore, an equation has been developed to transform the experimental pressure data, including three other corrections, into the true vapor pressure of the ice. If the valve is closed for only a very short time, < 3 s, and the pressure is measured and documented 60 to 100 times/s, these data can be recorded as shown in Fig. 1.78.1. The automatic pressure rise measurements (1) can then be plotted

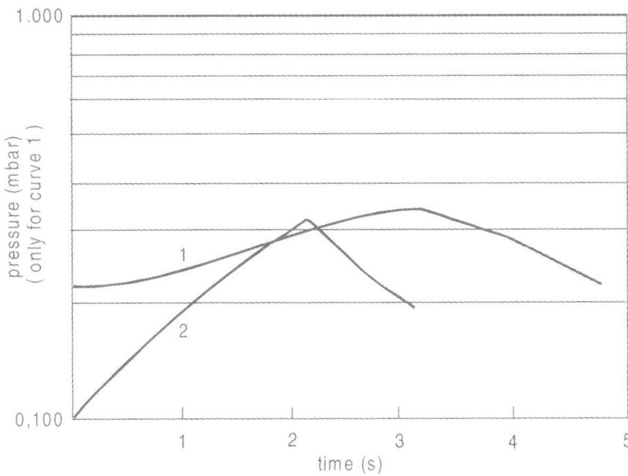

Fig. 1.78.1. Pressure rise as a function of time.
1, Pressure rise in the chamber after the valve is closed; 2, first derivation of 1. The maximum of 2 is reached at 2.14 s, the related equilibrium vapor pressure is p_s = 0.286 mbar corresponding to T_{ice} = –32.7 °C (Fig. 2 from [1.138]).

Table 1.11: Equilibrium water vapor pressure of ice and the related specific density of the vapor (from [1.109]).

t °C	p_s mbar	ρ_D g/m^3	t °C	p_s mbar	ρ_D g/m^3
–100	$1.403 \cdot 10^{-5}$	$1.756 \cdot 10^{-5}$	– 50	39.35	38.21
– 99	1.719	2.139	– 49	44.49	43.01
– 98	2.101	2.599	– 48	50.26	48.37
– 97	2.561	3.150	– 47	56.71	54.33
– 96	3.117	3.812	– 46	63.93	60.98
– 95	3.784	4.602	– 45	71.98	68.36
– 94	4.584	5.544	– 44	80.97	76.56
– 93	5.542	6.665	– 43	90.98	85.65
– 92	6.685	7.996	– 42	102.1	95.70
– 91	8.049	9.574	– 41	$114.5 \cdot 10^{-3}$	106.9
– 90	9.672	11.44	– 40	0.1283	0.1192
– 89	11.60	13.65	– 39	0.1436	0.1329
– 88	13.88	16.24	– 38	0.1606	0.1480
– 87	16.58	19.30	– 37	0.1794	0.1646
– 86	19.77	22.89	– 36	0.2002	0.1829
– 85	23.53	27.10	– 35	0.2233	0.2032
– 84	27.96	32.03	– 34	0.2488	0.2254
– 83	33.16	37.78	– 33	0.2769	0.2498
– 82	39.25	44.49	– 32	0.3079	0.2767
– 81	46.38	52.30	– 31	0.3421	0.3061
– 80	$0.5473 \cdot 10^{-3}$	$0.6138 \cdot 10^{-3}$	– 30	0.3798	0.3385
– 79	0.6444	0.7191	– 29	0.4213	0.3739
– 78	0.7577	0.8413	– 28	0.4669	0.4127
– 77	0.8894	0.9824	– 27	0.5170	0.4551
– 76	1.042	1.145	– 26	0.5720	0.5015
– 75	1.220	1.334	– 25	0.6323	0.5521
– 74	1.425	1.550	– 24	0.6985	0.6075
– 73	1.662	1.799	– 23	0.7709	0.6678
– 72	1.936	2.085	– 22	0.8502	0.7336
– 71	2.252	2.414	– 21	0.9370	0.8053
– 70	$2.615 \cdot 10^{-3}$	$2.789 \cdot 10^{-3}$	– 20	1.032	0.8835
– 69	3.032	3.218	– 19	1.135	0.9678
– 68	3.511	3.708	– 18	1.248	1.060
– 67	4.060	4.267	– 17	1.371	1.160
– 66	4.688	4.903	– 16	1.506	1.269
– 65	5.406	5.627	– 15	1.652	1.387
– 64	6.225	6.449	– 14	1.811	1.515
– 63	7.159	7.381	– 13	1.984	1.653
– 62	8.223	8.438	– 12	2.172	1.803
– 61	9.432	9.633	– 11	2.376	1.964
– 60	$10.80 \cdot 10^{-3}$	10.98	– 10	2.597	2.139
– 59	12.36	12.51	– 9	2.837	2.328
– 58	14.13	14.23	– 8	3.097	2.532
– 57	16.12	16.16	– 7	3.379	2.752
– 56	18.38	18.34	– 6	3.685	2.990
– 55	20.92	20.78	– 5	4.015	3.246
– 54	23.80	23.53	– 4	4.372	3.521
– 53	27.03	26.60	– 3	4.757	3.817
– 52	30.67	30.05	– 2	5.173	4.136
– 51	34.76	33.90	– 1	5.623	4.479

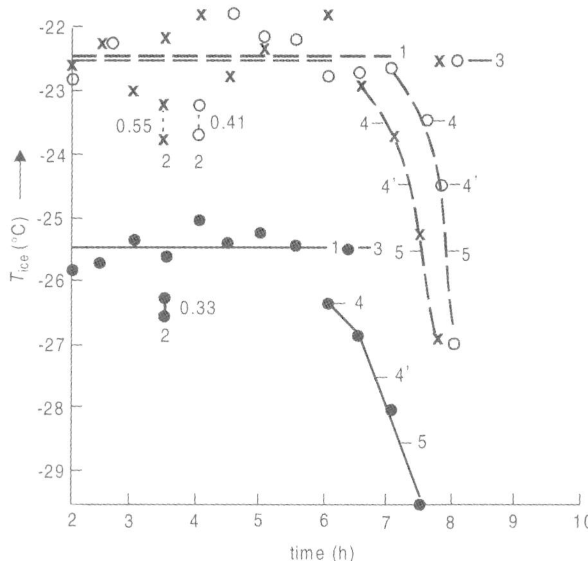

Fig. 1.78.2. T_{ice} as a function of drying time. O—O, X—X measurements and repetition with the same product and the same process data, •—• Another product with other process data.
1, average T_{ice}; 2, standard deviation of $T_{ice,}$ 3, maximum $T_{ice/n}$ for O and X identical –22.53 °C; 4, maximum $T_{ice/n}$ –1 °C; 4', maximum $T_{ice/n}$ –2 °C; 5, maximum $T_{ice/n}$ –3 °C (Fig. 6 from [1.138]).

and the computer calculates the first derivative (2). The peak time represents the time when p_s has been reached. In the example this time was 2.14 s after the valve has been closed. The related equilibrium pressure is $p_s = 0.286$ mbar corresponding to $T_{ice} = -32.7$ °C. Table 1.12 shows the results of such measurements. The average temperature from 2.0 to 4.8 h is –23.66 °C, the standard deviation 0.39 °C. The table shows also T_{ice}/n which is the sum of all measurements divided by the number of measurements. After 5 h T_{ice} seems to drop abruptly by 1.7 °C below the highest T_{ice}/n (–23,4 °C). As shown in Fig. 1.78.2, this decrease in ice temperature marks the end of main drying and can be used to switch over by hand or by automatic operation to secondary drying.

As disadvantage of the BTM method Bardat et al. [1.64] described the danger of collapse or melting of the product during the measurement. This can only happen if T_{ice} and thereby p_s are larger than the maximum tolerable T_{ice}. The measurement is not the reason for the collapse, it is the too high T_{ice} shown by the measurement. Bouldoires [1.66] pointed out that the BTM method can only be used for discontinuous installations having a valve drying chamber and condenser.

For the use of BTM, two conditions have to be fulfilled:

- The leak rate of the chamber has to be so small that the pressure rise in the chamber during the time of valve closure due to leak rate is small compared with the pressure rise due to the water vapor.

Table 1.12: Protocol of T_{ice} during main drying.

t_{MD} (h)	T_{ice} (°C)	$T_{ice/n}$ (°C)	t_{MD} (h)	T_{ice} (°C)	$T_{ice/n}$ (°C)
02.0	–23.5	–23.5	02.0	–2.3	–23.4
02.5	–23.5	–23.5	02.8	–23.6	–23.5
03.0	–23.3	–23.4	03.3	–23.4	–23.4
03.5	–23.4	–23.4	03.8	–23.4	–23.4
04.0	–23.7	–23.5	04.3	–24.1	–23.5
04.5	–24.4	–23.6	04.8	–24.3	–23.7
05.0	–25.1	–23.8	05.3	–25.9	–23.9

$T_{ice/n}$: sum of all n T_{ice} measurements divided by n, after thermal equilibrium has been reached (in this run after 2 h).

The main drying in this run could have been terminated after 5 h, but it is useful to wait for the next data, such that the decision is not dependent on one measurement.

Example: The pressure in the chamber rises in 3 s from 0.28 mbar to 0.41 mbar. With a chamber volume of 200 L this is 8.7 mbar L/s of water vapor. This vacuum chamber should have a leak rate not larger than 0.08 to 0.09 mbar L/s (1 % of 8.7 mbar L/s). Such a leak rate has no measurable consequences. Even if it were ten times larger (0.8 to 0.9 mbar L/s) the pressure in the chamber would rise in 3 s to 0.42 mbar. Converted into temperature, this would be an error of 0.1 °C. A leak rate of 0.8–0.9 mbar L/s is already larger than could be pumped off by a reasonable pumpset in this size of a freeze drying plant. The partial pressure of air, p_{air}, must be small compared with the water vapor pressure. At 0.28 mbar total pressure p_{air} should be 0.03–0.04 mbar. The vacuum pump which can pump 0.8–0.9 mbar L/s at 0.03–0.04 mbar must have a pumping speed of approx. 100 m^3/h, which is unusually large for a 200 L chamber. A vacuum pump with 40 m^3/h pumping speed will theoretically evacuate a chamber and condenser (total of 500 L) in approx. 6 min down to 0.1 mbar. Even if it takes 10 min with the loaded chamber, the pumping speed of the pump is sufficient. With this pump, the leak rate of the plant should not exceed 0.4 mbar L/s, which would be pumped at approx. 0.04 mbar.

The necessary leak-tightness of a plant can be summarized as follows:

- To insure an undisturbed water vapor transport (see Section 1.2.4) the leak rate of a freeze drying plant must allows BTM with sufficient accuracy. This applies for vapor pressure in the ice temperature range between –45 °C to –10 °C, corresponding to 0.07 mbar to 2.5 mbar.
  The pressure range for DR measurements is normally one decade below the above data, and this has to be considered in the specification of the plant. All measurements discussed above have to be carried out by capacitance vacuum gauge, because these instruments measure pressure independently of the type of gas. All vacuum gauges based on the change of heat conductivity as a function of pressure show a result which depends

not only on pressure from the gas mixture but also on the type of gas. Leybold AG [1.67] indicates that for instruments based on heat conductivity 'Thermovac' in the pressure range from 10^{-3} to 1 mbar measurements in pure water vapor (carried with an instrument calibrated in air) the reading value must be multiplied by 0.5 to give the correct water vapor pressure. If the mixture of water vapor and air changes, the reading value of e. g. 0.28 mbar water vapor pressure in pure water vapor corresponds to 0.14 mbar. At 80 % water vapor, the reading value must be corrected to 0.17 mbar. In freeze drying plants during main drying, the water vapor content can vary between 60 and 95 %. An average correction factor 0.65 can be used, as can be seen in Fig. 1.63: Here 0.34 mbar Thermovac corresponds to 0.16 mbar, showing a correction factor of approx. 0.5, with the progress of drying after 8 h 0.11 mbar Thermovac corresponding to 0.08 mbar measured by the Capacitron with a correction factor 0.73. Given that the reproducibility of the heat conductivity manometer in the pressure range of 10^{-2} to 1 mbar is approx. 10 % of the reading [1.67] while capacitance gauges are rated at 0.5 % of the reading [1.68], the advantages of the capacitance method are clear. The difference in price for both instruments is small compared with the investment cost, even of a pilot freeze drier. BTM should therefore always be carried out by a capacitance instrument.

- The second conditions of a reliable BTM is that, in the tolerable measuring time so much ice can sublime as to fill the chamber with saturated water vapor. The measuring must be chosen in such a way that the temperature of the ice during closing of the valve between chamber and condenser does not rise to a disturbing degree. Assuming extreme conditions, one can estimate that the temperature of the ice increases by approx. 0.25 °C/s under the following stipulations:

  - K_{tot} is high, e. g. 84 kJ/m^2 h °C
  - T_{tot} large, e. g. 50 °C
  - the product with 10 % solids has been dried so far that only 15 % of the water is ice and the layer thickness is 0.7 cm.

The measuring time should be < 3 s; this is possible, as can be seen from Table 1.12 and Fig. 1.78.1.

A chamber volume and the amount of ice to be sublimed during MD must satisfy the following conditions:

$$V_{ch}\, dp/dt << m_{H2O}/t_{MD} \tag{15}$$

V_{ch} = chamber volume (L)
dp = $p_s - p_{H2O,ch}$ (mbar) pressure rise during measuring time
dt = time (s) until p_s is reached
m_{H2O} = mass of water to be sublimed during the time MD
t_{MD} = time of MD (s), secondary drying not included

An example of Fig. 1.64:
V_{ch} = 160 L
dp = (0.51–0.13) mbar = 0.38 mbar
dt = 3 s
M = 2.24 kg water, 85 % to be sublimed = 1.90 kg
t_{MD} = 5 h = $18 \cdot 10^3$ s
(1g H_2O is converted into $1.24 \cdot 10^3$ mbar L)

results in
$$\frac{160 \cdot 0.38}{3} << \frac{1.9 \cdot 10^3}{18} \frac{1.24 \cdot 10^3}{10^3} \qquad (15a)$$
$$20.27 << 131$$

With the data of the example condition (17) is satisfied, BTM can be applied.

If in this example a chamber of 1000 L were used. the left side of the equation would become 126.7; in this case, the chamber is too larger or the amount of water to be sublimed too small. Thus 2–4 kg of product in a 160-L chamber or 30 to 80 kg in a 1000-L chamber will satisfy Eq. (15).

If these conditions are met, curves as shown in Fig. 1.78.1 are measured. Figure 1.78.2 shows one measurement and one repetition of this measurement, and a third measurement with another product and other process data.

Towards the end of main drying the data of T_{ice} will systematically decrease; this effect can be used for an automatic change from main to secondary drying (see Section 2.2.8).

The temperature measurement during secondary drying with Th or RTD is possible, as shown in Fig. 1.63, with an accuracy of approx. ±2 °C.

The change from main drying to secondary drying is difficult to determine by the product temperature, as shown in Fig. 1.79. This can also be seen in Fig. 1.80.2 and 1.80.3 [1.63]. Nail and Johnson [1.62] compared (in Fig. 1.81) the pressure measured by a heat conductivity vacuum gauge (TM) with pressure rise measurements during secondary drying and indicated the related RM. In Fig. 1.82.1, the pressure measured by TM is compared with the p_{H2O} measured by a mass spectrometer. The signal of the mass spectrometer is reduced during the first two time units, but changes very little between the 3 and 7 time unit. Connelly and Welch [1.137] also used a mass spectrometer to determine the end of main drying and of secondary drying. They found also that the change in output signal changed by ten to one between main drying and the end of secondary drying. It is suggested to use the water vapor pressure measured by the mass spectrometer not directly, but to divide these data by the total pressure measured by the mass spectrometer. As shown in Fig. 1.82.2 these normalized values show a plateau during the first approx. 7 h of main drying for 5 % bovine serum albumin, and afterwards a decay between 7 and 23 h. A further suggestion is not to plot these normalized values, but their natural logarithms. By this method, the shape of the plot is meaningful while the absolute value of the y-axis is more difficult to interpret. The authors concluded that the main drying is terminated at about 7 h of the cycle. However there was no measurable indicator whether to use the exact end of

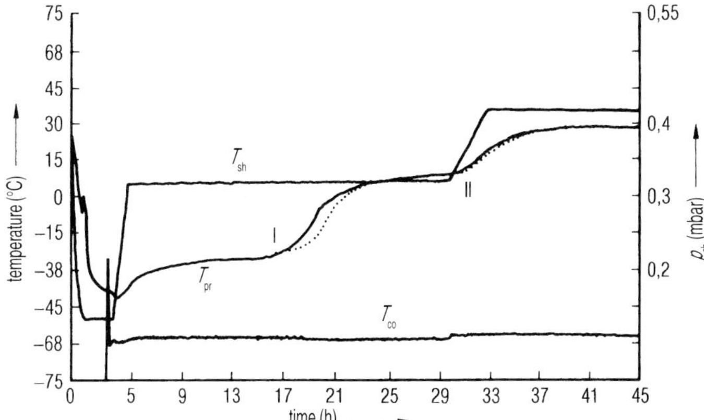

Fig. 1.79. Increase of the product temperature T_{pr} as a function of process time.
The increase (I) starts at approx. 16 h and reaches T_{sh} at approx. 22 h. Secondary drying (SD) has been started (II) at approx. 30 h. Between 16 h and 22 h is no measurable indicator of when to start SD. Also, the safety margin between 22 h and 30 h cannot be connected with the measured product temperature (Fig. 2 from [1.64]).

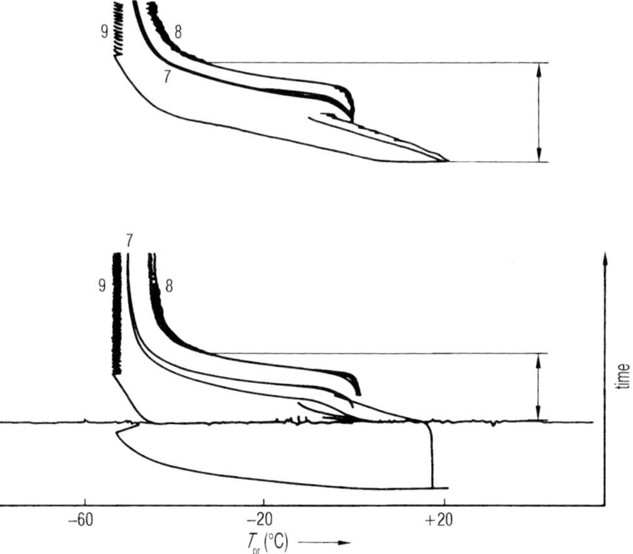

Fig. 1.80.1. Enlarged freezing phase of the two complete processes shown in Fig. 1.80.2 and 1.80.3.
Top: Vials with mannitol solution are cooled with 0.6 °C/min, while the shelf temperature is lowered from room temperature to –50 °C.
Bottom: The vials are loaded on the precooled (–50 °C) shelves (the temperature rises to approx. –40 °C), and frozen at ≈ 1 °C/min. The subcooling is more pronounced due to the quicker cooling.
7, T_{pr} measured by Th; 8, T_{pr} measured by RTD; 9, T_{sh}. Temperatures measured by RTD are 5–7 °C higher than those measured by Th (Fig. 5 from [1.63]).

the plateau or 1–2 h later. For example the decline of the curve in Fig. 1.82.3 changes again at approx. 12 h (this can also be found in the curve of Fig. 1.82.2). The end of secondary drying is suggested to be established by the following procedure:

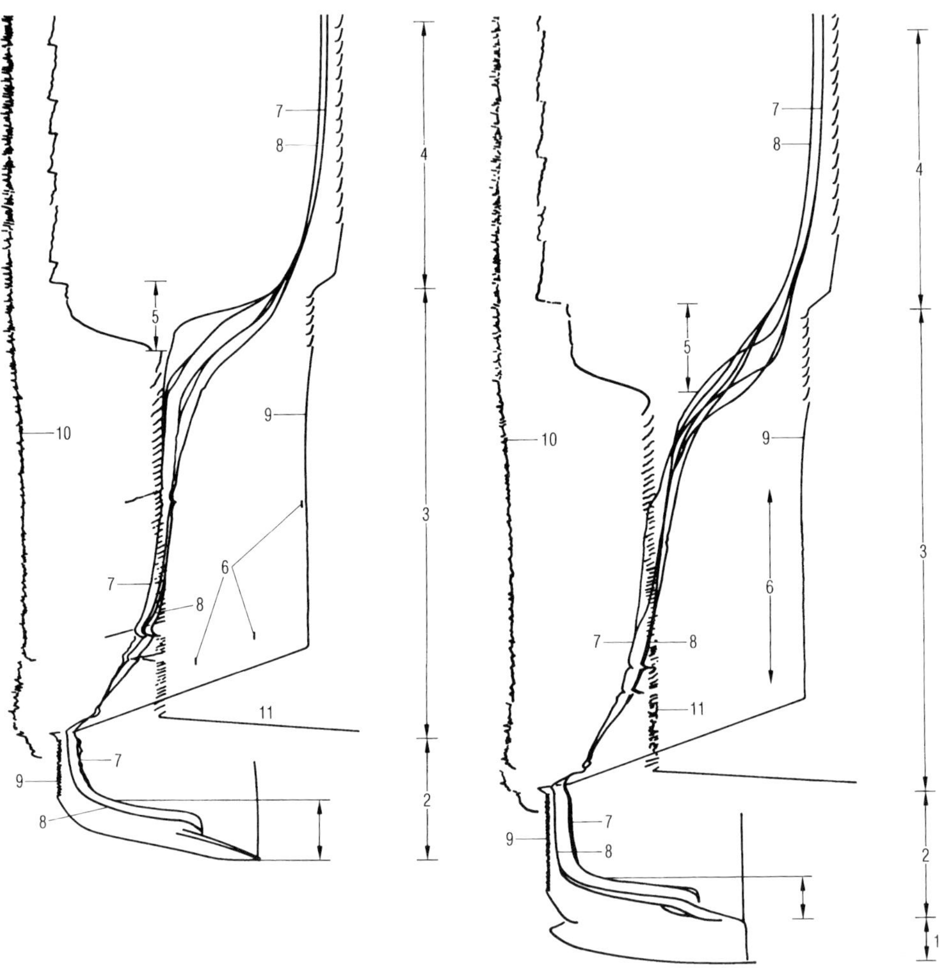

Fig. 1.80.2. Course of the freeze drying after the product has been frozen as per Fig. 1.80.1 (top). 2, Feezing; 3, MD; 4, SD; 5, DR measurements to define the end of MD; 6, some BTM; 7–9, as in Fig. 1.80.1; 10, T_{co}; 11, p_{ch}. At the beginning of DR measurements the pressure control in this example is deactivated. When the DR value has reached a predetermined number, T_{sh} (in this case) is risen to the maximum tolerable temperature. The optimum time frame for the change from MD to SD cannot be estimated from the T_{pr} plot (Fig. 6 from [1.63])

Fig. 1.80.3. Course of the freeze drying after the product has been frozen as per Fig. 1.80.1 (bottom). Nomenclature as in Figs. 1.80.1 and 1.80.2. The rise of T_{pr} is different. The optimal time frame for the change from MD to SD can not be estimated from the T_{pr} plot (Fig. 7 from [1.63]).

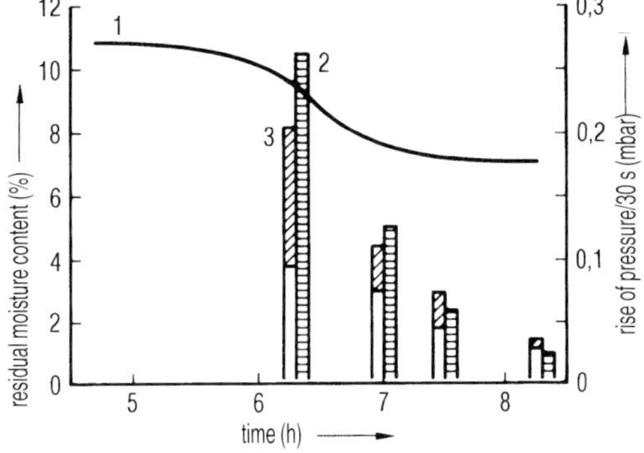

Fig. 1.81. Plot of the pressure measured by heat conductivitY vacuum gauge (TM) during SD. In addition pressure rises in 30 s and related RM data are shown.
1, p_{ch} measured by TM; 2, pressure rise in 30 s; 3, RM in % of solids (Fig. 5 from [1.62]).

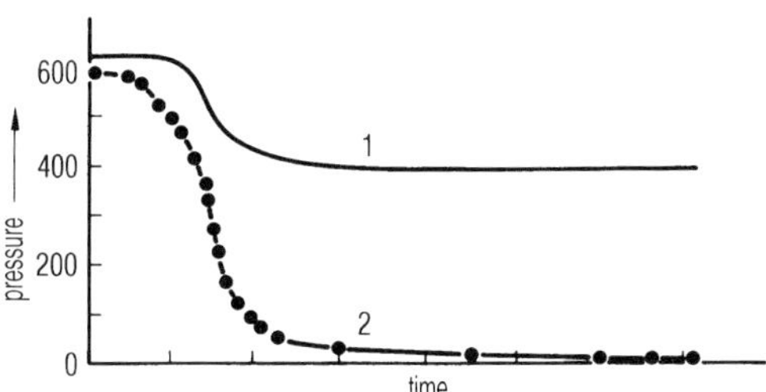

Fig. 1.82.1. Comparison of p_{ch} data: TM measurements and signals of mass spectrometer for mass 18 during freeze drying.
1, p_{ch} by MT; 2, mass spectrometer signal at mass 18 (Fig. 10 from [1.62]).

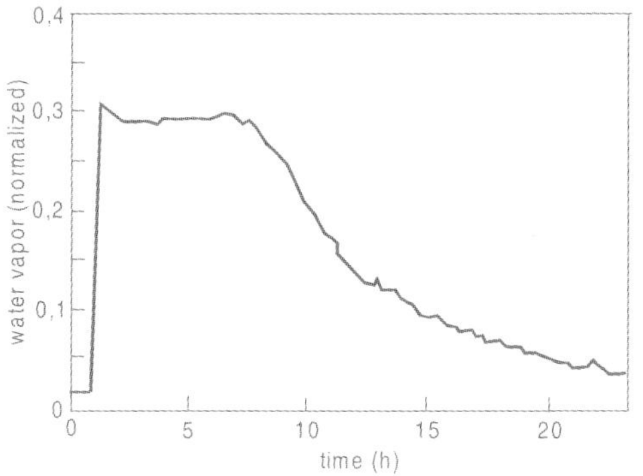

Fig. 1.82.2. Water vapor partial pressure divided by total pressure as a function of time of a 5 % bovine serum albumin (Fig. 3 from [1.137]).

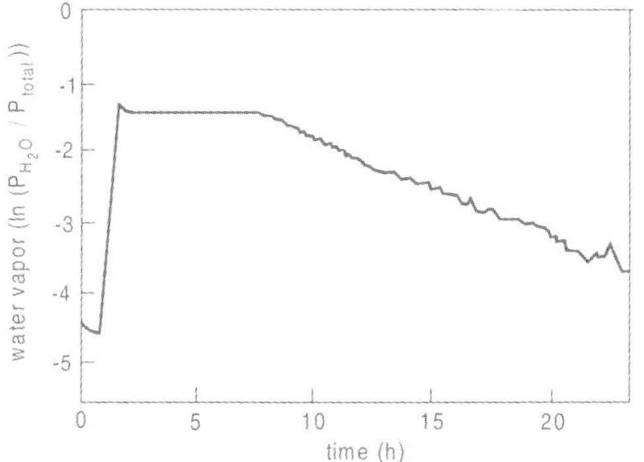

Fig. 1.82.3. Natural log of water vapor partial pressure divided by total pressure as a function of time for 5 % bovine serum albumin (Fig . 5 from [1.137]).

1. Taking a baseline measurement of the partial pressure of the ice on the condenser in the empty plant,
2. Measuring the partial pressure during the run and terminating the secondary drying if the two values are close together. In certain cases this might be too insensitive; in this case it is suggested to close the valve between chamber and condenser and measure the increase in water vapor pressure in a certain time. (*Note*: it is surprising that water vapor pressure at the beginning of main drying is only 40 % of the total pressure during the sublimation of distilled water and 30 % during the sublimation of bovine serum albumin).

The pressure rise measurements in Fig. 1.81 change during the final hours from 0.26 mbar to 0.05 mbar, showing that this method is more sensitive than p_{H2O} measurements with the mass spectrometer alone.

Figure 1.83 shows the comparison between measurements made with TM, CA and a hygrometer, and demonstrates that the hygrometer data are not much more sensitive to the change in vapor pressure than the data of the two other instruments. The end of main drying can be between 2.5 and 5 h, depending on which change of inclination is chosen. From the BTM measurements, one can conclude (see discussion of Table 1.12) that the main drying is terminated at approx. 3.5 h.

Figure 1.84 [1.69] summarizes the measurements of three runs of the product temperatures with RTD, T_{ice} with BTM and of the pressures by CA. The plots show that the difference in pressures during main and secondary drying is largest with no pressure control and still clearly recognized with p_c at 0.2 mbar in relation to an ice temperature of approx. –30 °C.

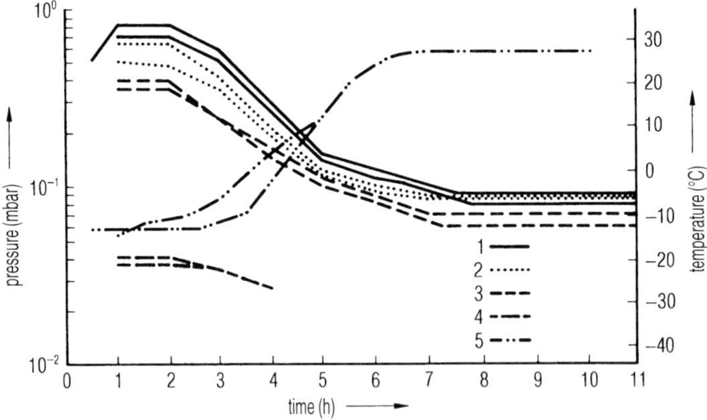

Fig 1.83. Course of two tests with identical products and identical process conditions, also using a hygrometer. The pressure drop at the end of MD, measured by the hygrometer is not more informative than the data provided by CA. Below a certain pressure (in this case below approx. 0.09 mbar) the hygrometer had to be recallibrated for the lower pressures.
1, p_{ch} (mbar) TM; 2, p_{ch} (mbar) CA; 3, hygrometer (System 3A, Panametrics GmbH, D-65719 Hofheim, Germany); 4, T_{ice} (BTM) (°C); 5, T_{pr} (°C) RTD (Fig. 7 from [1.110]).

The water vapor desorption can be measured as shown in scheme in the Fig. 1.85.1, and be calculated by Eq. (16):

$$D = \frac{\mathrm{d}p V_{ch}}{\mathrm{d}t} \ (\text{mbar L/s}) \qquad (16)$$

V_{ch} = chamber volume (L)
dp = pressure increase (mbar)
dt = measuring time for pressure increase (s)

By using $22.4 \cdot 10^3$ L mbar corresponding to 18 g H_2O, mbar L can be converted into g. This relationship is accurate enough as the temperature of the water vapor depends on several factors and will also be modified by a change of $T_{sh.}$ The desorption process can be best illustrated by using the desorption rate (DR), which measures the desorbed amount of water in % of the solids of the product per hour.

$$\mathrm{DR} = 2.89 \cdot 10^2 \, (V_{ch}/m_{fest}) \, (\Delta p/\Delta t)$$
(desorption of water vapor in % of solids per h) (16a)

V_{ch}, dp, dt as in Eq. (16)
m_{fest} = mass of solids (g)

Measurements of the desorption rate (DR) require three conditions:

- For a product a reproducible desorption isotherm exists, and the product does not change at the end temperature during secondary drying.
- The end temperature has to be applied for some time depending on the cake thickness in order to minimize the temperature gradients in the product
- The leak rate of the plant must be so small that a pressure rise due to the leak rate is also small compared with the pressure rise resulting from the desorbed water.

To measure DR values, one has to use measuring times of approx. 30 s. The prolonged time (compared with BTM) can be used, since the product temperature during this time is almost constant. On the other hand, the absolute pressures are approx. one decade smaller than during BTM (Fig. 1.64 $p_{MD} = 0.36$ mbar, $p_{SD} = 0.03$ mbar). To measure e. g. 1 %/h in the run of Fig. 1.64, one must calculate:

65.5 g solids in a chamber volume of 160 L by Eq. (16b)

$$\mathrm{d}p/\mathrm{d}t = D/V_{ch} = 1.4 \cdot 10^{-3} \ \text{mbar/s} \qquad (16b)$$

This pressure range can be measured by a CA. If the leak rate (qL) has to be small compared with the value to be measured follows:

$qL << 1.4 \cdot 10^{-3}$ 160 $<< 2.2 \cdot 10^{-1}$ mbar L/s or the maximum qL
$qL_{max} = 2.2 \, . \, 10^{-2}$ mbar L/s

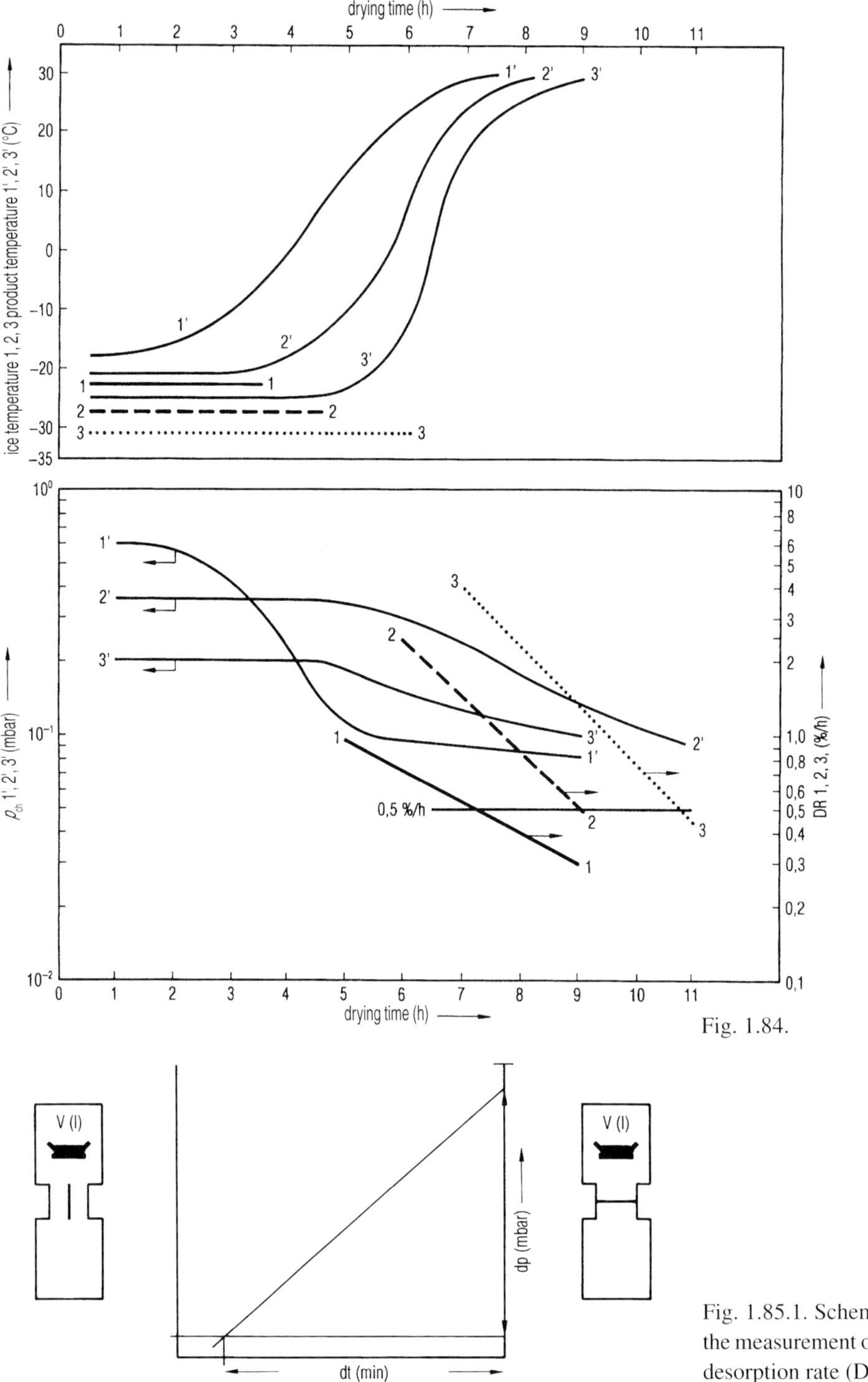

Fig. 1.84.

Fig. 1.85.1. Schema of the measurement of the desorption rate (DR).

Fig. 1.84. Summary of the results of three runs, which are differentiated by the control pressure: 1, No pressure control; 2, pressure controlled at 0.36 mbar; 3, pressure controlled at 0.20 mbar. The graphs show: T_{ice}, marked as 1, 2, 3, and T_{pr} marked as 1', 2', 3' in the upper drawing. In the lower drawing, the DR are marked as 1, 2, 3 and the p_{ch} as 1', 2', 3'. The increase in the product temperature (T_{pr}) and decrease of chamber pressure (p_{ch}) depend on the chamber pressure, because K_{tot} is pressure-dependent and T_{sh} has been programmed up to +30 °C in such a way that the control pressure has never been exceeded. In the test 1, T_{sh} has been raised to +30 °C as quickly as technically possible. The end of MD and SD are difficult to define by T_{pr} and/or by p_{ch}. The DR values measure the amount of water desorped from the product per hour in % of solids. The end of drying has been determined by DR:
1, after 7 h, DR = 0.55 %/h; 2, after 8.5 h, DR = 0.65 %/h; 3, after 11 h, DR = 0.45 %/h. As shown in Fig. 1.73, the dW (RM) can be calculated from the DR data and the end of drying can be expressed as residual moisture content in % of solids (based on Fig. 1 from [1.69]).

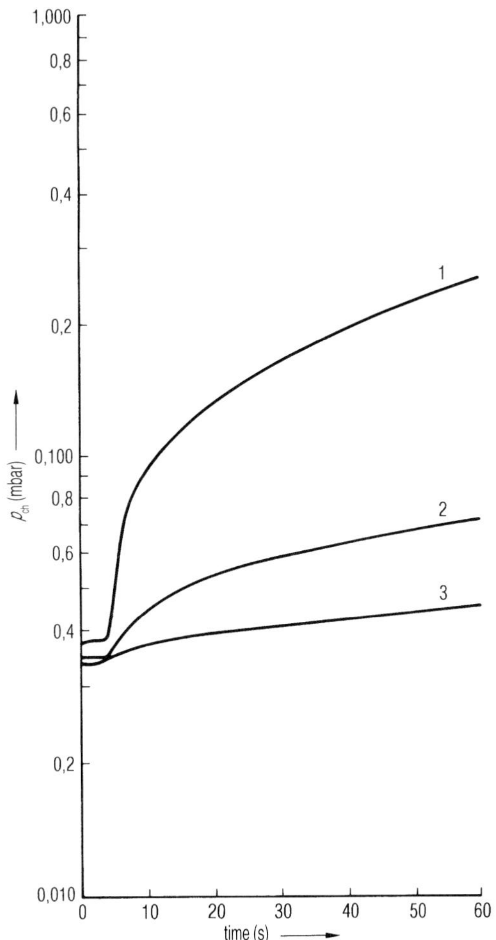

Fig. 1.85.2. Automatically measured and recorded pressure rise as a function of time after the valve between the chamber and the condenser has been closed. Three measurement have been selected from all made during SD.
1, Shortly after SD has been started; 2, 2.5 h later; 3, 5.75 h later. A computer can calculate dp/s from the measured $p_t - p_0/t \cdot p_0$ pressure after closing of the valve; P_t pressure after the measuring time t (based on measurements of AMSCO Finn-Aqua, D-50345 Hürth, Germany).

With this qL_{max} the calculated DR is approx. 10 % too large, DR = 1 %/h is calculated as 1.1 %/h or a DR 0.1 %/h is calculated as 0.2 %/h. For many freeze drying plants one can expect that a leak rate will be in the range of 10^{-3} mbar L/s. If the leak rate of a plant is stable and known, it can be accounted for in the DR-value. In a normal operation one would expect that a 100-L chamber is loaded with 2.5 kg of liquid product, containing 250 g solids, the leak rate could than be four fold larger, as mentioned above.

The pressure rise measurement can be made automatically, as shown in Fig. 1.85.2. The leak rate becomes critical if the solid content is small, e. g. 1 %, then qL_{max} has to be approx. $2 \cdot 10^{-3}$ mbar L/s, all other conditions being equal. In such cases, the leak rate of the chamber should be measured before charging the product

1.2.4 Water Vapor Transport during Drying

The water vapor transport in a freeze drying plant can be described schematically with the aid of Fig. 1.86: The ice (1) is transformed into vapor and has to flow out of the container (2) into the chamber (4). Between the chamber wall or any other limitation an area (3, F1) is necessary. The vapor flows then through the area (F2) into the condenser (7), having surface of (F3) on which the water vapor will mostly condenses. A mixture of remaining water vapor and permanent gas is pumped through (8), (9) and (10) by a vacuum pump (11).

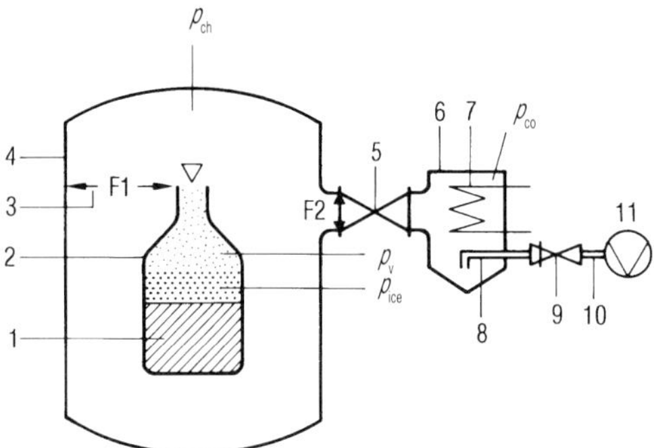

Fig. 1.86. Schema for the estimation of the water vapor transport in a freeze drying plant. 1, Frozen product; 2, vial or the end of a shelf; 3, open surface (F1) for the water vapor flow between 2 and 4; 4, chamber wall; 5, valve with an open area F2; 6, condenser chamber; 7, cooling and condensing surface in the condenser chamber having a surface of F3; 8, vacuum pipe with the diameter *d*; 9, stop valve; 10, vacuum pipe with the length l (from 8 to 11); 11, vacuum pump; p_{ice}, water vapor pressure at the sublimation front of the ice; p_v, pressure in the vial; p_{co}, pressure in the condenser.

Example: p_s at the sublimation front is 0.937 mbar (–21 °C) (see example in Table 1.9), in the chamber a p_{H2O} = 0.31 mbar has been measured, resulting in a pressure difference of approx. 0.6 mbar. With these data, the water vapor permeability $b/\mu = 1.1 \cdot 10^{-2}$ kg/h m mbar is calculated. With this data known, it is possible to calculate dp for different conditions, if the mass of frozen water m_{ice}, the time t_{MD}, the thickness (d) and the surface (F) are known. This dp depends from the amount vapor transported and thereby from the heat transfer (Table 1.9). In the examples given it changes between 0.17 mbar in a slow drying process (6 h) to 0.6 mbar for a shorter drying time, 2.5 h.

Transport out of the container (2) into the drying chamber produces no measurable pressure drop if the product surface is equal to the opening of the container (e. g. with trays). Vials without stoppers in the vial neck do not produce a measurable pressure drop, if e. g. 1 g/h water at a T_{ice} of –20 °C and a pressure difference of 0.6 mbar are transported. In this example the velocity of water vapor is a few m/s.

If stoppers are in the vials, in the freeze drying position, the situation is different: Depending on the type of stopper (Fig. 1.87), the drying performance, the can be reduced to 66 % or 77 %, generally speaking to 60 to 80 %. To achieve the same performance, the temperature would have to be increased from –20 °C to –17 °C, resulting in 30 % higher pressure. If the temperature increase is not tolerable, the pressure in the drying chamber must be reduced and a slower drying process be accepted.

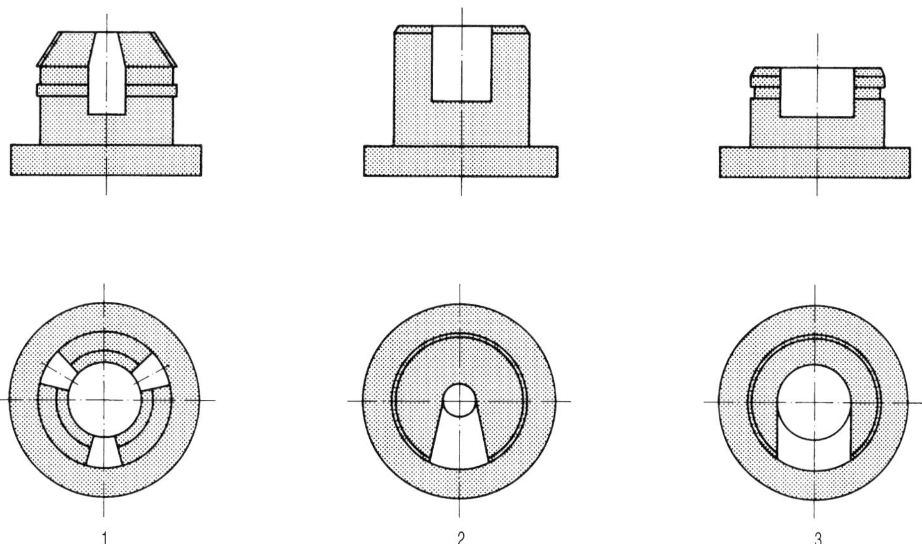

Fig. 1.87. Influence of different forms of stoppers on the water vapor transport out of the vials into the chamber. At a pressure of 1 mbar in the vial the following relative amounts of water vapor are transported into the chamber in 3 h:

Stopper	None	1	2	3
	100 %	77 %	75 %	66 %

These conditions remain comparable down to an ice temperature of approx. –40 °C, and drying performances can be expected to be reduced by a third by the stoppers. If the water vapor streams are smaller (the example used shows an upper limit), the reduction in performance may be smaller.

Vapor transport into the condenser depends strongly on the geometric design of the plant. Under favorable conditions, and including a valve between chamber and condenser, a vapor speed of 60–90 m/s can be expected, resulting in a pressure drop between the chamber and condenser of a factor of 2, as an order of magnitude.

With these estimated conditions and a condenser e. g. –45 °C = 0.07 mbar, the following pressure can assumed:

- p_{co} times 2 = 0.144 mbar (p_{ch})
- p_{ch} times 1,5 = 0.216 mbar (p_{F1})
- p_{Fl} +0.2 to 0.6 mbar = 0.4 to 0.8 mbar

this results in: p_s approx. –29.5 °C to approx. –22.5 °C

At this condenser temperature, and in this plant, products could be dried at ice temperatures between –29 °C and –23 °C. As shown in Table 1.9 an ice temperature of –22.3 °C (test run in Fig. 1.62) has been successfully operated at a condenser temperature of –45 °C and a pressure difference $p_{ch} - p_{co}$ = 0.5 mbar (see also Table 1.10, columns 1 and 2).

If the freeze drying conditions are extreme namely small solid content, low sublimation temperature, e. g.:

- solid content 1.7 %
- T_{ice} during MD –40 °C
- layer thickness 3.8 mm
- vials with stoppers

one has to consider that the water vapor permeability b/μ will be larger as in the earlier example. If $b/\mu = 6.9 \cdot 10^{-2}$ kg/m h mbar and ice temperature is –41 °C = 0.115 mbar are measured, the water vapor pressure in the chamber will be 0.065 mbar. The condenser temperature should therefore represent a pressure of approx. 0.035 mbar, which would require a condenser temperature of approx. –51 °C.

If stoppers with more favorable channels are used, the vapor pressure in the vials could have been 0.09 mbar, leaving a $\Delta p = p_{ice} - p_{F1}$ = 0.025 mbar which is in agreement with $b/\mu = 6.9 \cdot 10^{-2}$ kg/h m mbar of this test.

The water vapor transport from the chamber to the condenser depends largely from the geometric design of the installation. Assuming that only one bottleneck exists between the chamber and condenser having a diameter (*d*), which is large compared with the length (*l*) of the connection (Fig 1.88 (a)) one could expect a jet flow, which follows the equation (1.3.9) in [1.70], For this case, the connection must have the shape of a jet with no obstacle in it (e. g. a valve lid). Technically this case is not possible. Even in a plant as shown in Fig. 2.48.2, the water vapor has to pass through a ring-type jet and is then deflected towards the

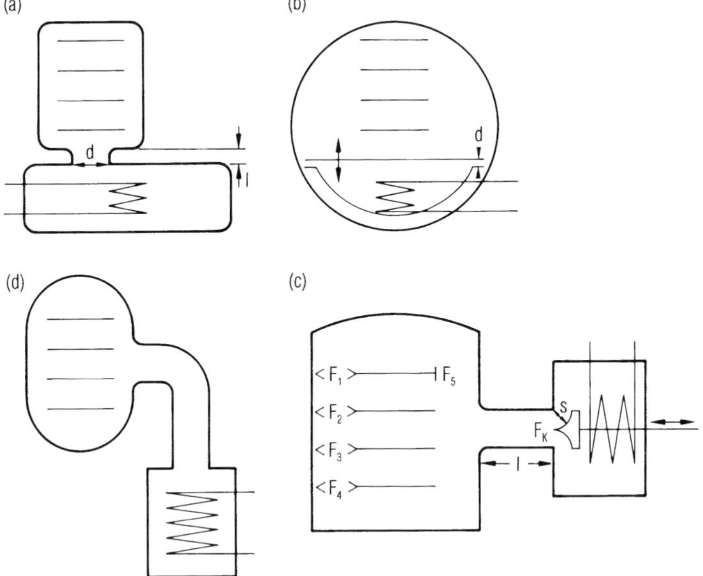

Fig. 1.88. Four different geometric layouts of condenser and drying chamber.
(a) Chamber and condenser in a housing not divided by a valve, water vapor transport into the condenser through an opening d with the length l. (d, maximum approx. condenser or chamber diameter). (b) Chamber and condenser located in the same housing, the condenser compartment can be separated from the drying chamber by a lid. In the open position, the water vapor flows through a clearance with the height d. (c) Drying chamber with four shelves, a connecting pipe to the condenser with the cross-section d (area F_K) and a length l. In the open position of the valve a circular clearance s can be an additional bottleneck after F_K. (d) Layout similar to (c), but chamber and condenser are connected by a 90° bend.

condenser surface. To estimate the influence of p_{ch}, d and l, the 'Günther-Jaeckel-Oetjen' equation (1.3.11) in [1.70] or its graphical plots (Fig. 1.3.4 or 1.3.6 in [1.70]) can be used. Figure 1.89 is an evaluation of the quoted equation and plot for the area of interest for freeze drying. It shows the specific flow of water vapor through tubes with a ratio of l/d = 1; 1.6; 2.5 and 5 as a function of pressure at the inlet of the tube.

In the pressure range from 0.66 to 1.32 mbar, the expected specific flow of water vapor is reduced for the mentioned ratios to 66 % or 55 % of that passing through an ideal jet. At 0.04 mbar pressure in the chamber the specific throughput is reduced to 33 % or 10 %. This becomes even more drastic, if the velocity of the water vapor is plotted as a function of p_{ch} (Fig. 1.90). In an ideal jet, the velocity of the vapor flow under the conditions chosen is approx. sound velocity. However even with l/d = 1 the velocity is strongly reduced as a function of pressure, and reaches at 0.04 mbar at only approx. one-third of this maximal speed.

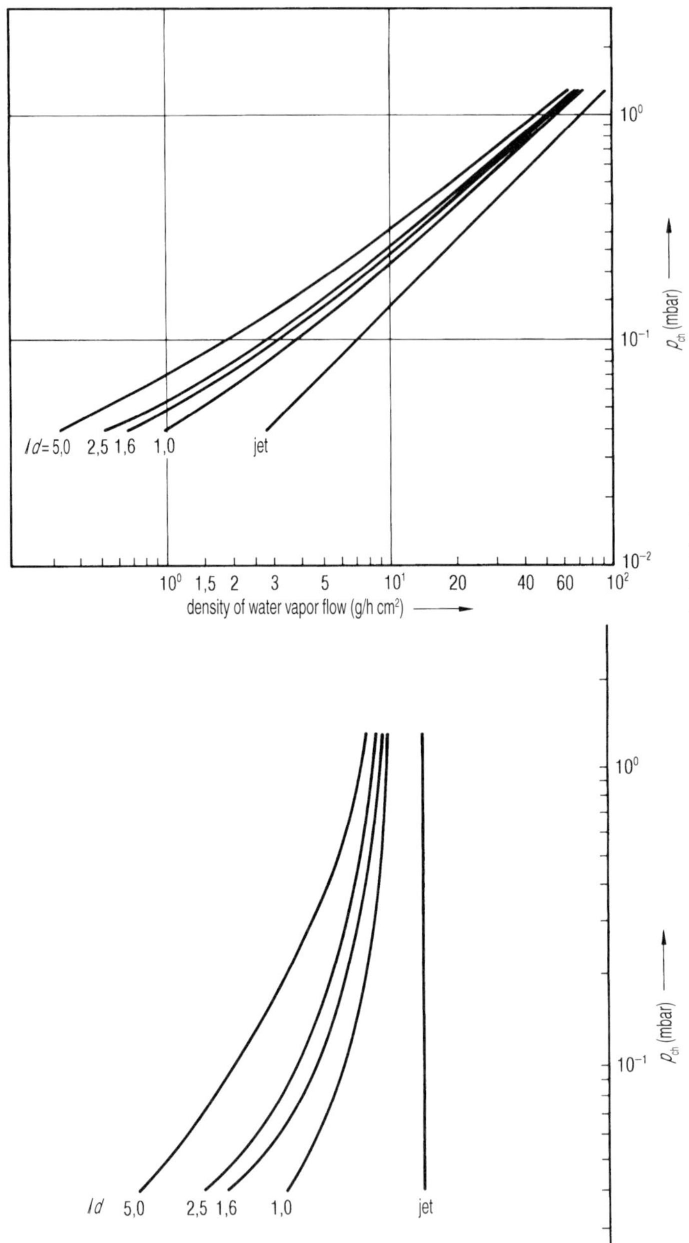

Fig. 1.89. Density of water vapor flow (g/cm^2 h) as function of p_{ch} with jet flow and different l/d as parameter.

Fig. 1.90. Rate of the water vapor flow (m/s) as a function of p_{ch} through a jet and different l/d pipe dimensions as parameter.

To summarize, one can say that: In the pressure range of 0.1 to 0.3 mbar and $l/d = 5$, a vapor velocity between 50 to 100 m/s can be expected. The graphs and figures are given to underline the influence of the pressure in the chamber and the geometry of the plant. A general guide line for the design of a plant can be:

- all cross-sections, through which the vapor has to flow, ahead of the smallest one must be large compared with smallest, e. g. F_k in Fig. 1.88 >> F_1 to F_5.
- if a valve is installed in F_k, its cap should influence the vapor flow as little as possible (see e. g. Fig. 2.48).
- the path of the vapor should be deflected as little as possible; a deflection of 90° (Fig. 1.88 (d) does not only prolong l, but it transpires that l has been increased by a multiple of l.

In an expediently designed plant, one can expect to reach in the pressure range above $8 \cdot 10^{-2}$ mbar a vapor velocity in the cross-section F_k of between 50 and 80 m/s ($l/d = 2.5$ to 5) However 90 m/s will be reached only, if the design uses special features, e. g. a funnel-like connection between the chamber wall and the location of the valve, slow changes in the outline, and smooth surfaces without sharp edges or holes. It is also recommendable, to clarify the maximal amount of vapor transportable at several pressures in a plant specification, e. g. at $p_{ch} = 1$ mbar a minimum of 3 kg/h and at $p_{ch} = 0.04$ mbar a minimum of 25 g/h flow of water vapor must be demonstrated during the acceptance test, while the condenser temperature is below –30 °C, respectively below –57 °C. Such measurements can be carried out practically with sufficient accuracy (see e. g. Fig. 2.19 and the related text).

1.2.5 Collapse and Recrystallization

A possible collapse of the product during main drying and a recrystallization during the drying can have a significant influence on the quality of the final product. Therefore these two events will be discussed again with regard to the drying process.

If, during freezing, not all freezable water has been frozen, the collapse temperature depends strongly on the amount of unfrozen water present. The highly concentrated, highly viscous, amorphous substance does not show at a temperature of e. g. –85 °C any measurable mobility. The water molecules can no longer migrate to the existing crystals and the unfrozen water is solidified. If the temperature is increased, the viscosity of e. g. 10^{14} Poise does not decrease with temperature, but with the difference of the temperatures $T - T_{g'}$ · $T_{g'}$ represents the highest possible T_g if all freezabale water is frozen. Incompletely frozen products have an unnecessary low T_g e. g. –85 °C, while $T_{g'}$ for this product is only –58 °C. If such an incompletely frozen product is freeze dried above –85 °C, the structure will soften and at T_c will collapse. In Fig. 1.91 devitrification can be recognized in the region of –40 °C ($T_{g'}$) and a collapse at approx. –36 °C (T_c). The upper picture is the enlarged zone at –36 °C. Figure 1.92 shows a metastable structure at –63 °C, which softens at –28 °C and

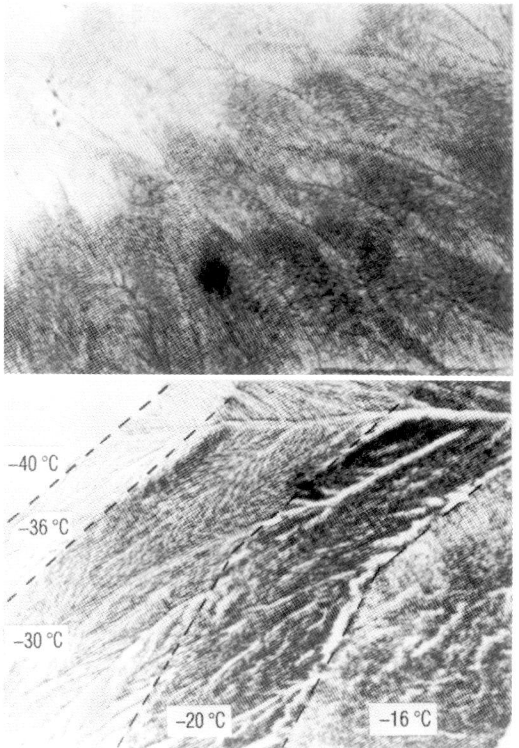

Fig. 1.91. Photographs taken by a cryomicroscope during rising temperatures in the sample. At –40 °C a devitrification and at –36 °C a collapse can be seen. The upper photograph is an enlargement of the –36 °C area (Fig. 9 from [1.111]).

Table 1.13: $T_{g'}$ data, related UFW (unfreezable water, %) T_r and T_c data ($T_{g'}$- and UFW data from [3.6]; T_r and T_c data from [1.113]).

Substance	$T_{g'}$ (°C)	UFW (%)	T_r (°C)	T_c (°C)
Dextran	– 9	–10	– 9	
Fructose	–42	49.0	–48	–48
Glucose	–43	29.1	–41	–40
Glycerine	–65	45.9	–60/–65	
Lactose	–28	40.8	–31	
Trehalose	–30	16.7		
Maltose	–30	20.0	–30/–35	
Ovalbumin	–10	–10	–10	
Polyethyleneglycol	–13	–65	–13	
Polyvinylpyrrolidone	–19.5	–24	–23	
Sorbitol	–43	18.7	–57	
Sucrose	–32	35.9	–32	–32

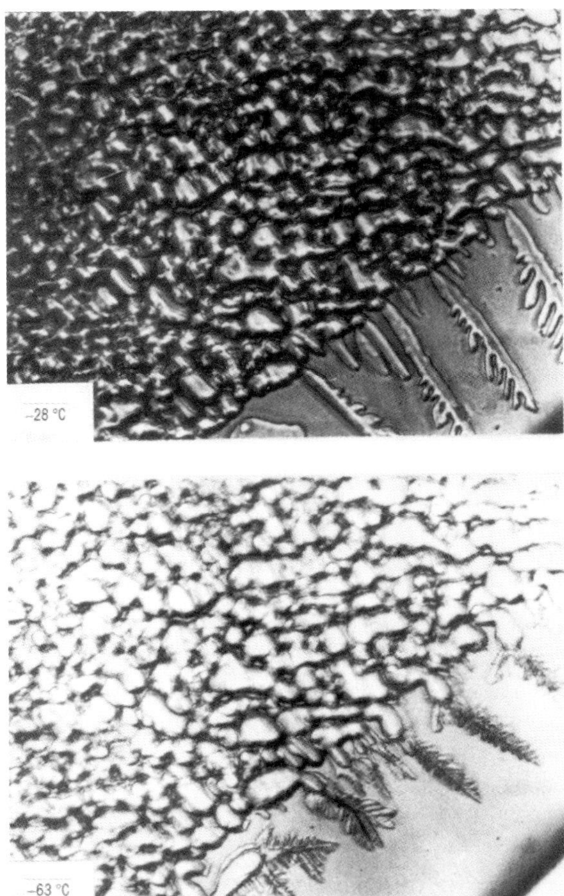

Fig. 1.92. Photographs by a cryomicroscope at –63 °C and –28 °C (Fig. 11 from [1.111])

the viscous, but liquid part can be seen (black). With only unfreezable water (UFW) in the product it can be dried closely below T_c, if T_{ice} is measured and controlled during MD.

Figure 1.93 shows how the more movable water (black) dissolves the already dry product into a mixture of ice crystals and a highly viscous syrup. At a different temperature (T_r) water molecules can move to existing ice crystals, which thereby grow by recrystalization.

$T_{g'}$, T_c and T_r can be close together, or are approx. in a 10 % range, as shown in Table 1.13. For sorbitol, $T_{g'}$ is shown at a higher temperature (–43 °C) than T_c (–57 °C), a fact which can not be explained.

All temperatures mentioned are influenced by the methods of their measurement [1.71], e. g. very thin samples in a cryomicroscope, very small amounts of product (mg range) in an installation for differential scanning calorimetry (DSC) and some temperature gradients in the sample during the measurement of the electrical resistance (ER). $T_{g'}$, T_c and T_r measured with pure substances can supply helpful information about the temperature range to

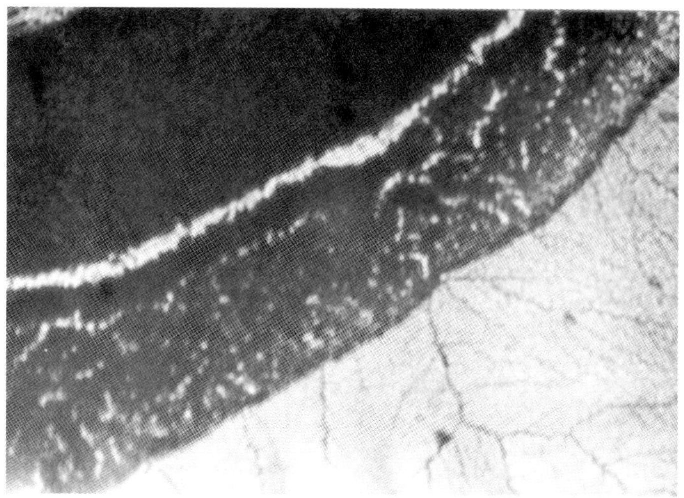

Fig. 1.93. Photograph taken by a cryomicroscope showing T_c (water, upper left corner, black); and T_r (white ice crystals and dried product (gray), lower right corner (from [1.112] unpublished).

be expected. For products containing two or more ingredients, the data must be measured for the specific mix since traces of additives or residues can change the data substantially, see e. g. Fig. 3.2, 3.3.1 and 3.3.2.

The question of aroma retention has been of special interest between 1968 and 1975 for the freeze drying of food. Thijssen and Rulkens [1.72, 1.73] are of the opinion that slow freezing and quick freeze drying provides good retention of the test substance, 0.1 % acetone in a dextrin solution, because the slow freezing produces large ice crystals, which include highly concentrated solutions between them. The pore size in a solution of 20 % dextrin frozen at 0.5 °C/min. is approx. 3 μm, at a freezing speed of 20 °C/min, is only 1.8 μm. The freeze drying speed with 3 μm pores and with 1.8 μm pores has a ratio of 0.17 : 0.07. Furthermore, the retention increases with increasing solid content: in a 10 % solution retention is practically 0, but between 20–30 % solids it increases to 45–60 %. Flink and Karel [1.74] showed (Table 1.14) that the loss in volatile substance, 1-butanol in a maltose solution, happened in the first 6 h of the MD, during SD from 6 to 24 hours the volatile content is practically unchanged.

Voilley et al. [1.75] confirmed the increasing retention with decreasing freezing speed and with the increasing number of carbon molecules of the alcohol (Fig. 1.94).

During MD the retention is unchanged with time, but decreases with increasing temperature during SD. Flink [1.76] proved, by additional tests, his model about the retention of volatile, called 'Microregion Entrapment'. A product frozen and ground up does not lose more volatiles than when frozen in a block. The microregions are smaller, as can be achieved by grinding, and there is no concentration of volatiles in the surfaces. If the microstructure is destroyed, e. g. by collapse, the retention decreases. Maltose, sucrose and lactose each have a better retention for volatiles than either glucose or dextran (Table 1 in [1.76]).

Table 1.14: Loss of 1-butanol during the freeze drying of a maltose solution (Table IV from [1.74]).

Drying time (h)	Average water content (g/100 g solid)	Average content of 1-butanol (g/100 g solid)
0	430	4
3	178	3.30
6	36	2.20
12	11	2.45
24	0.7	2.50

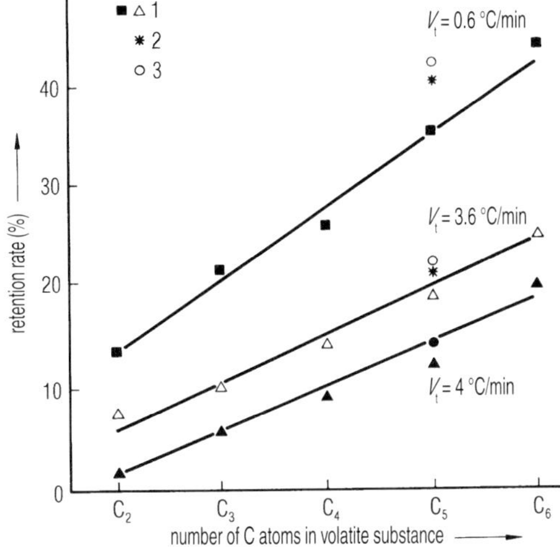

Fig. 1.94. Percentage of alcohol retained as a function of the number of carbon molecules in the alcohol molecule three freezing rates (v_t) as parameter. The solution consists of: 30 g saccharose, 15 g glucose, 15 g fructose, 15 g citric acid, 5 g $CaCl_2$, 15 g pectin, 5g freeze dried albumin, 900 g water and 100 ppm volatile substance.
1, Homologous series; 2, 3-methyl-1-butanol; 3, cyclopentanol (Fig. 1 from [1.75]).

Gero and Smyrl [1.77] showed the retention from formic acid to butyric acid as a function of the dextran concentration and a special behavior of the acid (Fig. 1.95) while Seger et al. [1.78] demonstrated that organic solvents used during the production of a formulation, e. g. methanol, ethanol and *n*-butanol cannot be completely removed by freeze drying as they influence the freezing structure and freeze drying process. During freezing, methanol and ethanol often form films on the surface, which makes drying difficult or impossible. The residues are pushed to the surface by the crystallizing ice and dry by evaporation from the liquid phase, thus forming skins. (*Note*: It is also possible that the mixture at the chosen conditions does not freeze completely and cannot be dried at all).

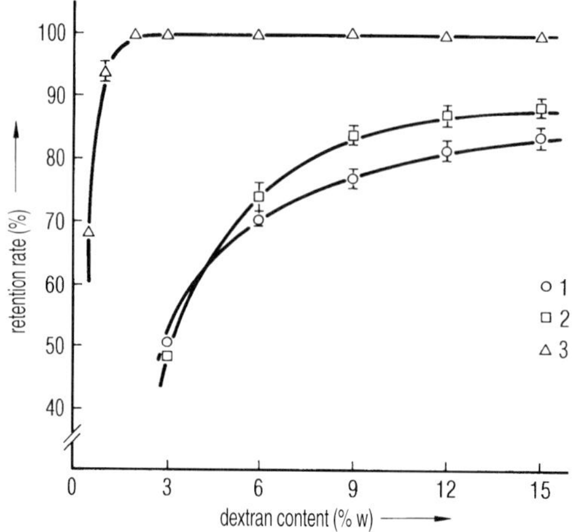

Fig. 1.95. Retention of acid in percent of the initial concentration (0.05 M) as the function of dextran concentration. Three retained acids:
1, n-butyric acid; 2, isobutyric acid; 3, lactic acid (Fig. 2 from [1.77]).

1.2.6 Drying Processes without Vacuum

From time to time, drying at low temperatures at atmospheric pressure has been discussed and tried experimentally, because vacuum installations are high-cost investments and expensive to operate. There are three basic problems which must be solved by such a low temperature drying process:

1. 1 kg of ice, when sublimated at 0.6 mbar, has a volume of approx. 2000 m^3. Since the atmospheric pressure is approx. 1700 times larger, approx. $3.4 \cdot 10^6$ m^3 of air must be transported to carry the water vapor (the vapor content is < 1 per thousand).
2. If only the diffusion of vapor in resting air is used to transport the vapor from the sublimation front to the condenser (or vapor absorber), only $4 \cdot 10^{-2}$ g/m^2 h can be transported over a distance of 100 cm. Even if the condenser could be positioned at a distance of 1 cm the result is only 4 g/m^2 h. Transport of vapor by diffusion cannot be used practically.
3. By mixing an absorbing granulate or powder with the product to be dried, the distances the diffusion can become very small, or the water molecules may move by surface diffusion. In both cases, the problem is the same: First to find an acceptable drying agent (absorber) and then to separate it quantitatively from the dried product.

In the last years, several publications have tackled these problems: Kahn-Wyler [1.79] lists four reasons which prove, that fluidized-bed drying (solving problem 2 above) is not suitable:

a. The structure of the frozen product is difficult to control.
b. The abrasion of the already-dried product is too large.

c. The separation of the carrier-substrate (glass-spheres) from the dried product is not complete enough
d. Abrasion of the installation results in product contamination

Labrude and Rasolomana [1.80] reported an atomizer- spray-drying system for oxyhemoglobin in a 0.25 M sucrose solution at temperatures between +80 and +100 °C, which resulted in an unchanged dry product if the relative humidity was kept below 3 %. When this dry product was compared with a freeze dried product, in both cases a met-oxyhemoglobin (met-HBO)-content of approx. 3 % was found. By ERP and spectrometric measurement, it was shown that the structure of the dried molecules had not changed measurably. However, with this process described, two problems remain: at +80 °C, water has a vapor pressure of approx. 470 mbar; 3 % of this value is approx. 14 mbar. Depending on the efficiency of the heat exchanger and water condenser, approx. 100 times more transport gas must be cooled than water vapor can be condensed. If the partial pressure of the vapor in the transport gas were to be e. g. 3–4 mbar (to allow in increase to 14 mbar during drying), condensation of the vapor must occur at approx. –5 °C. The transport gas must be cooled to that temperature and reheated to +80 °C. Absorption systems to remove water vapor are technically feasible, but the temperature of +80 °C would still very likely have to be lowered, and the dust produced by abrasion becomes a problem (see problems 1 and 3).

Wolff and Gibert [1.81] described the freeze drying of small pieces (maximum 5 mm) in a fluidized bed process at –5 °C to –15 °C. The absorber was granulated corn starch added to the product in an amount ten times the amount of water to be absorbed. The operation pressure was 0.5 mbar. Whether the enumerated advantages, including low investments, 35 % saving of energy and shorter drying time outweigh the disadvantages of the above- mentioned point 3, was not discussed by the authors.

Mumenthaler [1.82] discovered similar problems, as already mentioned: freezing in a fluidized bed with CO_2 clogs the filters, reducing the yield to only 80–90 %, with an additional loss of 10 % fines.

1.3 Storage

The storage of a freeze dried product starts with the end of the secondary drying and its transfer into a suitable packing. In the drying plant a certain residual moisture content (RM) is achieved as a function of the product temperature and the drying time (Section 1.2.2).

The desorption isotherm describes, under equilibrium conditions, the amount of water absorbed on the product at a given temperature as a function of water vapor pressure, as shown in Fig. 1.96. To approximate the equilibrium at a given temperature in a short time, the pressure during SD should be small compared with the equilibrium vapor pressure, e. g. at +40 °C and a desired RM < 1 %, p_{ch} should several times 10^{-2} mbar. If the product

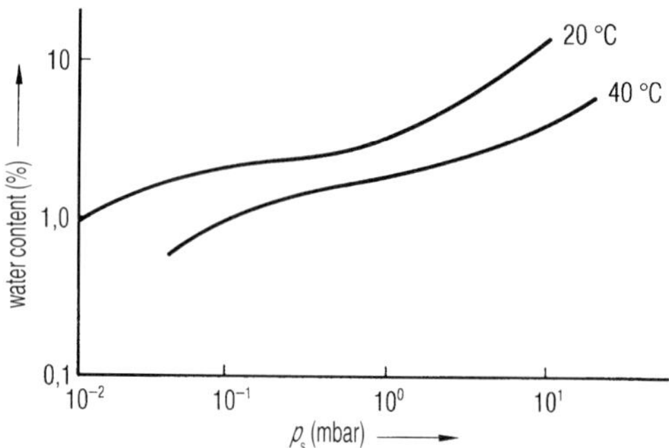

Fig. 1.96. Desorption isotherm of blood plasma (Fig. 9 from [1.63]).

(blood plasma) is to be exposed only to +20 °C, the pressure has to be small compared with 10^{-2} mbar. As shown in Section 1.2.2 a prolonged drying time does not result in a lower RM, only a higher temperature will achieve this. To maintain a low RM, a hygroscopic product has to be protected against the reintroduction of moisture already in the drying chamber. If vials are used, they can be sealed in the chamber, as shown in Section 2.3.3. If bulk material or food has been dried, the chamber has to be vented with dry air or inert gas. At +20 °C and 70 % relative humidity air contains approx. $1.3 \cdot 10^{-2}$ g H_2O/L. During the venting of a 200-L chamber with this air, 2.6 g water vapor are introduced. If the chamber is filled with 300 vials, each containing 1 cm^3 with a solid content of 10 %, the RM will be increased by approx. 9 %. If the solid content is only 1 %, the RM rises to approx. 90 %. The dew point of the venting gas should correspond to the end pressure of SD in the chamber, e. g. if end pressure is $2 \cdot 10^{-2}$ mbar, the dew point of the gas should be –55 °C, minimum –50 °C.

1.3.1 Measurement of the Residual Moisture Content (RM)

For all measurements of RM the product must be handled in such a way as to exclude water absorption from the surroundings. Filling a freeze dried product into another container and/or weighing it, should only be done in boxes or isolators filled with dry gas (see above).

The boxes can contain e. g. P_2O_5 or be rinsed with dry gas. Handling in the isolator should be done wearing rubber gloves fixed to the isolator. Balances used in such a dry gas need some modifications to avoid electrostatic charges, which could lead to substantial errors.

1.3.1.1 Gravimetric Method

Until a few years ago this method has been obligatory, as shown in Title 21 of the *Code of the Federal Regulations for Food and Drugs*, section 610.13 [1.83]. The weighed sample is stored at temperatures between +20 °C and +30 °C in a chamber, together with P_2O_5, and repeatedly weighed until the weight becomes constant. The smallest sample should be larger than 100 mg, if necessary taken from several vials. Higher temperatures lead to shorter times before the weight is constant, but they may desorb more strongly any bound water or even change the product. With this method, at 20–30 °C, water can be detected which is weakly bound to the solid.

1.3.1.2 Karl Fischer method

By this method the weighed dry product is dissolved in methanol and titrated with the Karl Fischer solution until the color changes from brown to yellow. The visual observation can be replaced by an ammeter, which shows an steep increase in current, when the titration is terminated (dead-stop-titration). The samples can be two to four times smaller than for the gravimetric method. To avoid the visual observation completely, iodine can be produced by electrolyzation and the water content is calculated by Coulomb's law. Such an apparatus (e. g. Fig. 1.97.1 and 1.97.2) is available commercially. The smallest amount of water to be detected by such instruments is 10 µg. Wekx and De Kleijn [1.84} showed, how the Karl Fischer method can be used directly in the vial with the dried product. The Karl

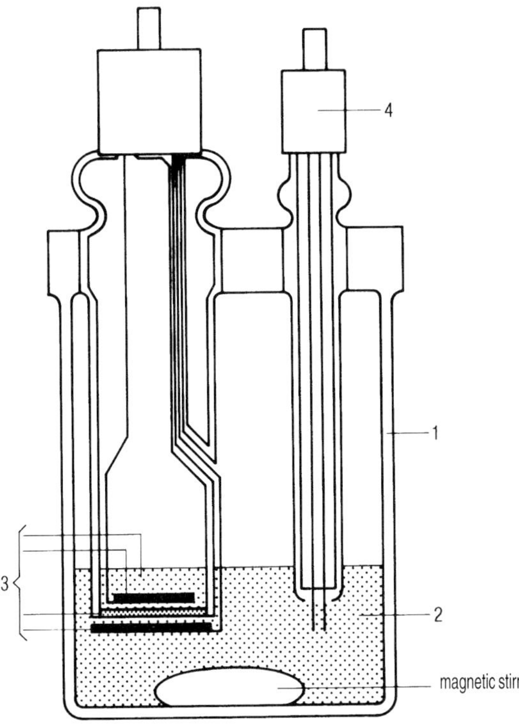

Fig. 1.97.1. Schema of the 'Coulometer Meßzelle DL 36' for measurement of residual moisture content (RM) after Karl Fischer. In the titration cell (1) iodine is electrolytically produced (3) from an iodine-containing analyt (2). Water in the titration cell reacts with the iodine. When the water is used up, a small excess of iodine is produced, which is detected by special electrodes, which leads to iodine production being stopped. The amount of water in the cell can be calculated from the reading of the coulometer, and the amount of electrical charge needed. The solids are introduced into the cell either by a lock, or the water is desorbed in an oven and carried by a gas stream into the cell. 10 µg in a sample can be detected with an accuracy of reading of 0.1 µg (KF Coulometer DL36, Mettler-Toledo AG, CH-8603 Schwerzenbach, Switzerland).

Fig. 1.97.2. KF Coulometer DL 37 with oven on the left.
(Photograph: Mettler-Toledo AG, CH-8603 Schwerzenbach, Switzerland.)

Fischer titration cannot be used if the product reacts with iodine in the Karl Fischer reagent, or does not dissolve in methanol, or the moisture cannot be extracted by the methanol.

1.3.1.3 Thermogravimetry

The weight loss of the product is measured by an electrical balance at constant temperature, or at a given temperature-time profile. For the balance and handling of the product, the rules given in Section 1.3.1 should be carefully observed, as the sample with such a balance can be as small as 2 mg. May et al. from the Center for Biologic Evaluation and Research, Food and Drug Administration [1.83] described the reading of a mass spectrometer during weighing to differentiate between desorbed water and volatile products, which might come from residual solvents or decomposed parts of the product.

1.3.1.4 Infra-red Spectroscopy

Kamat et al. [1.85] described a process by which the water content in freeze dried sucrose is measured by infra-red spectroscopy ($\lambda = 1000$ to 2500 nm) and a newly developed fiber optic (Fig. 1.98). However, whether an interpretation of the absorption lines with respect to the water content is possible must be investigated from product to product. The location of the lines and their relative intensity can prevent their necessary discrimination. For pure sucrose, the authors show that RM can be measured between 0.72 % and 4.74 %, with an error of 0.27 % (Fig. 1.99). By this process many vials of one charge could be measured, since a single scan of approx. 20 s duration is expected. Also in automated unloading systems (see Section 2.41) perhaps only a small proportion (e. g. 1 %) of vials need be checked. Of equal importance is the fact that vials could be re-measured again after days or even years. The authors consider that a minimum sample of 0.5 mg is sufficient for the measurement.

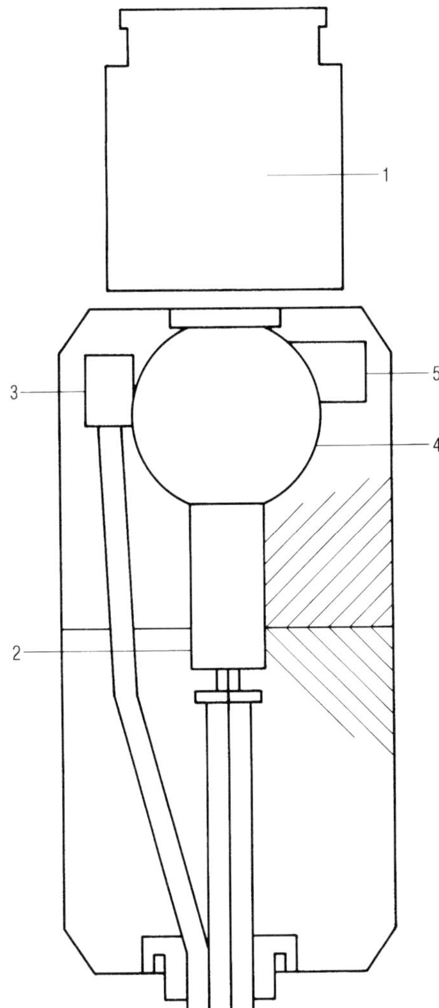

Fig. 1.98. Schema of the fiber-optic sensor to measure the residual moisture content (RM) through the bottom of the container with the product.
1, Product to be measured; 2, fiber optic, which concentrates the light through a safir window on the vial bottom; 3, reference fiber optic; 4, gold coated hollow sphere, in which the two light bundles are integrated; 5, analyzer of the light in the sphere (Fig. 1 from [1.85]).

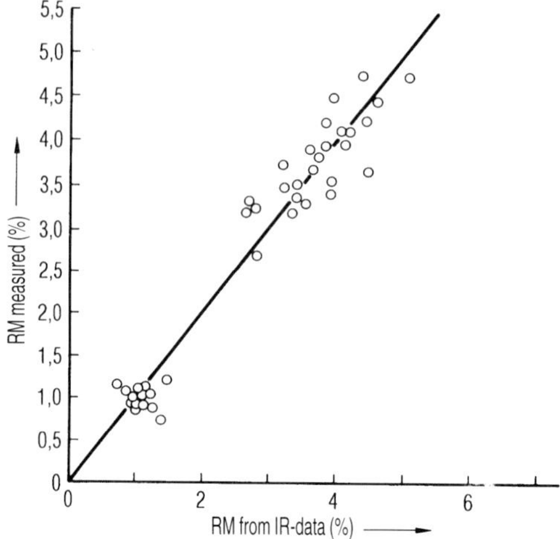

Fig. 1.99. Measured residual moisture content (RM) as a function of the RM data calculated from IR measurements. Within the range from 0.72 % to 4.74 % RM, the error was 0.27 % (Fig. 8 from [1.85]).

Summary of Section 1.3.1:

Water in the dried product can be bound in many different forms: as surface water, as water bound more or less to the dry substance, or as water of crystallization. Therefore each method can lead to different results for different substances. There are products for which the RM values by gravimetry and by Karl Fischer show very little differences. May et al. [1.83] presented four examples of such substances, but as shown in Table 1.15 the RM by gravimetry can be 0.3 to 0.6 % smaller than by Karl Fischer, while the thermogravimetric data, within the given errors, are close to the Karl Fischer data.

Table 1.15: Residual moisture content (RM) of different vaccines measured by four different methods.

Test method	% RM ± standard deviation of the vaccines			
	Rubella virus	Mumps virus	Rubella and mumps virus	Measles, mumps and rubella virus
Gravimetric *	0.42 ± 0.18	1.10 ± 0.40	0.41 ± 0.26	0.18 ± 0.14
Karl Fischer**	1.03 ± 0.14	1.54 ± 0.20	0.72 ± 0.16	0.80 ± 0.14
TG-profile **	1.26 ± 0.16	1.54 ± 0.15	0.76 ± 0.12	0.76 ± 0.11
TG-60 °C hold**	1.17 ± 0.20	1.53 ± 0.17	0.74 ± 0.13	0.70 ± 0.08

* Average of 5 to 12 determinations.
** Average.

TG-profile = Thermogravimetric profile by a given course of temperatures.
TG-60 °C hold: by a constant temperature of +60 °C (part of Table 1 from [1.83]).

1.3.2 Influence of Vial Stoppers on the Residual Moisture Content

The stoppers for vials contain a certain amount of water, which depends on the composition of the stoppers. De Grazio and Flynn [1.86] showed, that the selection of the polymer, the additives for the vulcanization, and the filler influence the adsorption and desorption of water. However even the best possible mixture increases the RM in 215 mg sucrose from 1.95 % to 2.65 % during 3 months storage time at room temperature. Other stopper mixtures show an increase up to 1.7 %. Pikal and Shah [1.87] demonstrated, that the desorption of water from the stopper and the absorption of water by the product depends, in the equilibrium state, on the mass and water content of the stopper and the water content and sorption behavior of the dry product.

If the stopper is as small as technically possible and its material optimized, the water content of the stopper depends on its prehistory: Steam-sterilized stoppers take up water (e. g. 1.1 % of their weight) which can only be removed by 8 h vacuum drying [1.87] or by 8 h recirculated hot air (110 °C) drying down to 0.1 % [1.88]. Figure 1.100 [1.87] shows that a steam-sterilized stopper, vacuum-dried for 8 h, releases a little less water to lactose than does an untreated stopper. A stopper, dried for only 1 h, increases the RM within 6 months stored at 25 °C by a factor of 2.4. Figure 1.100 [1.87] also shows that an equilibrium is reached which, practically, does not change later. The time to reach the equilibrium depends strongly on the temperature. For a given product the time to reach half-maximum increases, from 4 days at +40 °C to 10 months at +5 °C. It is surprising that the absorption isotherms for lactose are found to be temperature-independent at +25 °C and +60 °C; this applies also to vancomycin at +25 °C and +40 °C. Figure 1.101 shows the equilibrium water content as a function of stopper treatment and amount of dry product independent of the storage temperatures of +25 °C and +40 °C for two different products; vancomycin is clearly more hygroscopic than lactose.

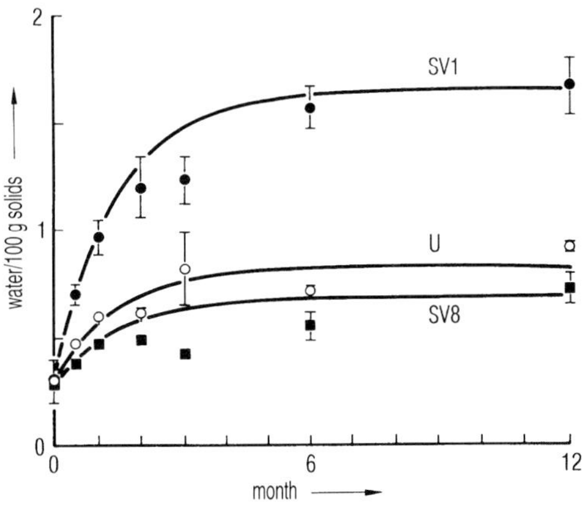

Fig. 1.100. Water content of 100 mg lactose at +25 °C as function of time. The vials have been closed with 13-mm stoppers of different pretreatment. Pretreatments: SV1 steam-Sterilized; U, untreated; SV8, steam-sterilized followed by vacuum-drying of minimum 8 h. The lines are calculated by a model system (Fig. 4 from [1.87]).

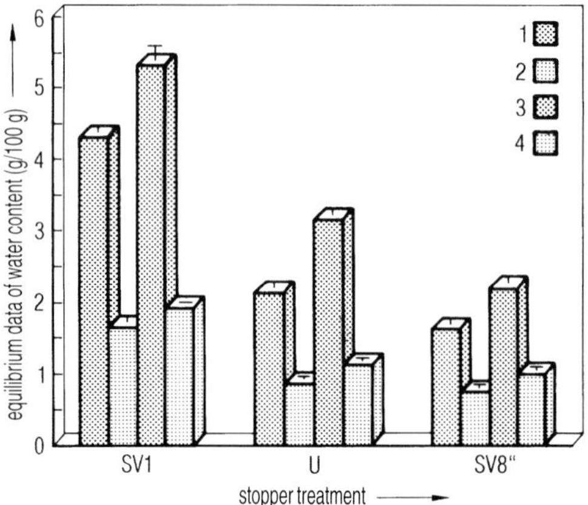

Fig. 1.101. Equilibrium water content in two different freeze dried products, each with two different amounts of product per vial. The equilibrium data are extrapolated from the +25 °C, respectively from the +40 °C values. SV1, U, SV8 see Fig. 1.100.
1, 25 mg lactose; 2, 100 mg lactose; 3, 25 mg vancomycin; 4, 100 mg vancomycin (Fig. 7 from [1.87]).

Earle et al. [1.89] showed, that the RM in the product Pedvax HIB TM did not change during storage at 2–8 °C for 24 months if the stoppers were steam-sterilized, vacuum dried for 6 h and finally dried at +143 °C for 4 h. If the vials were closed with stoppers which had not been dried, the RM increased in 12 months to approx. 5.3 %. Danielson [1.90] warned against toxic components which could diffuse or migrate from the stopper to the product. A protective coating does not avoid the extraction of these substances, but Teflon coating is better than none.

1.3.3 Qualities of the Dry Substances and their Changes

During storage of a freeze dried product, its qualities can change under the influence of at least three conditions: Residual moisture content, storage temperature, and gas mix in the packing assuming, that the freeze drying itself has been carried out under optimum conditions, and the product had the intended qualities at the end of the freeze drying process. The changes which can be related to one or mostly to a combination of the three conditions, can be more simply divided in four groups:

1. Changes in the reconstitution with water and/or the solubility.
2. Chemical reactions in the dried product.
3. Deterioration of the biological-medical activity of the product.
4. Changes of the physical structure of the product, e. g. from an amorphous to a partially or totally crystalline one.

Often changes occur which have to be accounted for as a combination of several of the four categories. In this chapter some typical examples are given, though some special changes are also mentioned in Chapters 3 and 4.

Liu and Langer [1.91] showed, that BSA, ovalbumin, glucose oxydase and β-lacto-Globulin rapidly lost their solubility at 37 °C and within 24 h, 97 % became insoluble if 30 % (w/w) of a buffered, physiological NaCl solution were added to the dried product. The aggregation induced by the moisture was attributed to intermolecular S–S binding. This aggegration can be minimized, if the RM is optimized for a given albumin.

Zhang et al. [1.139] studied the effect of the reconstitution medium on aggregate formation in recombinant keratonocyte growth factor (KGF). Several additives reduced aggregation substantially, but similar effects were observed by adjusting the ionic strength of the reconstitution medium. Optimization of the reconstitution conditions increased the recovery of soluble active proteins; for KGF, the recovery of the soluble active proteins corresponds to the native, monomeric form. Furthermore Zhang et al. [1.140] demonstrated that interleukin-2 (I) and RNase (II) show significant aggregation upon storage at +45 °C when pure water was used for reconstitution. The extent of aggregation can be substantially reduced, if e. g. heparin or phosphates are added to the reconstitution water. Yoshika et al. [1.141] studied the inactivation of β.-galactosidase (I) during storage in relation to the water mobility by O^{17} NMR. An increase in water content also produced an increase of the spin-lattice relaxation time, T_1, the inactivation being more dependent on T_1 than on the pH value. It was assumed that the water increased the mobility of the water around the enzyme, thus enhancing enzyme inactivation. The freeze dried samples with little water also showed an greater inactivation rate than was expected from details of pH value and water mobility, this inactivation being induced by the salts used as additives for freeze drying. Yoshika et al. [1.142] also used NMR but with H_1 spin-spin relaxation time T_2. T_2 has been measured on BSA and γ-globulin (BGG) as a function of hydration level. Freeze dried BSA and BGG became sensitive to aggregation, if the water content exceeded approx. 0.2 g/g protein. T_2 of the protein protons began to increase already at lower water contents and followed the increase in aggregation in time with increasing water content. For freeze dried BGG, as well the aggregation as the T_2 of the protein proton decreased at water contents > 0.5 g/g protein.

The stability of moisture-sensitive drugs has been studied by Vromans and Schalks [1.143] using amorphous vecuronium bromide. Its decomposition in a formulation depends more on the water activity a_w (see Chapter 4, point 2) than on the water content. Glass-forming excipients may not only be cryoprotective but also stabilizing.

Hsu et al. [1.92] demonstrated, that decomposition can take place in a packaged product. It was assumed that the freeze drying had ended with a monomolecular layer of water, which may not have been evenly distributed, but could be attached as cluster to certain locations of the molecule. This water was an optimum protection against denaturation during drying as well as during storage. This has been demonstrated by two very different products, made by gene technology: Too little water, less than one monolayer, makes tPA and met-hemoglobin physically unstable while a higher water content leads to biological instability during storage.

Examples of the fourth type of changes were described by To and Flink [1.93] and Van Scoik and Carstensen [1.94]: According to To and Flink the change from an amorphous to a crystalline phase is induced either by to high storage temperatures T ($T > T_c$) or by water absorption. (*Note:* More water increases the mobility of the amorphous solid, promoting nucleation of crystals and their subsequent growth.)

Van Scoik and Carstensen [1.94] differentiated with their experiments between nucleation of sucrose crystals and their growth. The nucleation parameters of temperature and residual moisture were discussed and additives suggested to stop nucleation, or to delay or speed it up. The influence of gases used for venting the chamber with vials or filled into the packaged bulk product, is difficult to summarize. Only O_2 should be excluded in most cases. Spieß [1.95] suggested dry air for storage of cauliflower and blueberry, while carrots and paprika should be stored in a gas with < 0.1 mg O_2/g dry substance. Less sensitive products such as peas, green beans or mushroom tolerate 0.2–0.5 mg O_2/g dry substance. For drugs, virus or bacteria, no general recommendation can be given, since the influence of CPAs, structural additives, buffer, etc. have to be taken into account.

The purity of gases used should also be specified, as certain impurities can have a decisive effect on storage behavior, e. g. gas desorption from stoppers. Greiff and Rightsel [1.96] have shown that influenza viruses without CPAs keep their infectivity best when stored with 1.6 % RM in helium. In argon, the infectivity decreases approx. 10 times and in O_2 20 times faster if the average of the storage temperatures is used.

References for chapter 1

[1.1] Riedel, L.: Enthalpie-water content diagram for lean beef (also valid for other meats with fat content below 4 %). Recommendations for the processing and handling of frozen foods, p. 28 and 29. International Institute of Refrigeration <IIR> 75 Paris 17e, France, 2. ed 1972

[1.2] Riedel, L.: s. [1.1] p. 32 and 34

[1.3] Duckworth, R. B.: Differential thermal analysis of frozen food systems. I. The determination of unfreezable water. Journ. Food Technol. 6, p. 317–327, 1971

[1.4] Steinbach, G.: Die Bedeutung des Einfriervorganges, Berechnungn des Gefriervorganges, Verein Deutscher Ingenieure, VDI-Bildungswerk, RW 1570, p. 3.

[1.5] Oetjen, G. W.: Industriel freeze-drying for pharmaceutical applications, Table 3, p. 288. Pharmaceutical freeze-drying, edited by Louis Rey, Joan C. May, copyright © 1999 by Marcel Dekker, Inc. New York, N. Y.

[1.6] Riehle, U.: Schnelleinfrieren organischer Präparate für die Elektronenmikroskopie. (Die Vitrifizierung verdünnter wässeriger Lösungen). Chem. Ing. Tech. 40, p. 213–218, 1986

[1.7] de Quervain, M. R.: Crystallization of water, a review. Freeze-Drying and Advanced Food Technology, Chapter 1, p. 3–15. Copyright ©1975 Academic Press, Inc., New York

[1.8] Oetjen, G. W.: Absorptionsmessungen an Lösungen von Neodymsalzen. Dissertation Universität Göttingen, Jan. 1947. Zeitschrift für Naturforschung Bd. 4a, Heft 1, 1949

[1.9] Dowell, L. G., Rinfret, A. P.: Low temperature forms of ice as studied by X-ray diffraction. Nature, Vol. 188, p. 1144–1148, Dec. 1960

[1.10] Luyet, B.: On various phase transitions occuring in aqueous solutions at low temperatures. Annals New York Academy of Science, p. 568, Fig. 14, 1960

[1.11] Thijssen, H. A. C., Rulkens, W. H.: Effect of freezing rate on rate of sublimation and flavour retention in freeze-drying, p. 99–114. International Institute of Refrigeration <IIR> (Comm. X, Lausanne, 1969)

[1.12] Reid, D. S.; Lim, M. H.; McEvoy, H. M.: Studies on the freesing processes in aqueous model systems. Paper 354 International Institute of Refrigeration <IIR> (Comm. C_1, Paris 1983)

[1.13] Burke, M.; Lindow, S. E.: Surface properties and size of the ice nucleation of active bacteria: Theoretical considerations. Cryobiology 27, p. 80–84, Copyright 1990 by Academic Press Inc. 1990

[1.14] Rasmussen, D., Luyet, B.: Contribution to the establishment of the temperature concentration curves of homogeneous nucleation in solutions of some cryoprotective agents. Biodynamica., Vol. 11, No. 225, p. 33–44, 1970

[1.15] Sutton, R. L.: Critical cooling rates to avoid ice cristallisation in solutions of cryoprotective agents. Journ. Chem. Soc. Faraday Trans. Vol. 87, p. 101–106, 1991

[1.16] Levine, H., Slade, L.: Principles of "cryostabilisation" technology from structure/property relationship of carbohydrate/water system-A Review. Cryo-Letters 9, p. 21–63, 1988. Published by Cryo-Letters, 7, Wootton Way, Cambridge. CB3 9LX, U. K.

[1.17] Reid, D., Kerr, W., Hsu, J.: The glas transition in the freezing process. Journ. of Food Eng. 22, p. 483–494, 1994

[1.18] Carpenter, J. F., Arakawa, T., Crowe, J. H.: Interactions of stabilizing additives with proteins during freeze-thawing and freeze-drying. Developments in biological Standardization Vol. 74, p. 225–239. Acting Editors: Joan C. May, F. Brown; S. Karger AG, CH-4009 Basel (Switzerland), 1992

[1.19] Timasheff, S. N., Lee, J. G., Pittz, E. P., Tweedy, N.: The interaction of tubulin and other proteins with structure stabilizing solvens. Journ. of Colloid and Interface Science, 55, p. 658–663, 1976

[1.20] Meryman, H. T.: The "Minimum Cell Volume" modes of freezing injury. Nature 218, p. 333, 1968 und International Institute of Refrigeration <IIR> (Comm. X, p. 897–900, Washington 1971)

[1.21] Pushkar, P. S., Itkin, U. A.: The Study of the intercellular and extracellular cristallization of the biological objects on freezing. International Institute of Refrigeration <IIR> (Comm. X, p. 861–868, Washington 1971)

[1.22] Nei, T.: Ice particles formed in various cells. International Congress of Refrigeration <IIR> (Comm. 1, p. 429–430, Paris 1983

[1.23] De Antoni, G. L., Perez, P., Abraham, A., Anon, M. C.: Trehalose, a cryoprotectant for lactobacillus bulgaricus. Cryobiology 26, p. 149–153, 1989. Copyright ©1989 Academic Press, Inc.

[1.24] Rey, L.: Influence of the preliminary freezing period and adsorption phenomina in freeze-drying. Vortrag auf der 5. Gefriertrockentagung Leybold, Köln, p. 3–19, 1962

[1.25] Mac Kenzie, A. P.: A current understanding of the freeze-drying of representative aqueous solutions. Fundamentals and applications of freeze-drying to biological materials, drugs and foodstuffs, p. 21–34, International Institute of Refrigeration <IIR> 1985–1

[1.26] Luyet, B.; Rasmussen, D.: Study by differential thermal analysis (DTA) of the temperatures of instability of rapidly cooled solutions of Glycerol, Ethylene Glycol, Sucrose and Glucose. Biodynamica., Vol. 10, Nr. 211, p. 167–191, 1968

[1.27] Willemer, H.: The condition of aqueous solutions during freezing demonstrated by E. R. measurements and low temperature freeze-drying, p. 463–472. International Institute of Refrigeration <IIR> (Comm. C1 und C2, Karlsruhe 1977)

[1.28] Cosman, M. D., Toner, M., Kandel, J., Cravalho, E. G.: An integrated cryomicroscopy system. Cryo Letters 10, p. 17–38, 1989. Published by Cryo Letters, 7, Wootton Way, Cambridge CB3 9LX, U. K.

[1.29] Dawson, P. J., Hockley, D. J.: Scanning microscopy (SEM) of freeze-dried preparationes: relationship of morphology to freeze-drying parameters. Developments in Biological Standardization Vol. 74, p. 185–192. Acting Editors: Joan C. May, F. Brown; S. Karger AG, CH-4009 Basel (Switzerland), 1992

[1.30] Gatlin, L. A.: Kinetics of a phase transition in a frozen solution. Developements in Biological Standardization, Vol. 74, p. 93–104. Acting Editors: Joan C. May, F. Brown; S. Karger AG, CH-4009 Basel (Switzerland), 1992

[1.31] Deluca, P. P.: Phase transitions in frozen antibiotic solutions, p. 87 bis 92, International Institute of Refrigeration <IIR> (Comm. C1, Tokyo, 1985)

[1.32] Takeda, T.: Cristallization and subsequent freeze-drying of cephalothin sodium by seeding method. Yakugaku Zasshi 109 (6), p. 395–401, 1989

[1.33] Roos, Y., Karel, M.: Thermal history and properties of frozen carbohydrate solutions. Paper 350. International Institute of Refrigeration < IIR> (XVIIIth Congress, Montreal 1991)

[1.34] Talsma, H., van Steenbergen, M. J., Salemink, P. J. M, Crommelin, D. J. A.: The cryopreservation of liposomes. 1. A differential scanning calorimetry study of the thermal behavior of a liposome dispersion containing mannitol during freezing/thawing. Pharmaceutical Research, Vol. 8, No. 8, p. 1021–1026, 1991. Copyright © 1991 Plenum Publishing Corp., New York, N. Y., USA

[1.35] Williams, N. A.: Differential scanning calorimetric studies on frozen cephalosporin I-solutions. Int. J. Pharm. 44 (1–3), p. 205–212, 1988

[1.36] Hanafusa, N.: The behavior of hydration water of protein with the protectant in the view of HNMR. Developments in Biological Standardization Vol. 74, p. 241–253. Acting Editors: Joan C. May, F. Brown; S. Karger AG, CH-4009 Basel (Switzerland), 1992.

[1.37] Nagashima, N., Suzuki, E.: Freezing curve by broad-line pulsed NMR and freeze-drying, p. 65–70. International Institute of Refrigeration <IIR> (Comm. C1, Tokyo 1985)

[1.38] Hatley, R. H. M.: The effective use of differential scanning calorimetry in the optimisation of freeze-drying processes and formulations. Developments in Biological Standardization Vol. 74, p. 105–122. Acting Editors: Joan C. May, F. Brown; S. Karger AG, CH-4009 Basel (Switzerland), 1992

[1.39] Harz, H.-P., Weisser, H., Liebenspacher, F.: Bestimmung des Fest-Flüssiggleichgewichtes in gefrorenen Lebensmitteln mit der gepulsten Kernresonanzspektroskopie. DKV-Tagungsbericht, p. 741–752, Hannover 1989

[1.40] Girlich, D.: Multikernresonanzuntersuchungen zur molekularen Dynamik wässeriger Saccharidlösungen. Dissertation. Naturwissenschaftliche Fakultät III, Biologie und vorklinische Medizin der Universität Regensburg 1991. Printed by: S. Roderer Verlag, Regensburg, 1992

[1.41] Mac Kenzie, A. P.: The physico-chemical basis for the freeze-drying process. Developments in Biological Standardization Vol. 74, p. 51–67. Acting Editors: Joan C. May, F. Brown; S. Karger AG, CH 4009-Basel (Switzerland) 1992

[1.42] Pikal, M. J.: Journ. Pharm. Sci. 66, p. 1312, 1977

[1.43] Pikal, M. J.: Journ. Pharm. Sci. 67, p. 767, 1978

[1.44] Kovalcik, T. R.: PDA, Journal of Parenteral Science & Technology, Vol. 42, S. 29, 1988. Copyright © 1988 PDA, Inc. Bethesda, Maryland, USA

[1.45] Gatlin, L. A.: Kinetics of a phase transition in a frozen solution. Developments in Biological Standardization, Vol. 74, p. 93–104. Acting Editors: Joan C. May, F. Brown; S. Karger AG, CH-4009 Basel (Switzerland), 1992

[1.46] Yarwood, R. J., Phillips, A. J.: Processing factors influencing the stability of freeze-dried sodium ethacrinate. Pharmaceutical Technology: Drug Stabilization, p. 40–48, 1989. Edited by Rubinstein, Michael, Verlag Ellis Horwood, Chichester, UK.

[1.47] Koray, D. J., Schwartz, J. B.: Effects of exipients on the cristallisation of pharmaceutical compounds during lyophilization. PDA, Journal of Parenteral Science & Technology, Vol. 43, No 2, p. 80–83, 1989. Copyright ©1989 PDA, Inc. Bethesda, Maryland, USA

[1.48] De Luca, P. P., Klamat, M. S., Koida, C.: Acceleration of freeze-drying Cycles of aqueous Solutions of Lactose and Sucrose with tertiary Buthylalcohol (tBA). Congr. Intern. Technol. Pharm. 5th Vol. 1, p. 439–447, 1989

[1.49] VDI-Wärmeatlas, 5. Auflage, VDI-Verlag Düsseldorf, p. Kb 5, 1988

[1.50] Oetjen, G. W., Eilenberg, H. J.: Heat transfer during freeze-drying with moved particles, p. 19–35 International Institute of Refrigeration <IIR> (Comm. X, Lausanne 1969)

[1.51] Steinbach, G.: Wärmeübertragung und Stofftransport bei der Gefriertrocknung. Berechnung von Gefriertrocknungsprozessen. VDI-Bildungswerk, BW 1610, 1974

[1.52] Steinbach, G.: Equations for the heat and mass transfer in freeze-drying of porous and non porous layers and bodiesm, p. 674 bis 683. International Institute of Refrigeration <IIR> (XIII, Washington 1971)

[1.53] Gehrke, H.-H., Deckwer, W.-D.: Gefriertrocknung von Mikroorganismen. II. Mathematische Beschreibung des Sublimationsvorganges. Chem. Ing. Tech. 62, Heft 9, 1990

[1.54] Sharon, Z., Berk, Z.: Freeze-drying of tomato juice and concentrate, studies of heat and mass transfer, p. 115 bis 122. International Institute of Refrigeration <IIR> (Comm. X, Lausanne, 1969)

[1.55] Magnussen, O. M.: Measurements of heat and mass transfer during freeze-drying, p. 65–74. International Institute of Refrigeration <IIR> (Comm. X, Lausanne, 1969)

[1.56] Gunn, R. D., Clark, J. P., King, C. J.: Mass transport in freeze-drying, basic studies and processing implications, p. 79–98. International Institute of Refrigeration <IIR> (Comm. X, Lausanne 1969)

[1.57] Kobayashi, M.: Vial variance of the sublimation rate in shelf freeze-drying. Paper 312. International Institute of Refrigeration <IIR> (Montreal 1991)

[1.58] Oetjen, G. W.: Vakuumtechnik. Ullmanns Enzyklopädie der technischen Chemie, 4. Edition, vol. 3, p. 104, Verlag Chemie, Weinheim/Bergstraße, 1973

[1.59] Pikal, M. J., Shah, S., Roy, M. L., Putman, R.: The secondary stage of freeze-drying: Drying kinetics as a function of temperature and chamber pressure. Intern. Journ. of Pharmaceutics, 60, p. 203–217. Elsevier Science Publishers B. V. (Biomedical Division) 1990

[1.60] Pikal, M.: Freeze-drying of proteins. Part I: Process design. Pharmaceutical Technology International, p. 37–43, 1991

[1.61] Hsu, C. C., Ward, C. A., Pearlman, R., Nguyen, H. M., Yeung, D. A., Curley, J. G.: Determining the optimum residual moisture in lyophilized protein pharmaceuticals. Development in Biological Standardization, Vol. 74, p. 255–271. Acting Editors: Joan C. May, F. Brown, S. Karger AG, CH-4009 Basel (Switzerland) 1992

[1.62] Nail, St. L., Johnson, W.: Methodology for in-process determination of residual water in freeze-dried products. Developments in Biological Standardization, Vol. 74, p. 137–152. Acting Editors: Joan C. May, F. Brown, S. Karger AG, CH-4009 Basel (Switzerland), 1992

[1.63] Willemer, H.: Measurements of temperature, ice evaporation rates and residual moisture content in freeze-drying. Developments in Biological Standardization, Vol. 74, p. 123–136. Acting Editors: Joan C. May, F. Brown, S. Karger AG, CH-4009 Basel (Switzerland), 1992

[1.64] Bardat, A., Biguet, J., Chatenet, E., Courteille, F.: Moisture measurement: A new method for monitoring freeze-drying cycles. PDA, Journal of Parenteral Science & Technology, Vol. 47, No. 6, p. 293–299, 1993. Copyright © 1993 PDA, Inc. Bethesda, Maryland, USA

[1.65] Neumann, K. H., Oetjen, G. W.: Meß- und Regelprobleme bei der Gefriertrocknung. First Intern. Congress on Vacuum Technology, Namur 1958

[1.66] Bouldoires, J. P.: Experimental study of heat and mass transfer during freeze-drying through dielectric and vapour pressure measurements, p. 189–206. International Institute of Refrigeration <IIR> (Comm. X, Lausanne, 1969)

[1.67] Leybold A G Köln: Vacuum Technology, its foundations, formulae and tables. Auflage 9, S. 52, 1987

[1.68] Welch, J.: Vacuum measurement in steam sterilizable lyophilizers. Journ. of Parenteral Sci. and Technol., Vol. 47, No. 1, Jan./Febr. 1993

[1.69] Willemer, H.: Water vapour pressure, its influence on the freeze-drying process and its control. 40th Annnual Congress of the International Association for Pharmaceutical Technology, Abstracts 1–67, Medpharm GmbH, Scientific Publishers, Stuttgart, März 1994

[1.70] Diels, K., Jaeckel, R.: Vakuum Taschenbuch, 2. Aufl., p. 22–24, Springer Verlag, Berlin, Göttingen, Heidelberg, 1962

[1.71] Pikal, M. J., Shah, S.: The collaps temperature in freeze-drying: Dependence on measurement methodology and rate of water removal from the glassy phase. International Journal of Pharmaceutics 62, p. 165–186, 1990

[1.72] Thijssen, H. A. C., Rulkens, W. H.: Effect of freezing rate on rate of sublimation and flavour retention in freeze-drying, p. 99–114. International Institute of Refrigeration <IIR> (Comm. X, Lausanne 1969)

[1.73] Thijssen, H. A. C., Rulkens, W. H.: Retention on aromas in drying food liquids. De Ingenieur, Chemische Techniek (Niederlande) 80, 47, p. 45–56, 1968

[1.74] Flink, J., Karel, M.: Retention of organic volatiles in freeze-dried solutions of carbohydrates. Reprinted with permission from Journal of Agricultural and Food Chemistry, Vol. 18, No. 2, S. 295, 1970, Copyright © 1970 American Chemical Society, Washington, DC 20005, USA

[1.75] Voilley, A., Sauvageot, F., Simatos, D.: Co,fficients de volatilit, r,lative et retention au cours de la lyophilisation de quelques alcools, p. 639–647. International Institute of Refrigeration <IIR> (Washington, 1971)

[1.76] Flink, J.: The retention of volatile components during freeze-drying: A structurelly based mechanism. Freeze-drying and Advanced Food Technology, p. 351–372, 1975. Copyright © 1975 Academic Press, Inc., New York, N. Y., USA

[1.77] Gero, L., Smyrl, T. G.: Behavior of low molecularweight Organic acids during freeze-drying. Journal of Food Science, Vol. 47, S. 954–957, 1982. Copyright <S010> 1982 Institute of Food Technologists, Chicago IL, USA

[1.78] Seager, H., Taskis, C. B., Syrop, M., Lee, T. J.: Structures of products prepared by freeze-drying solutions containing organic solvents. PDA, Journal of Parenteral Science & Technology, Vol 39 (4), p. 161–179, 1985. Copyright © 1985 PDA Inc., Bethesda, Maryland, USA

[1.79] Kahn-Wyler, A.: Kaltlufttrocknung von pharmazeutischen Präparaten und gefrorenen Lösungen in der Wirbelschicht. Dissertation, Philosoph.-Nat. Fakultät der Universität Basel, 1987

[1.80] Labrude, P., Rasolomana, M.: Atomization of oxyhemoglobin in the presence of sucrose. Study by circular dichroism and electronic paramagnetic resonance, comparison with freeze-drying. S. T. P. Pharma 4 (6), p. 472–480, 1988

[1.81] Wolff, E., Gibert, H.: La lyophilization en lit fluidise d'adsorbant, optimisation et applications. Paper 313, International Institute of Refrigeration <IIR> (Montreal 1991)

[1.82] Mumenthaler, M.: Sprühgefriertrocknung bei Atm.-Druck: Möglichkeiten und Grenzen in der pharmazeutischen Technologie und der Lebensmitteltechnologie. Dissertation Universität Basel, 1990

[1.83] May, J. C., Wheeler, R. M., Etz, N., Del Grosso, A.: Measurement of final container residual moisture in freeze-dried biological products. Developments in Biological Standardization Vol. 74, p. 153–164. Acting Editors: Joan C. May, F. Brown, S. Karger AG, CH 4009 Basel (Switzerland), 1992

[1.84] Wekx, J. P. H., De Kleijn, J. P.: The determination of water in freeze-dried pharmaceutical products by performing the Karl-Fischer-titration in the glass container itself. Drug Dev. Ind. Pharm. 16,(9) p. 1465–1472, 1990

[1.85] Kamat, M. S., Lodder, R. A., De Luca, P. P.: Near-infrared spectroscopic determination of residual moisture in lyophilized sucrose through intact glass vials. Pharmaceutical Research 6, p. 961–965, 1989. Copyright ©1989 Plenum Publishing Corp, New York, N. Y., USA

[1.86] De Grazio, F., Flynn, K.: Lyophilization closures for protein based drugs. PDA, Journal of Parenteral Science & Technology Vol 46 (2), p. 54–61, 1992, Copyright ©1992 PDA Inc., Bethesda, Maryland, USA

[1.87] Pikal, M. J., Shah, S.: Moisture transfer from stopper to product and resulting stability implications. Developments in Biological Standardization Vol. 74, 1991, p. 165–179. Acting Editors: Joan C. May, F. Brown, S. Karger AG, CH-4009 Basel (Switzerland), 1992

[1.88] Brinkhoff, O.: Primärpackmittel für Lyophilisate, p. 145 Essig-Oschmann, Lyophilisation, Bd. 35. Copyright: Wissenschaftliche Verlagsgesellschaft., Stuttgart 1993

[1.89] Earle, J. P., Bennett, P. S., Larson, K. A., Shaw, R.: The effects of stopper drying on moisture levels of haemophilus influenzae conjugate vaccine. Development in Biological Standardization, Vol. 74, . 203–210. Acting Editors: Joan C. May, F. Brown, S. Karger AG, CH-4009 Basel (Switzerland), 1992

[1.90] Danielson, J. W.: Toxicity potential of compounds found in parenreral solutions with rubber stoppers. PDA, Journal of Parenteral Science & Technology, Vol 46 (2), p. 43–47, 1992, Copyright © 1992 PDA Inc., Bethesda, Maryland, USA

[1.91] Liu, W. R., Langer, R.: Moisture induced aggregation of lyophilized proteins in the solid state. Biotechnol. Bioeng. 37 (2), p. 177–184, 1991

[1.92] Hsu, C. C., Ward, C. A., Pearlman, R., Nguyen, H. M., Yeung; D. A., Curley, G.: Determining the optimum residual moisture in lyophilized protein pharmaceuticals. Development in Biological Standardization, Vol. 74, p. 255–271. Acting Editors: Joan C. May, F. Brown, S. Karger AG, CH-4009 Basel (Switzerland) 1992

[1.93] To, E. C., Flink, J. M.: "Collapse", a structural transition in freeze-dried carbohydrates. Journ. Food Technology 13, p. 583–594, 1978

[1.94] Van Scoik, K. G., Carstensen, J. T.: Nucleation phenomina in amorphous sucrose systems. International Journal of Pharmaceutics 58 (3), p. 185–196, 1990

[1.95] Spiess, W.: Verfahrensgrundlagen der Trocknung bei niedrigen Temperaturen. VDI-Bildungswerk BW 2229, p. 5, 1974

[1.96] Greiff, D., Rightsel, W. A.: Stabilities of dried suspensions of inflenza virus sealed in vacuum or under different gases. Applied Microbiology 17, Table 3, p. 830–835, 1969

[1.97] AMSCO, Finn-Aqua, D-50354 Hürth.
[1.98] Umrath, W.: Kurzbeitrag für die Tagung Raster-Elektronenmikroskopie in Medizin und Biologie, unpublished, Umrath, W. D-50321 Brühl
[1.99] Umrath, W.: Cooling bath for rapid freezing in electron microscopy. Journal of Microscopy, Vol. 101, Pt 1, p. 103–105, 1974. Copyright ©1974, Blackwell Scienticfic Publications Ltd, Oxford, UK
[1.100] Umrath, W.: unpublished results, W. Umrath, D-50321 Brühl
[1.101] Luyet, B.: On the amount of water remaining amorphous in frozen aqueous solutions, Biodynamica. Vol 10, Nr. 218, p. 277–291, Dec. 1969
[1.102] Willemer, H., D-50668 Köln. Unveröffentliche Messungen.
[1.103] Rasmussen D., Luyet B.: Complementary study of some non equilibrium phase transitions in frozen solutions of glycerol, ethylene, glycol, glucose and sucrose. Biodynamica. Vol 10, No 220, Dec. 1969
[1.104] Nunner, B.: Gerichtete Erstarrung wässriger Lösungen und Zellsuspensionen. Dissertation März 1993, Rheinisch-Westfälische Technische Hochschule Aachen. Copyright 1994, Verlag Mainz, Wissenschaftsverlag, D-52072 Aachen.
[1.105] Knowles, P. F., Marsh, D., Rattle, H. W. E.: Magnetic resonance of biomolecules. John Wiley & Sons, Ltd., New York, Chichester
[1.106] Kochs, M., Körber, Ch., Nunner, B., Heschel, I.: The influence of the freezing process on vapor transport during sublimation in vacuum-freeze-drying. Journal of Heat and Mass Transfer 34, p. 2395–2408, 1991
[1.107] Willemer, H.: Moderne Anlagen der Lyophilisation in Essig-Oschmann: Lyophilisation, Wissenschaftliche Verlagsgesellschaft mbH, Stuttgart, vol. 35, 1993
[1.108] Lentges, G., Oetjen, G. W., Willemer, H., Wilmanns, J.: Problems of measurement and control in freeze-drying down to –180 °C, p. 707–715. International Institute of Refrigeration <IIR> (XIII Cong. Washington 1971)
[1.109] Smithsonian Metrological Tables, 6th ed, 1971 and VDI-Wasserdampftafeln, 6. Ed., 1963
[1.110] Willemer, H.: Influence of product temperature and gas composition on the lyophilisation process. International Congress, p. 63–77, Basel 1994, Copyright © 1994 PDA, Inc., Bethesda, Maryland, USA
[1.111] Willemer, H.: Determination of freezing and freeze-drying data based on light optical micrographs and electrical resistance measurement, p. 9–15, 7th International Freeze Drying Course, Lyon 1997
[1.112] Willemer, H.: unpublished works
[1.113] MacKenzie A. P.: Collapse during freeze-drying, qualitative and quantitative aspects. Freeze-drying and advanced food technology, p. 282. Copyright © 1974 Academic Press Inc. New York, N. Y., USA
[1.114] Sutton, R. B.: Critical cooling rates for aqueous cryoprotectants in the presence of sugars and polysaccharides. Cryobiology 29, p. 585–598, 1992 Copyright © 1992, Academic Press, Inc.
[1.115] Correleyn, S., Remon, J. P.: The use of maltodexin in the lyophilization of a model protein, LHD (Lactate dehydrogenase) . Pharm. Res., 13 (1), p. 146, 1995. Publisher APGI, Chatenay Malabry, Fr.
[1.116] Shalaev, E. Yu., Kaney, A. N.: Study of the solid-liquid state diagram of the water-glycin-suchrose system. Cryobiology, 31, p. 374–382, 1994. Copyright © 1994 Academic Press Inc.

[1.117] Jang, J. W., Kitamura, S., Guillory, J. K.: The effect of excipients on the glass transition temperatures for FK 906 in the frozen and lyophilized state. PDA Journal of Pharmaceutical Science & Technology, Vol. 49, No. 4, p. 166–174, 1995

[1.118] Gordon, M., Taylor, J. S.: Ideal copolymers and the second-order transitions of synthetic rubbers in non-crystalline copolymers. J. Appl. Chem, p. 493, 1952

[1.119] Nicolajsen, H., Hvidt, A.: Phase behavior of the system trehalose-NACl-water. Cryobiology 31, p. 199–205, 1994

[1.120] Carpenter, J. F., Prestrelski, S. J., Anchordogy, T. J., Arakawa, T.: Interaction of stabilizers with proteins during freezing and drying. ACS Symposium Ser. 567 (Formulation and delivery of proteins and peptides), p. 134–147, 1994

[1.121] Hsu, C. C., Walsh, A. J., Nguyen, H. M., Overcashier, E. D., Koning-Bastiaan, H., Bailey, R., Nail, S. L.: Design and application of a low-temperature peltier-cooling microscope stage. J. Pharm. Sci. 85 (1) p. 70–71, 1996

[1.122] Kanaori, K., Nosaka, A., J.: Studies on human calcitinin fibrillation by protonnuclear magnetic resonance spectroscopy: Characterization of the lyophilized fibril. Proceedings of the International Society of Magnetic Resonance, XIIth meeting, part 1, p. 274–275. Bull. Magn. Reson., 17, p. 1–4, 1995

[1.123] Prestelski, S. J., Arakawa, T., Carpenter, J. F.: Separation of freezing- und drying-induced denaturation of lyophilized proteins using stress-specific stabilization. II. structural studies using infrared spectroscopy. Arch. Biochem. Biophys. 303 (2), p. 465–473, 1993

[1.124] Carrington, A. K., Sahagian, M. E., Goff, H. D., Stanley, D. W.: Ice crystallization temperatures of sugar/polyasaccharide solutions and their relationship to thermal events during warming. Cryo-Letters, 15, p. 235–244, 1994

[1.125] Williams, N. A., Gugliemo, J.: Thermal mechanical analysis of frozen solutions of mannitol and some related stereoisomers: Evidence of expansion during warming and correlation with vial breakage during lyophilization. PDA, Journal of Parenteral Science & Technology, vol. 47, No 3, p. 119–123, 1993. Copyright © 1993 PDA, Inc. Bethesda, Maryland, USA

[1.126] Morris, K. R., Evans, S. A., Mackenzie, A. P., Scheule, C., Lordi, N. G.: Prediction of lyophile collapse temperatures by dielectric analysis. PDA, Journal of Parenteral Science & Technology, vol. 48, No 6, p. 318–329, 1994. Copyright © 1994 PDA, Inc. Bethesda, Maryland, USA

[1.127] Smith, G., Duffy, A. P., Shen, J., Ollliff, C. J.: Dielectrc relaxation spectroscopy and some applications in the pharmaceutical science. J. Pharm. Sci., 84, 9, p. 1029–1044, 1995

[1.128] Kasrajan, K., De Lucca, P. O.: Thermal analysis of tertiary butyl alcohol-water system and its implication on freeze-drying. Pharm. Res. 12 (4), p. 484–90, 1995

[1.129] Wolff, E., Gibert, H. Rudolf, F: Vacuum freeze-drying kinetics and modeling of a liquid in a vial. Chem. Eng. Process., 25 (3), p. 153–158, 1989

[1.130] Ybema, H., Kolkmann – Roodbeen, L., te Booy, M., P., W., M., Vromans, H.: Vial lyophilization: Calculation on the rate limitation during primary drying. Pharm. Res. 12 (9), p. 1260–63, 1995

[1.131] Chang, B. S., Fuscher, N. L. The development of an efficient single-step freeze-drying cycle for interleukin-1 receptor antagonist formulation. Pharm. Res. vol. 12, No. 6, p. 831–837, 1995

[1.132] Kasraian, K., DeLuca, P. P. The effect of tertiary butyl alcohol on the resistance of the dry product layer during primary drying. Pharm. Res., 12 (4), p. 491–95, 1995

[1.133] Schellenz, G., Engel, J. Rupprecht, H.: Sublimation during lyophilization detected by temperature profile and X-ray technique. International Journal of Pharmaceutics 113, p. 133–140, 1995. Copyright © 1995 Elsevier Science B. V.

[1.134] Drummond, J. N., Day, L. A.: Influence of vial construction and material on performance, uniformity and morphology during freezing and freeze drying. PDA proceedings of the International Congress, p. 401–427, Osaka 1997

[1.135] Willemer, H., Spallek, M., Auchter-Krummel, P., Heinz, J.: Freezing and freeze drying of pharmaceuticals in tubing, vials with quartz-coated surfaces and resin vials, PDA, Proceedings of the International Congress, p. 99–108, Basel 1998

[1.136] Milton, N., Pikal, M. J., Roy, M. L., Nail, S. L.: Evaluation of manometric temperature measurement as a method of monitoring product temperature during lyophilization. PDA Journal of Pharmaceutical Science & Technology, vol. 51, No. 1, p. 7–16, 1997

[1.137] Connelly, J. P., Welch, J. V.: Monitor lyophilization with mass spectrometer gas analysis. J. of Parenteral Science & Technology, vol. 47, No. 2, pages 70–75, 1993

[1.138] Haseley, P., Oetjen, G. W.: Equipment data, thermodynaeumic measurements, and in-process control quality control during freeze-drying. PDA International Congress, p. 139–150, Basel 1998

[1.139] Zhang, M. Z., Wen, J., Arakawa, T., Orestrelsky, S. J.: A new strategy for enhancing the stability of lyophilized protein: The effect of the reconstitution medium on the keratinocyte growth factor. Pharm. Res. 12 (10), p. 1447–1452, 1995

[1.140] Zhang, M. Z., Pikal, K., Nguyen, T., Arakawa, T., Prestrelski, S. J.: The effect of reconstitution medium on the aggregation of lyophilized recombinant interleukin-2 and ribonuclease A. Pharm. Res. 13 (4), p. 643–646, 1996

[1.141] Yoshika, S., Aso, Y., Izuutsu, K., Terao, T.: Stability of beta-galactosidase, a model protein drug, is related to water mobility as measured by oxygen-17 nuclear magnetic resonance (NMR). Pharm. Res. 10 (1), p. 103–108, 1993

[1.142] Yoshika, S., Asu, Y., Kojima, Sh.: Determination of molecular mobility of lyophilized bovine serum albumin and gamma-globulin by solid-state H^1 NMR and relation to aggregation-susceptibility. Pharm. Res. 13 (6), p. 926–930, 1996

[1.143] Vromans, H., Schalks, E. J. M.: Comparative and predictive evaluation of the stability of different freeze dried formulations containing an amorphous, moisture-sensitive ingredient. Drug Dev. Ind. Pharm., 20 (5), p. 757–768, 1994

[1.144] Oesterle, J., Franks, F., Auffret, T.: The influence of tertiary butyl alcohol and volatile salts on the sublimation of ice from frozen sucrose solution: Implications for freeze drying. Pharmaceutical Developments and Technology, 3 (2), p. 175–183, 1998. Copyright © 1998 by Marcel Dekker, Inc., New York, N. Y., USA

2 Installation and Equipment Technique

2.1 Freezing Installation

2.1.1 Cooling by Liquids: Shell-Freezing and Spin-Freezing

The freezing of liquids in vials, bottles or flasks in a liquid bath is the most common freezing method used in laboratories. As the liquid must have a low melting point, alcohol (ethanol, melting point –114 °C) cooled by CO_2 (boiling point –80 °C at 1 atm) is frequently used. The bath can also be cooled by refrigerated coils.

In the cooled bath the container can be rotated slowly (shell-freezing) or quickly (spin-freezing), as shown in Fig. 2.1. The aim of both methods is to reduce the thickness of the liquid product before freezing to e. g. 15 or 20 mm. For production purposes, this method can not be used, since the liquid must be removed from the surfaces before loading the vacuum plant. This can be done by hand for a limited number of containers, but not in a production scale.

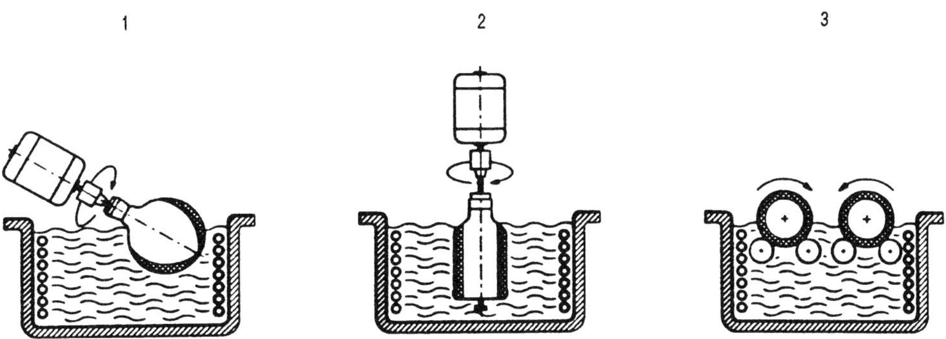

Fig. 2.1. Freezing methods.
1. Shell-freezing; a flask is placed in cold bath in such a way, that the neck of the flask is covered by the liquid. A motor turns the flask and the product freezes on the wall. 2. Spin-freezing; one or more bottles are fixed to a jig and immersed in the bath. The jig turns the bottle(s) so fast around its (their) axle(s), that the liquid is distributed evenly on the wall(s). 3. Shell-freezing: the bottles are placed on cylinders in the bath, the cylinders turn in the bath. The bottles are turned by the cylinders around their axes (Fig. 3 from [2.20]).

Fig. 2.2. Automatic filling of a product to be freeze dried in vials, which are cooled by LN_2 . In the installation the following steps are carried out with a speed of up to 9000 vials per hour: Precooling of the vials down to –180 °C, filling of the vials, shock freezing of the product, placing the stoppers on the vials (photograph: Groninger & Co. GmbH, D-7180 Crailsheim).

If the content of vials has to be frozen in a production process as quickly as possible in liquids, LN_2 must be used. Figure 2.2 shows an automatic filling plant for vials, which are subsequently cooled in LN_2. The product can be filled into precooled vials with high freezing speeds. Depending on the size of the vials and the amount of product filled, freezing speeds up to 100 °C/min can be achieved. Such freezing speeds minimize freeze concentration- and separation effects, but lead to very small crystals. They will dry more slowly and can contain unfrozen (not crystallized) water. The disadvantages of the unfrozen water are described in Section 1.2.5. By a thermal treatment of the product (see [1.45] and Section 1.1.6) these problems can be overcome, but the procedure on a production scale is complex. It is necessary to warm the product for a given time at a given temperature, recool it to low temperatures, and then begin the drying process. Temperature and time must be kept within small tolerances. The data can be developed, e. g. by studies with a cryomicroscope or by DSC measurements.

2.1.2 Cooled Surfaces

In most laboratory-, pilot- and production plants, in which the content of vials and trays are freeze dried, the shelves can be cooled to –40 °C or –50 °C, while in special plants –60 °C or a little less can be reached. The containers can be loaded onto precooled or room temperature shelves (Fig. 1.80.1–1.80.3). The possible freezing speeds can be estimated per

Eq. (3) and by methods described in Section 1.1.1. If the shelves are precooled, the loading must be done quickly to minimize the condensation of water vapor from the air on the shelves. For production plants with tens or hundreds of thousand of vials, special loading installations (see Section 2.4.1) are necessary to minimize this problem.

For the freezing of food, stainless steel belt conveyors are used, which are cooled with a spray of cold brine. The design of such conveyors is somewhat difficult, since sealing of the moving belt between the brine and food can cause leaks and abrasion.

2.1.3 Product in the Flow of Cold Air, Foaming and Freezing of Extracts and Pulps

The freezing of pharmaceutical products is practically always done in vials, bottles or sometimes trays. Food or similar products can be frozen in a flow of cold air in a fluidized bed freezer, if the product is granulated or in small pieces. These conveyor belt – or fluidized bed freezers are available commercially in various forms. Figure 2.2 shows a plant in which vials with product can be cooled in LN_2 or cold N_2 gas. Whether and when such a process provides enough advantages of quality to justify the cost can only be decided from case to case. Two advantages are: (i) Freezing in a sterile gas with little O_2 and (ii) very rapid freezing.

Figure 2.3 illustrates a process in which the outer layer of a product is quickly frozen as a congealed crust. These 'CRUSToFREEZE'™ plants have a capacity between 1500 and 5000 kg/h and require 0.5 to 0.8 kg LN_2 pro kg of product, which has to be frozen totally on a conveyor belt. Figure 2.4 shows the product exit of the plant in Fig. 2.3. The freeze drying of coffee and tea extracts, fruit pulps or small pieces of meat require a multi- stage pretreatment. The granulated end product from coffee and tea extracts should have a defined grain size, a desired color, and a predetermined density. Fruit pulps should become granulated, with the appearance of fruit pieces, while meat pieces should not stick together like a small meat ball, but be recognized as single pieces when presented in a meal.

Coffee and tea extracts are therefore foamed by N_2 or CO_2 during cooling and partially freezing (e. g. to –5 °C) in a type of ice-cream machine. This foam must have a desired Density, with the inclusion of certain amount of small ice crystals. The foam is cooled on a conveyor belt not to –18 °C, but to –40 °C or colder, as this product must pass a grinding and sieving system to achieve a desired grain size and density.

A typical grinding and sieving system (Fig. 2.5) produces dust which has to be collected. If in addition a CIP (cleaning in place) installation is installed, the total preparation equipment becomes a major part of the whole freezing and freeze drying installation. Figure 2.6 shows only the collecting container for the dust, which is sorted by size while below the container is the vibration transport system.

The color of the end product is influenced by the freezing speed (fine crystals show a lighter color). Furthermore, the color is influenced by the structure of the foam and the surface of the dried product. To freeze meat in single pieces, special temperatures have to be used during cutting and preparation of each type of meat.

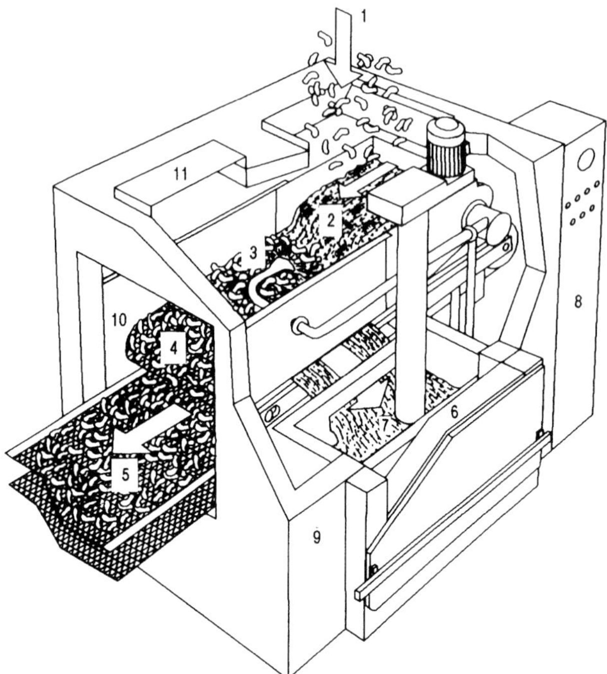

Fig. 2.3. Schema of the IQF-freezing process 'CRUST o FREEZE'®, AGA AB, Frigoscandia Equipment AB, S-25109 Helsingborg.
1, product entrance; 2, IQF-mixer; 3, injection of LN_2; 4, conveyor belt; 5, belt for frozen product; 6, LN_2 pump; 7, storage container; 8, control unit; 9, Isolated housing; 10 and 11, access openings.

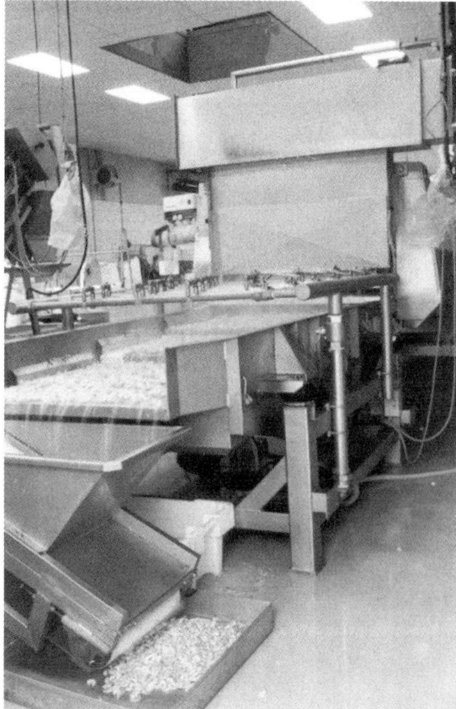

Fig. 2.4. Exit lock of the 'CRUST o FREEZE'® as shown in Fig. 2.3.

Fig. 2.5. Granulating and sieving installation to produce a granulate of specified dimensions from frozen plates or large lumps. The mills can be seen in the background on top, the sieves can be seen in the lower part in front. The photograph was taken before installation in the cold room (photograph: Vibra Maschinenfabrik Schultheis GmbH & Co, D-6050 Offenbach am Main).

Fig. 2.6. Containers for the dusts which are produced during milling and sieving. The dust is exhausted and collected in the containers. The photograph is taken before installation in the cold room (photograph: Vibra Maschinenfabrik Schultheis GmbH & Co, D-6050 Offenbach am Main).

2.1.4 Droplet Freezing in Cold Liquids

The process sounds simple, but becomes difficult if droplets of uniform size are to be produced. The other problem is the formation of a gas veil, which is produced if the liquid, e. g. LN_2, evaporates (see Table 1.5. and Fig. 1.5)

Figure 2.7.1. shows the schema of a process for freezing pellets for freeze drying and Fig. 2.7.2 illustrates the details of the freezing chamber. Such installations are offered under the trade name Cryopel™ with throughputs between 10 and 250 kg/h.

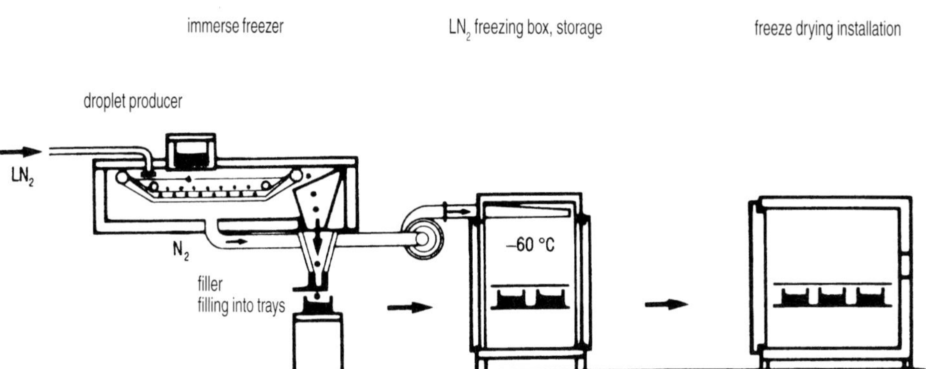

Fig. 2.7.1. Course of processes in a freezing plant, in which the liquid product is frozen dropwise in LN_2 (Cryopel'® Firma Messer Griesheim GmbH, D-47809 Krefeld).

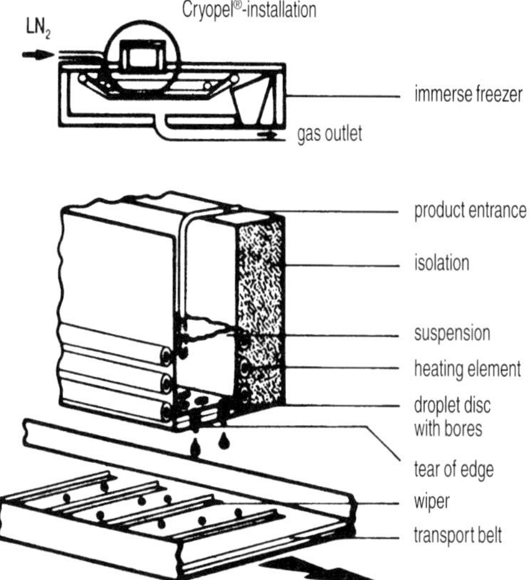

Fig. 2.7.2. Details of the freezer shown in Fig. 2 77.1.

This process is of special interest if a product has to be frozen more quickly than is possible on belts or in trays: A pellet of 2 mm diameter is cooled from 0 °C to –50 °C in approx. 10 s, or at a rate of approx. 300 °C/min. The advantages are: minimum freeze concentration, free-flow product, small ice crystals (which are acceptable in this case of small transport distances for energy and water vapor). It is likely that some pellets (those too large or too small) will need to be removed by sieving.

Yokota [2.1] sprays the liquid to be frozen in to a film of cold n-hexane, which flows down the inner wall of a conical vessel. The frozen particles are sieved off. With this method, two problems need to be solved: (i) The droplet size cannot be influenced, product parts can be extracted, and the product and n-hexane must to be completely separated, and (ii) the process must be sterile.

2.1.5 Freezing by Evaporation of Product Water

This method (see Section 1.1.1) is only mentioned in order to highlight the problems associated with it no applications are given. The withdrawal of water from an aqueous product under vacuum is a vacuum-drying process with known consequences namely structure changes and shrinkage. Depending on the viscosity of the product it is difficult to dissolve skin surfaces, or sticking lumps are formed. The remaining water may no longer be freezable. This method cannot be recommended as a freezing step for products which are to be freeze dried.

2.2 Components of a Freeze Drying Plant

2.2.1 Installations for Flasks and Manifolds

Figure 2.8.1 shows a typical installation for flasks and other containers in which the product is to be dried. The condenser temperature for this plant is offered either as –55 °C or as –85 °C. For this type of plant, a condenser temperature of –55 °C is sufficient as this temperature corresponds with a water vapor pressure of approx. $2.1 \cdot 10^{-2}$ mbar, allowing a secondary drying down to approx. $3 \cdot 10^{-2}$ mbar. This is acceptable for a laboratory plant, in which the limitations are not the condenser temperature but the variation of heat transfer to the various containers, the rubber tube connections and the end pressure of the vacuum pump (2 stage pump, approx. $2 \cdot 10^{-2}$ mbar). Figure 2.8.2 shows that these units are designed for very different needs. The ice condenser in this plant can take up 7.5 kg of ice at a temperature down to –53 °C.

Fig. 2.8.1. Freeze drying plant for flasks and bottles, 16 connections 1/2", titan condenser for maximum 3 kg of ice in 24 h, condenser temperature –55 °C or –85 °C (Flexi-Dry® MP, FTS Systems, Inc., Stone Ridge, New York).

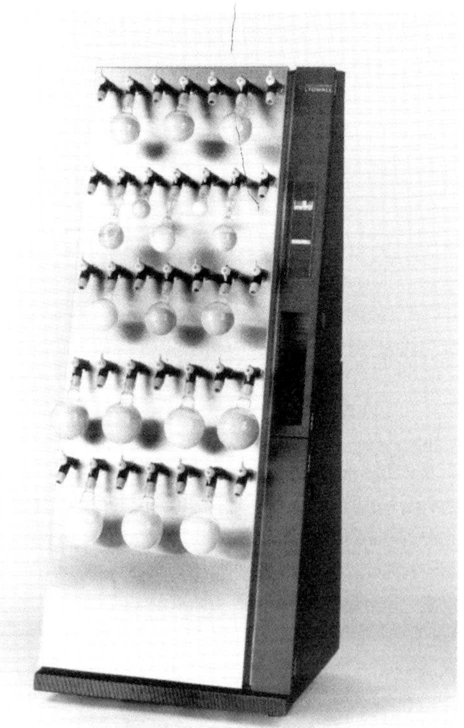

Fig. 2.8.2. Freeze drying plant for flasks or bottles, 35 connections NS 29/32, maximum 7.5 kg ice in 24 h, T_{co} down to –53 °C (Lyowall® Firma AMSCO Finn-Aqua GmbH, D-50354 Hürth).

These installations are relatively easy to handle, their disadvantage being the control of T_{ice}. The heat transfer from the air is difficult to adjust, since it depends on the geometry of the containers and their location at the plant. If e. g. –30 °C has to be the ice temperature, the containers have to be cooled by a bath of approx. –10 °C. It is preferable to dry such products in a chamber with shelves, which can be cooled and heated.

2.2.2 Drying Chambers and Forms of Trays

Drying chambers for freeze drying are built in three basic configurations (Fig. 2.9):

- Bell with baseplate
- Rectangular or cylindrical chamber
- Tunnel with round cross-section

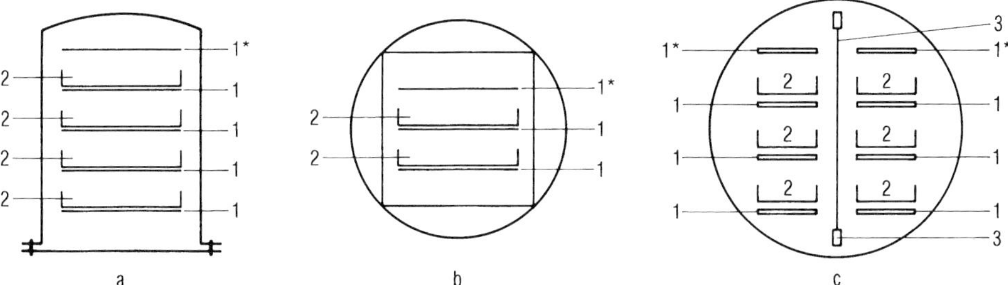

Fig. 2.9. Basic types of freeze drying chambers. A, bell jar or vertical cylinder; b, rectangular or cylindrical chamber with one (or 2) door(s); c, tunnel dryer, in which the trays are transported in and out by a system (shown as a carrier on a monorail).
1, temperature controlled shelves; 1*, temperature controlled plate, to expose the product on the upper shelf to the same condition as on the other shelves; 2, trays or vials; 3, transport system for trays.

Bells with baseplate are used for laboratory plants; they are cost-effective, but cannot be sterilized by steam. Figure 2.10 shows a typical bell installation in which the shelves are usually heated, but cooling can be provided as well as a closing mechanism for vials.

In Fig. 2.11, a cylindrical drying chamber is shown with a Plexiglas™ door and a hydraulic closing system for the vials (not to be sterilized by steam). Figure 2.12 represents a rectangular production chamber, sterilizable by steam, designed for automatic loading and unloading (see Section 2.4.1). The shelves are loaded with vials through a small door (see Section 2.4) which can be closed by a hydraulic system (see Section 2.4.1).

Fig. 2.10. Freeze drying plant of the type in Fig. 2.9 (a). 1600 cm^2 temperature-controlled shelf area stoppering device for vials on four shelves, valve between chamber and condenser, for BTM and DR-measurements, freezing is possible between the condenser coils or in the shelves if they are cooled and heated by brine from a thermostat, T_{co} down to –55 °C (LYOVAC® GT 2, AMSCO Finn-Aqua, D-50354 Hürth).

Fig. 2.11. Freeze drying plant of the type in Fig. 2.9 (b). 4000 cm^2 shelf area, T_{sh} from –50 °C to +70 °C, stoppering device for vials, T_{co} down to –65 °C (Lyoflex 04® Firma BOC Edwards Calumatic B. V., 5107 NE Dongen).

Fig. 2.12. Rectangular chamber, steam-sterilized, with a small loading door in the main door, which is opened at an angel of approx. 30° (lower left corner) (AMSCO Finn-Aqua, D-50354 Hürth).

Tunnels as drying chambers are used for luxury and food products or other products prepared on a large scale, e. g. collagen. For example the rapid loading and unloading of e. g. 500 kg product in 15 min is a typical requirement. The product, in trays, is placed on cars hanging on an overhead rail. The cars can either be quickly moved between the heated shelves (Fig. 2.13.1) and later unloaded the same way or, they can be moved continuously moved through the tunnel. In the other method (Fig. 2.13.2), larger trays are pushed through the distances between the heating plates (see Section 2.5.2). All chambers must be easy to clean (see Section 2.3.4), i. e. the surfaces must be smooth, and all corners rounded, leak tight and with no measurable resistance to the flow of water vapor. If the water evaporation rate is high (e. g. up to 3 kg/h m^2 in food freeze drying) or the operation pressure is low (p_c 0.08 mbar during MD for pharmaceuticals), the transport path for water vapor have to be carefully calculated (see Section 1.2.4). If at all possible, bottles or vials should be placed directly on the shelves, as the heat transfer is approximately when there are no trays between the vials and shelves. For pilot and small production plants, trays can be used with a bottom that may be removed before evacuation. If trays are used, they should have a machined bottom, as can be seen from Table 1.9; the heat transfer coefficient for machined bottoms can be up to twice that of trays with uneven bottoms.

For granulated luxury products and food, two basic forms of tray are used (Fig. 2.14):

a) Large, rectangular or square trays with low walls (e. g. 400 · 500 · 30 mm); or
b) ribbed trays (e. g. 500 · 160 · 50 mm)

Trays of type (a) are pushed through the plant between the heated shelves without contact (System Atlas Industries A/s, DK 2750 Ballerup, Danmark). The ribbed trays (b) are made

Fig. 2.13.1. Freeze drying tunnel plant. Upper part in front: monorail of the transport system. In the tunnel: transport rail for the carrier with trays. In the tunnel: heated shelves in between which the carrier with trays is moved. When the carrier is position, the trays are lowereed on to the shelves by lowering the carrier (System "CQC", ALD Vacuum Technologies GmbH, D-63526 Erlensee).

Fig. 2.13.2. Freeze drying plant, in which the trays are placed on to slide rails in between two heating plates by a lift and moved in position by pushing the trays by the last one from the lift (System "CONRAD"®, ATLAS INDUSTRIES A/S, DK-2750 Ballerup).

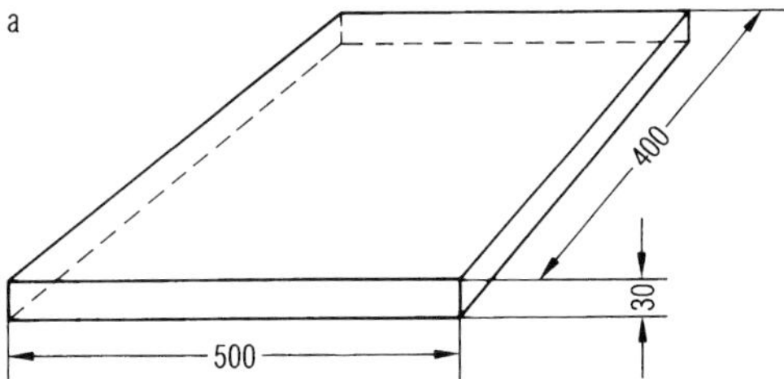

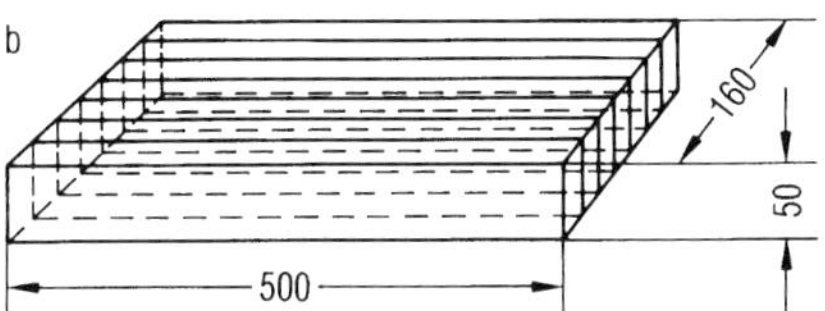

Fig. 2.14. Type of tray for freeze drying small cubes or granules of food.
Tray a: e. g. 400 · 500 mm, 30 mm height; tray b: e. g. 500 · 160 mm, 50 mm height.

from extrusion-molded aluminum with a machined bottom (Fig. 2.15). During the drying process they are placed on the heated shelves, but for transport they are elevated slightly and lowered onto the shelves again in the new position (System Leybold AG; now ALD Vaccum Technologies GmbH, D-63526 Erlensee, Germany). The distances between the ribs can be modified to meet the dimensions of the granulate. For certain products with small granulates (mm range) ribbed trays are used with inserts to facilitate vapor flow from the product into the chamber, e. g. V-shaped sieves as shown in Fig. 2.16.1. This is especially important if the water evaporation rate is high, e. g. 2–3 kg water/m^2 h.

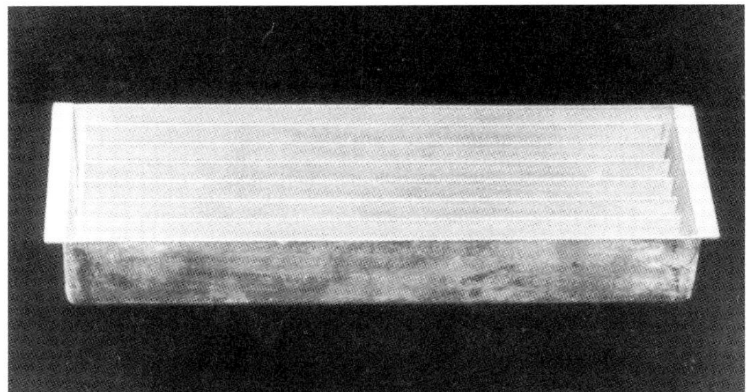

Fig. 2.15. Ribbed trays with seven compartments (type b in Fig. 2.14) (ALD Vacuum Technologies GmbH, D-63526 Erlensee).

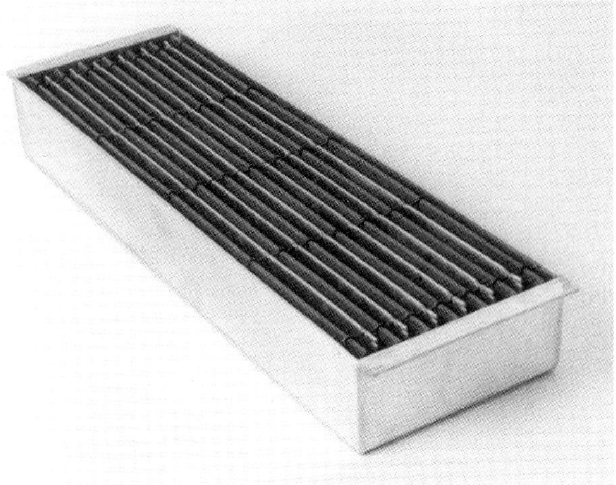

Fig. 2.16.1 Ribbed tray with a V-shaped sieve between the ribs (ALD Vacuum Technologies GmbH, D-63526 Erlensee).

Rolfgaard [2.2] compares the types of trays and heating systems: The ribbed trays are said to have an uneven temperature distribution, because the distances between shelf and tray vary between 0.1 mm and 1 mm. The ribs could compensate this only partially. The variation in distances is correct, but Rolfgaard overlooks that the thermal conductivity in the bottom of the tray is so effective that practically no temperature differences are established in the bottom. Even with an evaporation of 3 kg ice/m^2 h and the assumption that all heat is Transmitted only in the center of the tray (8 cm from the border of the tray), the temperature difference between border and center is approx. 5 °C. During the drying under actual conditions, no measurable temperature differences can exist.

However there is a major difference between the two forms of trays and heating systems. As shown in Fig. 1.57, approx. 1.3 kg ice/h m^2 can be sublimated by radiation heat, if the shelves have a temperature of +100 °C and the product temperature is –20 °C. The main difference is the method of heat transfer: With a flat tray and mostly radiation energy, the density of the heat flow is limited, and it can be substantially larger with ribbed trays standing on the heated shelf. Using the temperatures as above and an average value K_{tot} = 100 kJ/h m^2 °C from Tables 1.9 and 1.10, approx. 4.3 kg ice/h m^2 can be sublimated.

Figure 2.16.2 [2.3] shows the ice evaporation/m^2 h for different flat and ribbed trays. The difference in sublimation rates is a factor of approx. 4, or 400 %. The ribbed trays are more expensive than flat trays, as Rolfgaard states. However as shown in Fig. 2.16.2, the ice sublimation rate of ribbed trays is approx. 3.5 times larger than that of flat trays. This is understandable from Eq. (15) in which the layer thickness *d*, is decisive for the drying time (if the maximum possible T_{ice} during MD and the maximum T_{pr} during SD are applied). If the rib distance is chosen similar to the layer thickness in a flat tray the drying time becomes similar, but the ribbed tray has a load per m^2 which is three to four times higher than a flat tray, and the necessary heat transfer is possible by contact and convection.

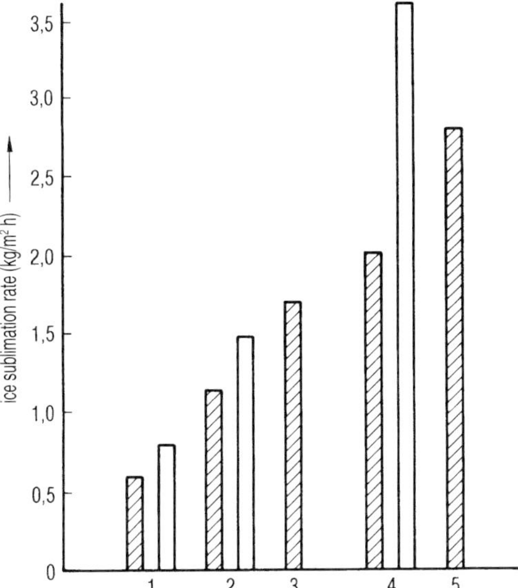

Fig. 2.16.2. Ice sublimation rate (kg/m^2 h) for five different types of tray.
1. Flat tray as in Fig. 2.14 (a), with 20 mm filling height.
2. Flat tray with one rib of 20 mm height.
3. Ribbed tray as in Fig. 2.15.
4. Ribbed tray as in Fig. 2.16.1.
5. Ribbed tray for cubes or granules of food, designed by Dr. Otto Suwelack, D-48723 Billerbeck.
All tests have been carried out with the same granulate.
Hatched columns: heating plates or shelves were heated in approx. 30 min to 100 °C, when the tray bottom reached +40 °C, the temperature was reduced in such a way that the +40 °C was kept constant. White columns: heating plate or shelves were heated to +140 °C in approx. 2 h and the temperature reduced after +40 °C was reached as above (measurements by Dr. Otto Suwelack, D-48723 Billerbeck).

2.2.3 Shelves and their Cooling and Heating

As shown in Section 1.2.1, heat transfer from the shelves to the container depends largely on planar shelves and trays. Stainless steel shelves in pharmaceutical plants are polished until the roughness height is RA 1.5 (corresponding to approx. 1.0–1.5 µm). The deflection should be smaller than 1 mm/m. The small roughness height also improves the cleaning and sterilization. In food installations using ribbed trays, the shelves are made from deep-drawn plates with tolerances of 0.1 mm. With radiation heating the roughness is not important for the heat transfer.

Shelves in plants for pharmaceuticals are mostly cooled and heated with brine. The amount of brine per time and its distribution in the shelf has to be guided in such a way, that the temperature difference between inlet and outlet of the shelf is < 1.5 °C during the maximum sublimation rate in MD. This maximum temperature difference and the maximum heat to be supplied at this difference should be written into a plant specification. It is recommended to use two cooling systems for the cooling of the brine and the condenser, otherwise the temperature control of the condenser and of the brine can influence each other. In large continuous plants, which need no cooled plates because the product is frozen outside and temperatures below +30 °C are normally not needed, the plates are heated by vacuum- or pressurized steam, or by a heat carrier.

2.2.4 Water Vapor Condensers

The volumes of water vapor are too large to be pumped by mechanical vacuum pumps in the pressure range of freeze-drying: 1 kg of ice at 0.4 mbar represents a volume of approx. 2800 m^3, or at 0.04 mbar approx. 25 000 m^3 (see Fig. 1.2). Only steam ejectors could do this, but these need large quantities of cooling water and steam, in addition to large areas for the multi-stage systems. In today's plants, water vapor is therefore condensed on cold surfaces, consisting of plates or mostly of tube coils.

Condensers must fulfil four essential requirements:

1. The surface area has to be large enough to condense the ice at a maximum thickness, which does not reduce the heat transfer from the tube surface to the condensing surface of the ice. The heat conductivity of ice depends on its structure. If this structure is solid and smooth, one can calculate with 6.3 kJ/m h °C, if the structure is more like snow, the heat conductivity is much lower, e. g. one decade; this happens typically if the air pressure is too high (see Fig. 2.17). To condense 1 kg of water in 1 h on 1 m^2 surface on top of an existing ice layer of 1 cm, the temperature difference between tube surface temperature and the ice surface temperature is approx. 4.5 °C. To reduce this temperature difference to 2 °C, the condenser would have to be defrosted every 30 min. Therefore, it is practical to design the condensing surface large enough to take up the total amount of ice of one charge in a layer of approx. 1 cm, e. g. if the total amount of ice to be condensed is 10 kg the surface should be 1 m^2 to form a layer of 1 cm, if the main drying time is long with a small amount of water per hour, half the size (0.5 m^2) may be sufficient. In this example the temperature difference during a main drying cycle of 5 h is limited to 1 or 2 °C.
2. The temperature difference between inlet and outlet temperature at the coil(s) of the refrigerant should be smaller than 1 °C ($\Delta T < 1$ °C), to ensure a uniform condensation on the total coil. On warmer areas no ice will condense until the temperature at the ice surface has increased to the warmer temperature on the coil. For large surfaces it is necessary to use several coils or plates in parallel, each of which must be separately temperature controlled. If the condenser is operated in an overflow mode, the weight of the liquid column should not change the boiling temperature of the liquid at the bottom of the column measurably.

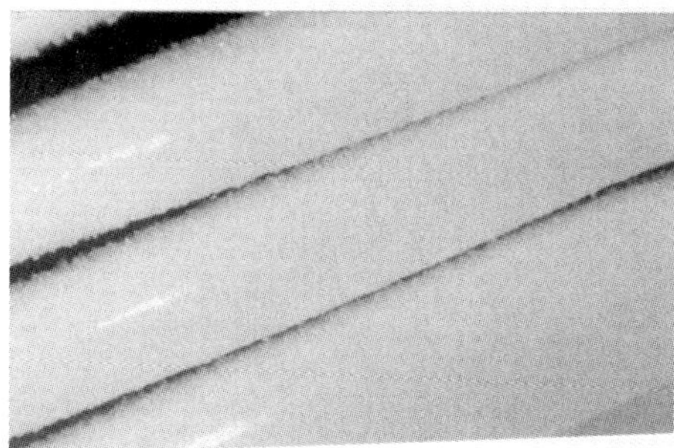

Fig. 2.17. Coils in a condenser covered by ice, observed through a window during two freeze drying processes. Left: Smooth, solid surface. Right: Porous, snow-like surface, which occurs typically, if the pressure of permanent gases during MD is high (photographs: Dr. Otto Suwalck, D-48727 Billerbeck).

3. The flow of water vapor should deviate as little as possible before the first condenser surface. The condenser design has to ensure that the water vapor is completely frozen and the remaining water vapor pressure is practically equal to the vapor pressure at the ice surface. This can only be achieved if the vapor passes over several condenser surfaces in series.
4. The permanent gases must be pumped off at the lowest position in the condenser. They are more dense than water vapor, concentrate at the bottom of the condenser and fill up the condenser housing in time. This permanent gas reduces the vapor transport to the cold surfaces and form a 'snow ice' as can be seen in Fig. 2.17. A condenser (Fig. 2.18.1) meets these requirements in general, but other designs are possible (see Fig. 2.52 B).

The qualities of a condenser can be judged in general terms as follows:

- It is important to know the leak rate of the valve D (Fig. 2.18.1) not only against the atmosphere, but also between the chamber and condenser for the pressure rise measurements for BTM and DR data. The leak rate can be measured as follows:
  Pressure rise measurement in the chamber and condenser with valve D open and valve G closed. The same measurement with valve D closed and G open and with valve D closed, but atmospheric pressure in the condenser. There should be no measurable difference between the 3 data and the absolute value should correspond to the criteria given in Section 1.2.3. The effect of the closing time and closing characteristic on the BTM and DR measurements can be checked by the following procedure: The shelves of the chamber and the condenser are at operation temperatures (e. g. +30 °C and –55 °C), and the chamber and condenser are evacuated by the pumping system. Air is injected through a needle valve until the pressure becomes similar to the operation pressure during SD, e. g. 0.04 mbar. Valve D is closed for 60 s as during DR measurements, the pressure

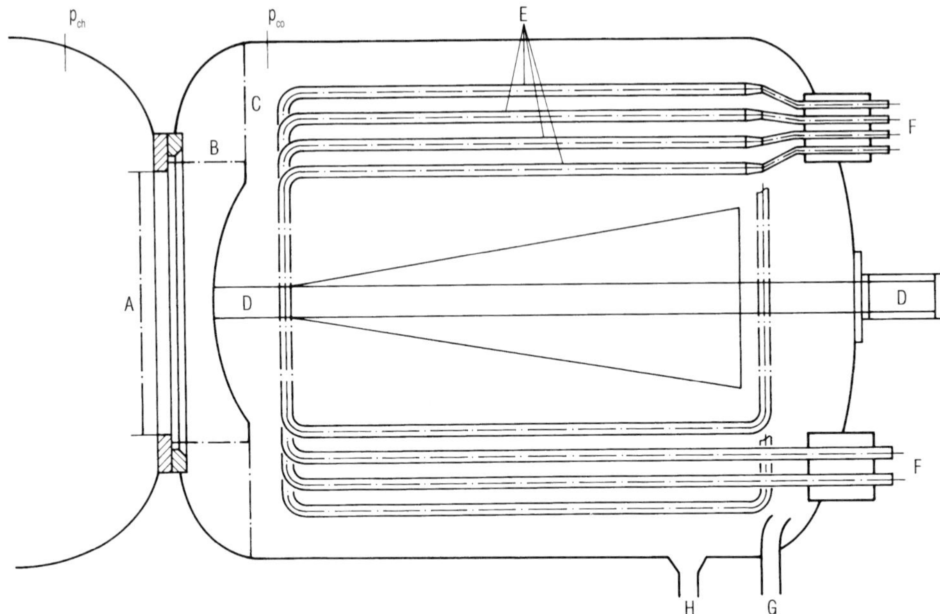

Fig. 2.18.1. Schema of a water vapor condenser for a freeze drying plant.
A, free diameter of the connection to the chamber; B, cylindrical opening by the movement of D; C, opening between condenser wall and valve plate; D, valve plate, hydraulic valve drive; E, condensation surface of the refrigerated coils; F, in- and outlet of the refrigerant; G, tube connection to the vacuum pump; H, water drain during defrosting of the condenser; p_{ch} and p_{co} pressure in the chamber and in the condenser, respectively.

raise recorded, and valve D opened at the end of 60 s. The is done over several hours to test the reproducibility of the valves operation. Figure 2.18.2 shows the result: In the upper plot the valve does not close correctly; in the lower plot the valve operates reproducibly. During the first 4 h the pressure rise per second (dp/s) has been an average $1.263 \cdot 10^{-4}$ mbar/s, SA $0.06 \cdot 10^{-4}$ mbar/s; in the second 4 h $1.203 \cdot 10^{-4}$ mbar/s, SA $0.04 \cdot 10^{-4}$ mbar/s and from 8 to 12 h $1.160 \cdot 10^{-4}$ mbar/s, SA $0.04 \cdot 10^{-4}$ mbar/s. dp/s changes in the first 4 h by ≈ 5 %, in the following 4 h by ≈ 3 %, and the next 4 hours by ≈ 2 % and is ≈ 1 % thereafter. This small effect is due to the decrease of water desorbed from the chamber surfaces.
The BTM measurements during MD do not show this effect since they are made under almost equilibrium conditions. However it is helpful to be aware of such an effect, which can vary with the design of the plant. The data reported are measured with a plant, as shown in Fig. 1.66.

- Maximum ice condensation per time by the following tests:
  Shelves of the chamber are loaded with ice, the chamber and condenser evacuated, and the shelves temperature raised until T_{ice} has reached the value to be tested. The tempera-

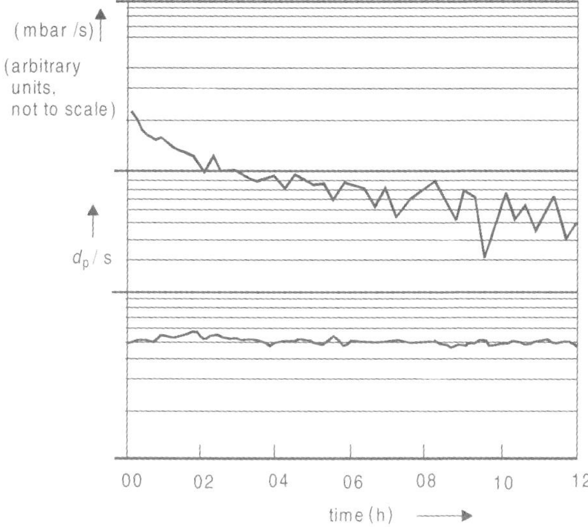

Fig. 2.18.2. Comparison of valve D (Fig. 2.18.1) functions: Upper plot: the valve does not operate reproducibly. Lower plot: measurements of the pressure rise per second (dp/s) are in the order of magnitude of $1.2 \cdot 10^{-4}$ mbar/s, standard deviation (SA) approx. $0.04 \cdot 10^{-4}$ mbar/s (measurements by AMSCO Finn-Aqua, D-50356 Hürth, Germany).

ture difference between inlet and outlet temperature of the refrigerant should remain unchanged or not exceed 1 °C or as a maximum of 2 °C. After a rise of the shelf temperature it is important to wait approx. 30–60 min to allow for an equilibrium status, before the temperature difference becomes meaningful. In this way the maximum p_{ch} can be determined at which the condenser operates as specified.

- To measure the absolute amount of water vapor transported per time at a certain pressure, the test described above can be carried out with a weighed amount of ice on the shelves, either as plates directly on the shelves or with water in trays frozen on the shelves (Fig. 2.19). Sublimation at the desired pressure should be continued for 5–6 h. A shorter time results in problems, since the time to reach approx. equilibrium conditions is 1 h or more, depending on the size and the design of the plant. If the chamber pressure rises with increasing shelf temperature, but the condenser temperature changes only very little, the condensation on the condenser surface is not the bottle neck, but the water vapor transport between chamber and condenser is the controlling part of the process. In this case, the pressure difference between chamber and condenser should be measured, best with a capacitance gauge designed to measure pressure differences. The pressure difference is expected to be small (e. g. 0.01 to 0.05 mbar). The pressure difference gauge avoids the inaccuracy of two instruments. From the difference and the absolute pressure p_{ch}, the amount of water vapor transported can be estimated by the principles discussed in Section 1.2.4.

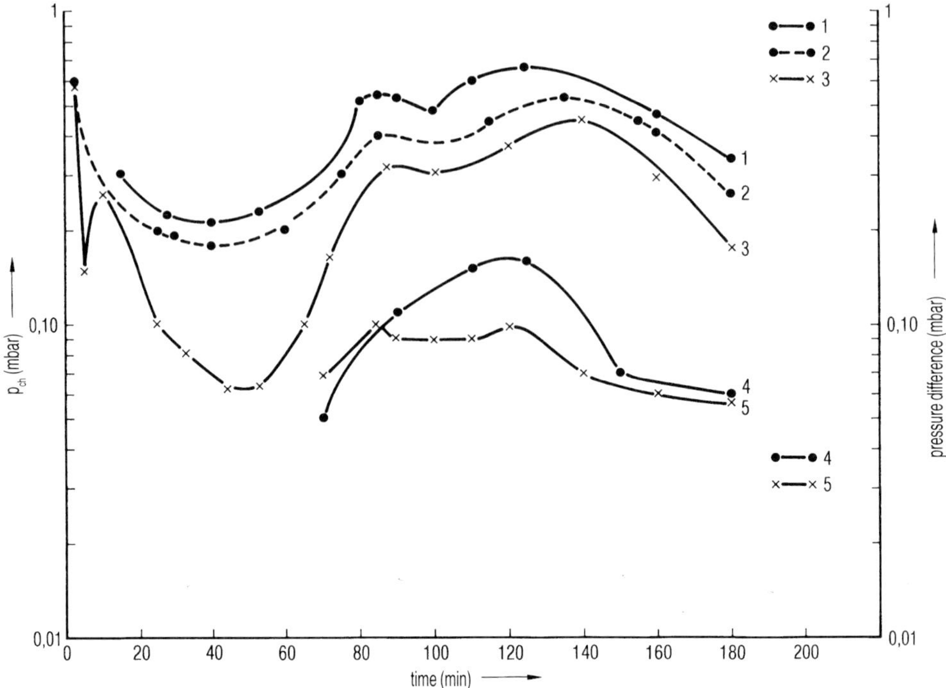

Fig. 2.19. Plots of a test to determine the specific water vapor flow or the water vapor speed in a production freeze drying plant with approx. 30 m^2 shelf area. For the tests 300 kg of distilled water were filled into ribbed trays, which were placed on the shelves. Six RTD were placed in different trays and frozen with the water. The RTD temperatures and BTM measurements were practically identical, because the RTD were always immersed in the ice, and during the test only 25 % of the ice was sublimated.
1, p_s, calculated from T_{ice} by Table 1.11; 2, p_{ch}; 3, p_{co}; 4, pressure difference ($p_s - p_{ch}$); 5, pressure difference ($p_{ch} - p_{co}$).
(Comment: The temperature of the shelves was hand controlled, because automatic control of the large plant would have been too slow for the short test time.)

For production plants, the loading of the plant with the required amount of ice (which should correspond to a full charge) is time consuming and several tests should be avoided. For a sufficient estimation of the water vapor transport and the bottle necks of it, one test can be carried out as described here:

A plant having a shelf area of ≈ 30 m^2 has been loaded with 300 kg of water in trays and frozen on the shelves. Water vapor transport and condenser temperatures have been measured in this case between 0.4 and 0.6 mbar, which is approx. two to three times higher than the normally expected operation pressure of the plant (to get an measurable quantity of ice sublimed in a reasonable test time). The data of the test are shown in Fig. 2.19 . Three Pt 100 (resistance thermometers) have been frozen in the ice. One CA each have been connected to the chamber and to the condenser at the places marked p_{ch} and p_{co} in Fig. 2.18.1. Furthermore, the surface temperature of the condenser coils are measured.

Approx. 50 min after the start of heating and evacuation the equilibrium conditions start to become visible, and after approx. 90 min they are effective. The pressure difference between T_{ice} (converted into pressure) and the chamber pressure depends on the absolute pressure, which corresponds to the amount of water vapor transported per time, but the difference $(p_{ice} - p_{ch}) < p_{ice}$. In the first 50 min, T_{ice} or (p_s) and the pressure in the condenser each drop, because ice is only sublimated after the shelf temperature has started to rise from the –30 °C seen at the start. After 2 h and 15 min the shelf temperature has been lowered to pass the pressure range of 0.3 mbar a second time in order to avoid a possible distortion by the non equilibrium conditions in the beginning.

The following conclusions can be drawn from this experiment for the water vapor transport and the working of the condenser:

- The water vapor flows from the sublimation front into the chamber and to the connection between chamber and condenser with a favorable small pressure drop; there are no measurable flow resistances, e. g. between the shelves or the shelves and the chamber walls.
- The pressure losses between chamber and condenser are surprisingly small (only 25–30 % of the chamber pressure). The reason for this becomes understandable by the last conclusion.
- The condenser design and surface can handle the vapor flow during main drying of this test. The possible low temperatures could be needed during secondary drying.
- The visual observation of the condenser coils shows, in the visible zone, a compact solid, glassy structure. Inclusions of permanent gasses resulting in 'snow like' surfaces were not seen.
- If the sublimated amount of water is calculated per surface area of diameter A (Fig. 2.18.1) and time, as shown in Fig. 2.20 for different quotients of l/d of the connection between chamber and condenser, the average data of this test is 4.7 g/h cm^2. With l/d = 5, the vapor flow density should be 10–14 g/h cm^2 in the pressure range between 0.3 and 0.4 mbar. Figure 2.20 shows, that 4.7 g/h m^2 can only be expected in the pressure range between 0.32 and 0.4 mbar, if l/d is much larger than 5. A rough extrapolation indicates, that the l/d value in this test had to be approx. 20. If one or two 90° bends are part of the connection, the actual length can be smaller, since each 90° bend does not contribute to the resistance by its physical length, but by a multiple of it. The actual flow resistance of a bend depends more on its design and surface structure than on its physical length. Therefore it is difficult to estimate the resistance, one has to measure it by tests as described above. In this test the average vapor speed has be calculated as approx. 50 m/s. From Fig. 1.90 one can see, that at l/d = 5, the vapor speed of approx. 100 m/s can be expected in the measured pressure range.

If the installation would have been designed with an l/d = 1.6, which would likely be the best possible technical solution, the same water vapor flow density can still be achieved at

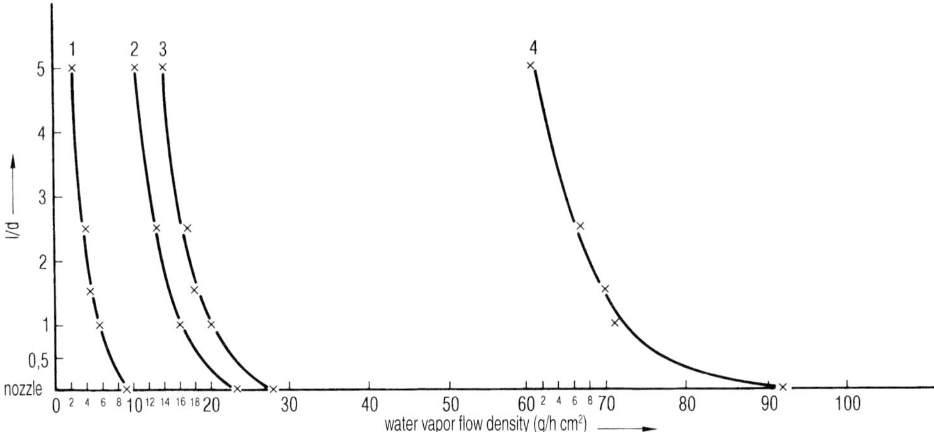

Fig. 2.20. Water vapor flow density (g/cm^2 h) as a function of the ratio *l*/*d* of the connecting tube from the chamber to the condenser at four different pressures p_{ch} as parameters (*l* length, *d* diameter of the tube).
1, 0.1 mbar; 2, 0.32 mbar; 3, 0.4 mbar; 4, 1.3 mbar.

a pressure one decade lower, at 0.04 mbar. The design of the connection between chamber and condenser is one of the most critical parts of a freeze drying plant. It should cleaerly be straight, and as short as possible. As Figs. 1.89 and 1.90 show, this becomes especially important for pressures below a few times 10^{-1} mbar.

2.2.5 Refrigerating Systems and Refrigerants

The special requirement of refrigerating machines for freeze drying plant is the capability to run with small load near the end temperature of the compressors. This can reduce the necessary cooling of the motor and the circulation of the lubricant. A safe solution is to limit the minimum suction pressure of the compressor, e. g. by a bypass, which guarantees a minimum flow of refrigerant. The temperature of the refrigerant in coils or plates cannot be controlled by a centralized sensor, but each group of coils or plates must have its own sensor and injection valve for the refrigerant. All of these must be well tuned with each other. The refrigerant for the condenser is normally directly injected into the coils or plates, while the shelves are mostly cooled and heated by a brine. Therefore, the condenser temperatures are normally lower than that of the shelves.

Haseley [2.4] differentiates between two categories of refrigerants: The first group can be exchanged in compressor systems without changing the compressor itself, but changing the injection valves. The second group with no 'Ozone Depletion Potential' (ODP).

To the first category belong e. g.:

HP 80/R 402	DuPont	R 125/R 290/R 22 0.03 ODP
Isceon 69 L	Rhone-Poulenc	R 290/R 218/R 22 0.03 ODP

For comparison, other ODP values: R 22 ODP 0.05, R 502 ODP 0.6, R 13 B 1 10.0 ODP.

Because HP 80 and Isceon 69 contain R 22, the use of them in existing installations is only for a limited time. For new installations, hydrofluorcarbons (HFC) are available with ODP values at 0, e. g.

HP 62	DuPont	R 143a/R 125/R134a
Genetron AZ 60	Allied Sign	R 143a/R 125
FX 40	Elf Ato chem	R 32/R 125/R 143a

These three refrigerants have a 'Coefficient of performance', which is 1- to 25 times smaller than that of R 502.

Willemer [2.5] shows in Fig. 2.21 the cooling time of 31 m^2 shelf area in a freeze drying plant by a 2 stage compressor operated with HP 80, R 13 B 1 and R22. –50 °C is reached with R13B1 in ≈ 60 min, with HP 80 in ≈ 85 min, and with R 22 in ≈ 90 min. At –40 °C, the time difference is only ≈ 15 min. Whether and how the use of LN_2 is technically advisable and economically justified has been studied by several authors. Snowman [2.6] sees the following advantages: No compressors, condenser temperatures between –70 °C and –120 °C, and a reserve of cooling medium during technical problems (the product can be kept cold with LN_2 in a tank). The disadvantages are: LN_2 is more expensive per kW cooling output than the electrical energy needed for the same cooling effect, and the installation has to be designed for LN_2, a change in an existing plant to LN_2 cannot be justified. Snowman describes a method to save LN_2 and to control a desired condenser temperature.

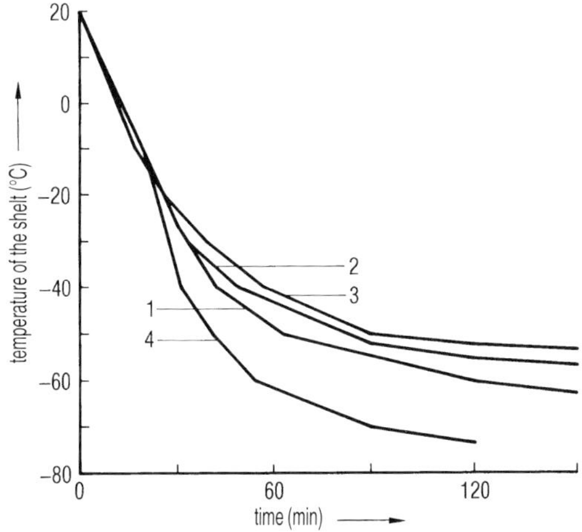

Fig. 2.21. Temperature of the shelf in a freeze drying plant as a function of cooling time, calculated for four different refrigerants.
1, R13B1; 2, HP80; 3, R22; 4, LN_2 (Fig. 1 from [2.5]).

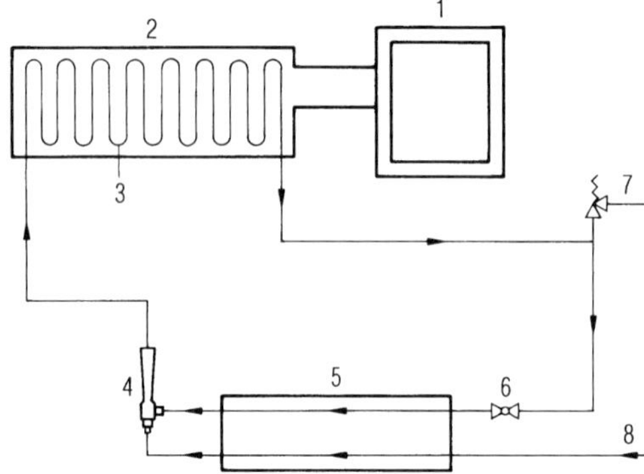

Fig. 2.22. Cycle cooled by LN_2 of a condenser in a freeze drying plant. A part of the N_2 is cooled in a heat exchanger and pumped back in the cycle by a jet pump.
1, drying chamber; 2, condenser; 3, condenser coil; 4, jet pump; 5, heat exchanger; 6, throttle valve; 7, pressure controlled N_2 outlet; 8, LN_2 inlet (from [2.6] page 342).

Figure 2.22 [2.6] demonstrates the method of a cooling circuit with recirculated flow: An injector pump operated with just evaporated LN_2 aspirates the warmer N_2 coming from the condenser and feeds the mixture back in the condenser. The desired condenser temperature can be controlled by a throttle valve. To achieve a uniform temperature distribution, the gas mixture is alternately fed to one or the other end of the condenser. No results of such a system are given.

Cully [2.7] estimates for a freeze drying plant with 30 m^2 shelf area that, starting with the third year of operation, LN_2 cooling is more economic than cooling with compressors. The main saving, in his opinion, comes from avoiding high maintenance costs of the compressors. The investment cost for compressor- or LN_2 plants is assumed equal in this calculation. A modification of an existing plant is also in the opinion of Cully uneconomical, the pay back time being approx. 9 years.

The company AMSCO Finn-Aqua [2.8] shows a different possibility of saving LN_2 and yet retain some of the advantages. In the condenser housing, besides the conventional coils, a LN_2-cooled plate is installed, which will be only operated under two conditions:

- during secondary drying, if an operation pressure below e. g. 10^{-2} should be reached quickly. The conventionally cooled condenser remains in operation at its end temperature.
  In this operation some ice sublimes from the condenser to the LN_2-cooled surface. However, the surfaces of the LN_2 plate can be controlled between –80 °C and –100 °C, that corresponds to a water vapor pressure of approx. $5 \cdot 10^{-4}$ to $2 \cdot 10^{-5}$ mbar.
- should the compressors fail temporarily, the LN_2 condenser can maintain a low pressure e. g. 0.1 mbar for e. g. 2 h, depending on the LN_2 on hand. This would allow a product temperature to be kept of approx. –40 °C in the chamber. This is supported by a heat exchanger in the brine circuit which can be also cooled by LN_2.
  It is possible that the maintenance problems with multi-stage screw compressors are substantially smaller than with piston compressors, but so far only one publication has been found [2.30] comparing the technical data and operation cost of piston- and screw compressors for two brands. The maintenance cost of both systems are not discussed.

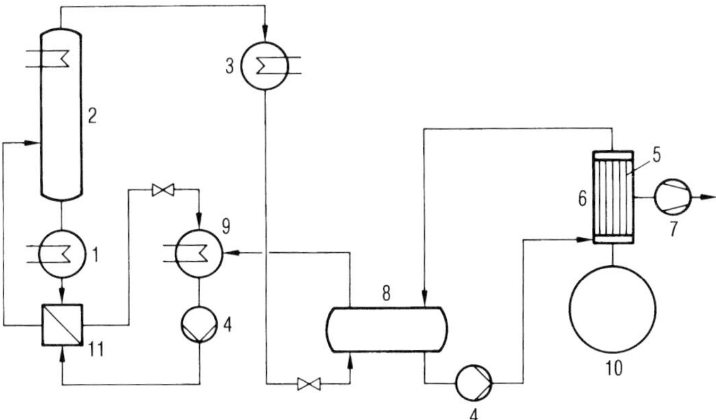

Fig. 2.23. Schema of an absorption refrigeration plant based on information of Deutsche Babcock-Borsig AG, D-13500 Berlin and ALD Vacuum Technologies GmbH, D-63526 Erlensee.
1, expulsion of NH_3 from the water-NH_3 solution; 2, rectification column for NH_3; 3, NH_3-condenser; 4, NH_3 pump; 5, condenser coils for water vapor; 6, condenser housing; 7, vacuum pump; 8, separation of the gaseous NH_3 from liquid NH_3; 9, absorption of the gaseous NH_3 in the water from the expulsion; 10, drying chamber; 11, heat exchanger.

In large freeze drying plants for food and other mass products it is economical to use ammonia absorption plants for cooling, which can be heated by steam or directly fired by oil or gas. If the required capacity is in the order of 500 kW at –55 °C or larger the low operation- and maintenance cost of the absorption plant should be studied and evaluated. The investment cost are shown in Fig. 2.24.1, and the total cost per year for a 1000 kW at –55 °C shows a substantial advantage for the absorption plant. In a NH_3 absorption plant (schema in Fig. 2.23) a water ammonia-mixture is evaporated in a steam heated or oil-fired boiler (1), the vapor is separated in a rectifier column (2) in NH_3 vapor and a remaining solution. The ammonia vapor is liquefied in a condenser (3) and pumped (4) into the water vapor condenser (5) (which is the NH_3 evaporator). In this schema, NH_3 is not injected into the water condenser but is pumped (4) through the condenser. In this case only a part of the ammonia (10–20 %) evaporates by the heat transmitted from the frozen ice on the surface of the condenser coils. The mixture of liquid and gaseous ammonia is conducted into a separator (8), the NH_3 vapor flows into an absorber (9), where it mixes with the remaining solution from the rectifier column and returns to the boiler (1).

As shown in Fig. 2.24.2, in large plants the absorption system is more economical than compressor installations, independently of the price of steam or electricity. The low maintenance cost are reflected in the calculation, but the high uptime and reduced production interruption should also be accounted for in an evaluation; no large, heavy moving machine parts are the reason for this advantage.

		absorption refrigeration plant			compressor refrigeration plant		
refrigerating capacity tot.	kW	1000	500	250	1000	500	250
evaporation temp.	°C	−55	−55	−55	−55	−55	−55
heat demand	kW	3681	1840	920	na	na	na
steam	kg/h	6700	3350	1675			
use of heat							
condenser	kW	1645	823	411	1552	776	388
absorber	kW	2180	1090	545			
solution cooler	kW	856	428	214			
cooling tower	kW	4681	2341	1170	1870	935	467
cooling water	m^3/h	410	205	102	165	83	41
fresh water demand	m^3/h	8.5	4.3	2.1	3.4	1.7	0.8
current demand							
engines	kW	35	18	9	900	450	225
cost estimate	TDM	3900	2700	2400	2500	1250	625

Fig. 2.24.1. Comparison of technical data and cost estimates of absorption- and compressor refrigeration plants (data provided by Deutsche Babcock-Borsig A. G., D-13500 Berlin).

			absorption refrigeration plant			compressor refrigeration plant	
capital costs yearly							
A investment		TDM	3900			A TDM	2500
p interest rate		%	6			p %	8
n amortization time		ano	25			n ano	10
q = q + p/100			1.06				1.08
		TDM/ano	305.1	305.1	305.1	TDM/a	372.6
energy costs		heating cost	DM 8/MWh	DM 15/MWh	DM 30/MWh		
steam		TDM/a	117.8	220.8	441.6	TDM/a	0.0
fresh water	DM 1/m^3	TDM/a	34.4	34.4	34.4	TDM/a	13.6
current	DM 150/MWh	TDM/a	21.6	21.6	21.6	TDM/a	540.0
operation and maintenance							
% of the first investment		TDM/a (1 %)	39.0	39.0	39.0	TDM/a (6 %)	150.0
staff costs		TDM/a	20.0	20.0	20.0	TDM/a	60.0
jährliche Gesamtkosten		TDM/a	537.8	640.9	861.7	TDM/a	1136.2

Fig. 2.24.2. Total annual cost of a refrigeration plant with a capacity of 500 kW at –55 °C and 8000 annual operating hours as well for an absorption – as a compressor plant.
$Kk = A \cdot ((q - 1) \cdot q^n / q^{n-1})$ (data provided by Deutsche Babcock-Borsig AG, D-13500 Berlin).

2.2.6 Vacuum Pumps

The vacuum pumping system in a freeze drying plant has to fulfil two tasks:

- To reduce the air pressure in the chamber and condenser to the necessary partial pressure of air (mostly 0.01 to 0.1 mbar).
- To pump off the gases from the product and the air entering the plant through leaks at a partial pressure of the permanent gases, which has to be small compared with the water vapor pressure.

Fig. 2.25 is a review of the working range of different vacuum pumps and pump combinations. The pressure range of interest for freeze drying is $5 \cdot 10^{-1}$ to $2 \cdot 10^{-3}$ mbar, for which single stage-vacuum pumps are not qualified. Two-stage pumps reach approx. 10^{-2} mbar with gasballast, which makes them applicable in the upper part of the pressure range. This type of pump is only available with a maximum pumping speed of 300 m^3/h. In a freeze drying plant the pumps have permanently to pump some water vapor with the permanent gases, and should therefore always be operated with gasballast to avoid condensation of water in the pumps.

During the operation with gasballast, such an amount of air will succeed into the pump house (after the pump house is separated from the vacuum chamber) that the water vapor at

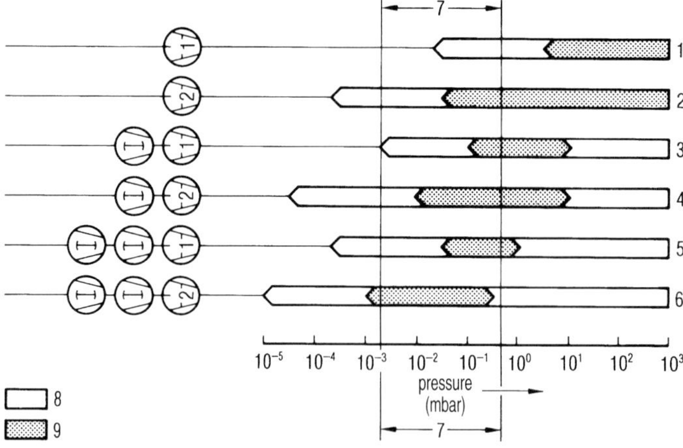

Fig. 2.25. Working range of vacuum pumps and multistage pumpsets.
1, single- tage pumps; 2, two-stage pumps; 3, two-stage pumpsets with single-stage backing pump; 4, three-stage pumpset with two-stage backing pump; 5, three-stage pumpset with single-stage backing pump; 6, four-stage pumpset with two-stage backing pump; 7, working range of pumpsets ($2 \cdot 10^{-3}$ to $5 \cdot 10^{-1}$) in a freeze drying plant depending on T_{co} and p_{ch} during NT; 8, start-up range of the pumpset; 9, working range of the pumpset (part of Fig. 2 from [2.21]).

the operation temperature of the pump cannot condense during the compression phase of the pump. An example: water vapor is pumped at a partial pressure of 0.5 mbar. The temperature of the pump is +70 °C. Under these conditions the water vapor will condense if the compression exceeds approx. 310 mbar. If the pressure in the pump house is enlarged by air from 0.5 mbar to e. g. 50 mbar, the compression needs only to be 1000/50 = 20. The original water vapor of 0.5 mbar is compressed by a factor of 20, the water vapor pressure reaches only 0.5 · 20 mbar = 10 mbar. No condensation can take place at 70 °C in the pump house.

Figure 2.26 shows, that a volume of 1000 L (chamber and condenser) will be evacuated to 0.01 mbar in approx. 8 min by a pump with the capacity of 100 m^3/h. For a volume of 100 L, a pump with a capacity of 10 m^3/h is sufficient. The pump for evacuation only can be relatively small. The pump with 100 m^3/h has this capacity also at a pressure of 0.05 mbar; however at this low pressure 100 m^3/h are only 1.4 mbar L/s or $1.1 \cdot 10^{-3}$ g/s. This pumping capacity is more than sufficient, if the leak rate is smaller than 0.01 mbar L/s, which can be expected for most plants. The critical dimension for the pump size can be the gas from the product. In a chamber of 700 L (plus 300 L condenser volume), there may be e. g. 10 kg of product, which may have a minimum of 10 g (but often 100 g) of air dissolved within, which may become free during the main drying of e. g. 8 h, which is 1.2 or up to 12 g/h or $0.3 \cdot 10^{-3}$ to $3 \cdot 10^{-3}$ g/s. Therefore, the pumping speed of 1.1 g/s might be sufficient, but it would be preferable to use a pump with a three-fold larger pumping speed. Two-stage pumps with such a capacity are expensive, or are not available.

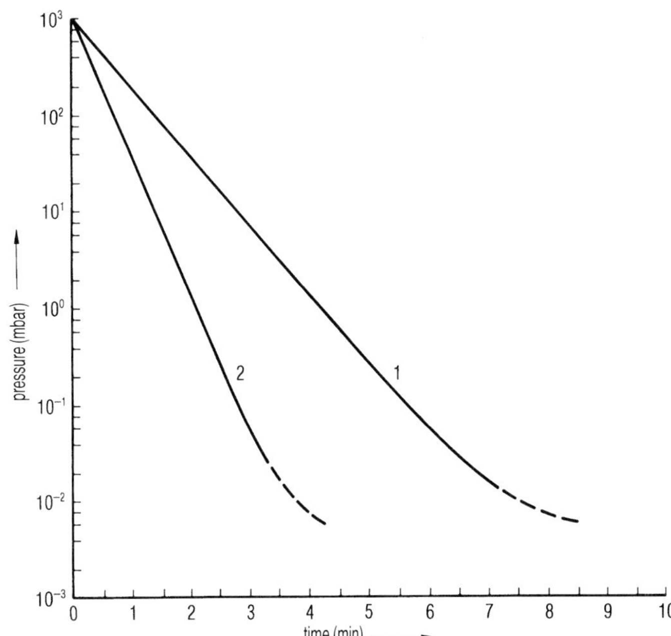

Fig. 2.26. Pressure as a function of the evacuation time of a 1000 L volume with: 1, a two-stage vacuum pump with a pumping capacity of 100 m^3/h, operated with gasballast; 2, a two-stage vacuum pump with a pumping capacity of 200 m^3/h, operated with gasballast (from [2.22]).

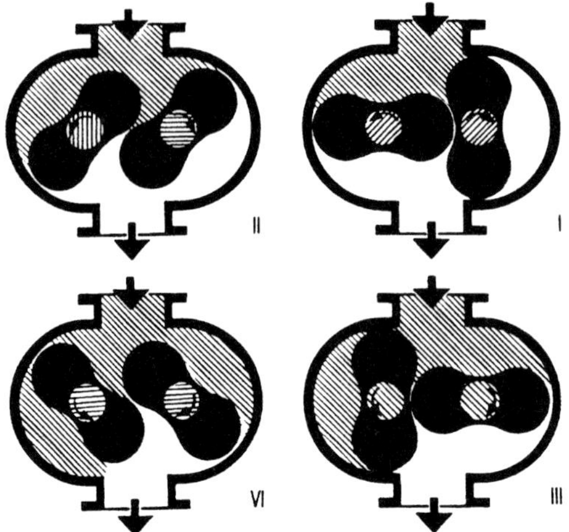

Fig. 2.27. Working principle of a one-stage roots pump (pumping direction vertical) (from [2.23]).

As shown in Figure 2.25, it is more efficient to use a backing pump combined with one or two roots blowers (Fig. 2.27). The backing pump requires only 40 m^3/h capacity, combined with a roots pump of 200 m^3/h. This system evacuates the 1000 L also in 8 min down to 0.01 mbar, but below 0.1 mbar the system has a capacity of 200 m^3/h or $2.2 \cdot 10^{-3}$ g/s at 0.05 mbar. Such a pumping system is preferable for freeze drying compared with a large two-stage pump alone.

If large pumping capacity is required in production freeze drying plants, a single backing pump with two blowers in parallel (Fig. 2.25) is an effective solution. Figure 2.28 shows the complete pumping system for a freeze drying plant. Two backing pumps are used for two reasons:

- during the evacuation both pumps run parallel, while during the freeze drying one pump can be sufficient
- the second pump can be shut off to save electricity, and acts as a stand-by.

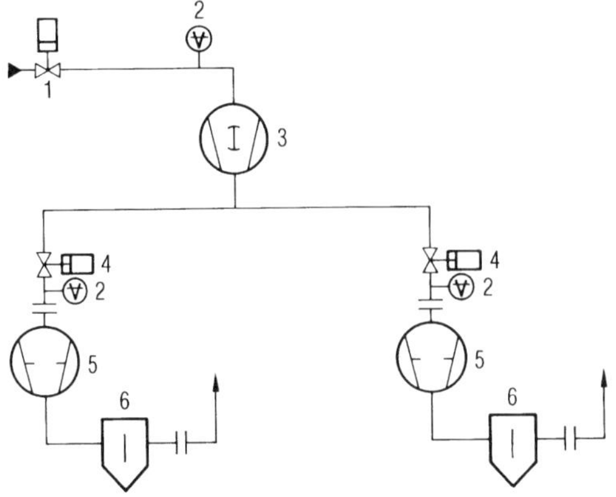

Fig. 2.28. Schema of a vacuum pump set, designed for a production freeze drying plant.
1, vacuum valve behind the condenser; 2, vacuum gauge; 3, roots pump; 4, vacuum valve between roots- and backing pump; 5, backing pump; 6, exhaust filter.

A small amount of oil vapor diffuses back from all oil sealed vacuum pumps in spite of the air flow to the pump. This amount is greatest if the air flow becomes small and the pump is running close to it final pressure. If a small amount of air is fed via a needle valve iinto the pump, so that the pump always runs at a pressure 5 to 10 times larger than its end pressure, 98 % of the back streaming can be avoided. It is possible to stop 99 % of the back streaming by using an trap filled with activated aluminum oxide. However this method is not recommended for practical reasons: The aluminum oxide has to be exchanged from time to time and it is difficult to decide the correct time for the exchange since some gases from the drying chamber can also be absorbed.

If the back streaming of oil has to be completely excluded, pumps without oil seals must be used as they are known from the semi-conductor industry. In these pumps two to four stages (depending on the manufacturer) of claw pumps (schema Fig. 2.29.1) are used. Figure 2.29.2 shows the working diagram of a four stage claw pump, which are built up to 100 m^3/h. For larger pumping capacity, an oil-free roots pump can be added in series. Such pumpsets have to be adopted for freeze drying conditions: Between the third and fourth stage a water-cooled condenser must be installed to avoid condensation of water vapor during compression. Depending on the operation pressure in the condenser of the freeze drying plant, a second condenser may have to be installed between the second and third stage.

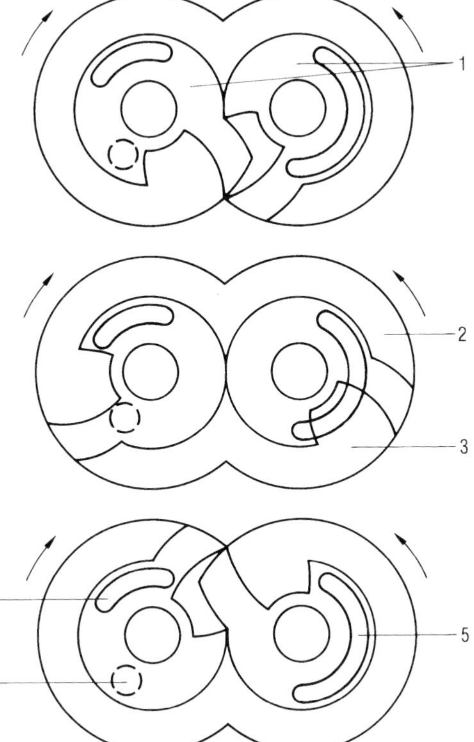

Fig. 2.29.1. Working schema of a dry vacuum pump operating on the so-called claw principle. 1 rotors; 2, compression chamber; 3, suction chamber; 4, exhaust slit; 5, suction slit; 6, purge between stages (Figure from [2.24]).

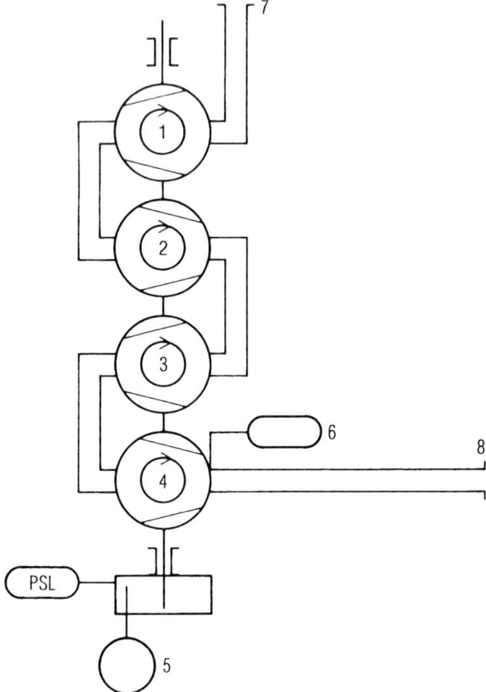

Fig. 2.29.2. Working schema of a four-stage dry vacuum pump, type Leybold DRYVAC®.
1–4, four stages in series each of the type shown in Fig. 2.29.1; 5, drive; 6, thermo switch, indicating deviations from normal operation; 7, flange of the suction line; 8, flange of the exhaust line (Figure from [2.25]).

2.2.7 Absorption Systems for Water Vapor

In the first freeze drying plant [2.9] phosphorous pentoxide has been used to absorb the water vapor, but to day this technique is outdated. The development of refrigeration technology with compressors or NH_3 absorption and the availability of LN_2 has replaced water absorption by silica gel or similar products. Some trials aiming to revive this technology are detailed in Section 1.2.6.

2.2.8 Measuring and Control Systems

The total pressure during freeze drying may be measured by several methods, though only two are mostly used: heat conductivity, and the membrane pressure difference gauge. Their operating principles and their advantages and disadvantages are described below.

The principle of design of a heat conductivity gauge (TM) is shown in Fig. 2.30. Electrical energy is fed into the wire (2) in such a way, that the temperature of the wire is kept constant. This amount of heat per time is in the area 2 of Fig. 2.31 and is approx. propor-

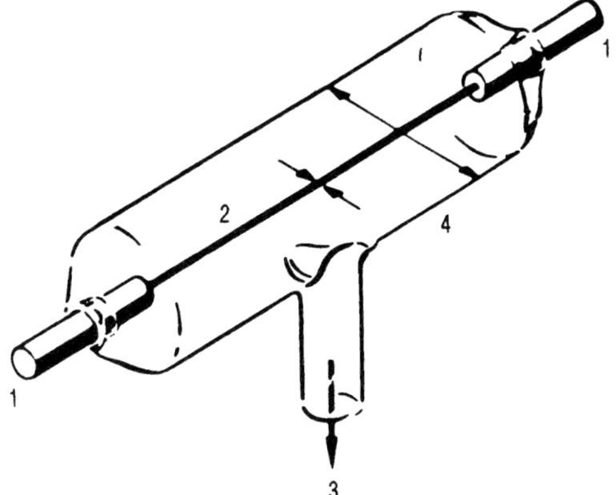

Fig. 2.30. Schema of a vacuum gauge on the principle of heat conductivity (TM).
1, wire support; 2, wire, d = 5 to 20 μm; 3, connection; 4, housing (Fig. 11.14 from [2.19]).

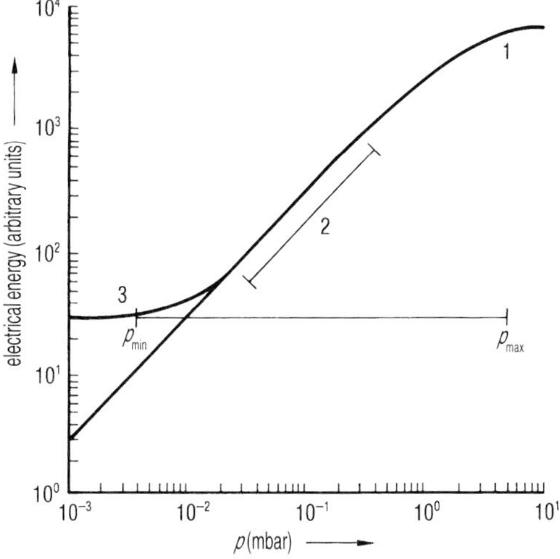

Fig. 2.31. Electrical energy fed to the wire (to keep it at constant temperature) as a function of pressure in the TM housing.
1, range of pressure independent conductivity; 2, range of pressure proportional conductivity; 3, range, in which the heat conductivity through the gas is negligible; p_{min} to p_{max} useful measuring range (Fig. 11.15 from [2.19]).

tional to the pressure. This range is between area 1, pressure independent and area 3, practically no heat conductivity. The measuring range of such a measuring tube is between 10^{-2} and approx. 3 mbar. The position of area 2 depends on design details; it can be moved within certain limits, e. g. the lower end can be moved below 10^{-2} mbar, but the upper limit is also reduced. The reproducibility of such an instrument is given e. g. by Leybold AG as smaller than 20 % of the observed value between 10^{-3} and 10^{-2} mbar and as smaller than 15 % of the observed value between 10^{-2} and 1 mbar; an observed value of e. g. 0.1 mbar can be 0.085 or 0.115 mbar. For Barometric Temperature Measurement (BTM; see Section 1.2.3), the conversion of p_s into ice temperatures (T_{ice}) with the data used above would lead to p_s 0.115 mbar = –41.0 °C, p_s 0.100 mbar = –42.2 °C, p_s= 0.085 mbar = –43.5 °C. The inaccuracy of more than ±1 °C is not acceptable. In addition, this measurement depends

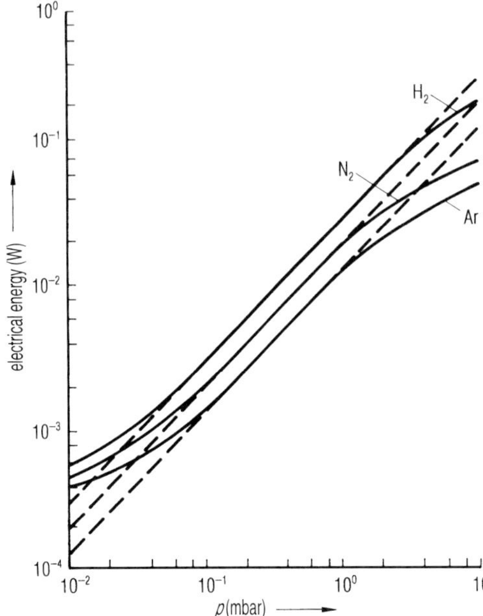

Fig. 2.32. Electrical energy to keep the wire temperature constant as function of pressure for different gases as parameter (Fig. 11.16 from [2.19]).

from the type of gas, as shown in Fig. 2.32. If the reading of a TM, calibrated in air, is 0.1 mbar, it would be 0.05 mbar partial pressure in pure water vapor. During the main drying a rough correction factor of 0.65 can be used to know the order of magnitude of pressure, e. g. the instruments reads 0.1 mbar in a freeze dryer during MD, the total pressure can be expected to be in the order of 0.065 mbar. In spite of these disadvantages TMs are only slowly replaced by membrane differential gauges (CA).

The membrane differential gauge measures the deflection of a membrane caused by the pressure difference between the two sides of the membrane. If one side of the membrane is evacuated down to 10^{-5} mbar and than sealed, the gauge can measure pressures down to 10^{-3} mbar with an maximum error of 1 %. The membrane is part of a capacitor, whose capacitance is changed by the deflection of the membrane. The pressure measured this way is independent of the type of gas and, since a capacitance can be measured very accurately, the instruments have a high resolution and reproducibility. For freeze drying, the CAs should have a measuring range either of 10^{-4} to 1 mbar or of 10^{-3} to 10 mbar. The resolution with these instruments is $1 \cdot 10^{-4}$ respectively $1 \cdot 10^{-3}$ mbar and the reproducibility at 0.1 mbar better than ±0.005 mbar. In the temperature range of 40 °C the vapor pressure changes by approx. 0.0014 mbar for 0.1 °C. Theoretically, temperature differences of ±0.3 or ±0.4 °C should be measurable. In the explanation of Table 1.12 a standard deviation of 0.38 °C has been calculated.

The membrane- or capacity gauges are more expensive than TMs, but the difference can be neglected even in the cost of a pilot plant. Sensors for such gauges are available for sterilization by steam.

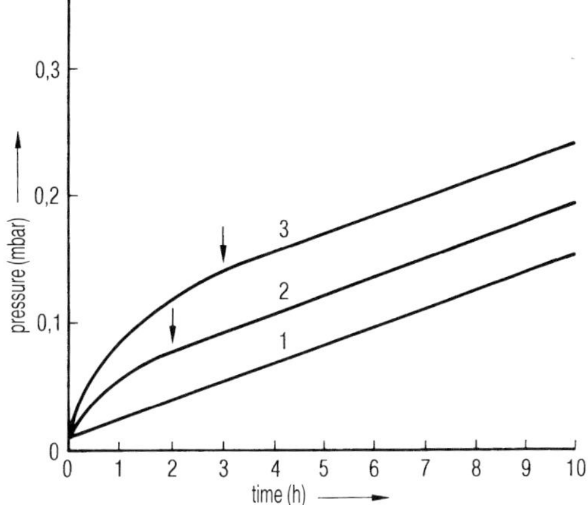

Fig. 2.33.1. Pressure in the chamber and condenser as a function of time after both have been shut off from the pumpset.
1, Plot of pressure, if no measurable gas is desorbed from the walls; 2, plot of pressure, if desorption of gas stops after 2 h at a pressure of 0.08 mbar; 3, plot of pressure if desorption of gas stops after 3 h at a pressure of 0.14 mbar.

Partial gas- or vapor pressures during freeze drying can also be measured by a mass spectrometer, and water vapor pressures by hygrometers, sensitive only for water vapor. Both systems are necessary for development- and analytical work, but in production plants they need only to be used to check or identify process data.

The leak rates of a freeze drying plant can be measured at the empty plant with the condenser cooled and the shelves heated by measuring the pressure rise per time multiplied by the installation volume in the dimension (mbar L/s). It should be noted, that the plant has to be evacuated for several hours, e. g. down to 10^{-2} mbar, before the pressure rise measurements, to avoid the influence of small amounts of ice and the desorption of gas from the surfaces. Furthermore, the pressure rise should be measured up to 0.2 or 0.4 mbar to detect possible gas desorption. Only if the pressure rise has been for some time proportional with time (Fig. 2.33.1), it represents a leak rate, which is defined as

$$\text{LR} = \mathrm{d}p/\mathrm{d}t\ \text{V} \quad (\text{mbar L/s}) \tag{17}$$

If the chamber and condenser have a volume of 1000 L, the leak rate (LR) can be calculated with the data from plot 1 in Fig. 2.33.1:

$$\text{LR} = \{(0.13 - 0.10) / 7200\}\ 1000 = 0.0042 = 4.2 \cdot 10^{-3} \text{ mbar L/s} \tag{17a}$$

If the pressure rise of plot 3 is used during the first 3 h, the leak rate would appear to be approx. $1.2 \cdot 10^{-2}$ mbar L/s.

Figure 2.33.2 shows two pressure rise measurements of the same plant: Plot 1 measured by TM, and plot 2 by CA. During the first approx. 20 h, the quotient plot 1/plot 2 ≈ 0.6, indicating that the gas desorbed is mostly water as can be expected at 30 °C. Other gases as e. g. CO_2 desorb only at much higher temperatures. The pressure in the first hour and also in the first 3 h (not shown) has been $8 \cdot 10^{-2}$ mbar/h, and this falls during the next 4 h to

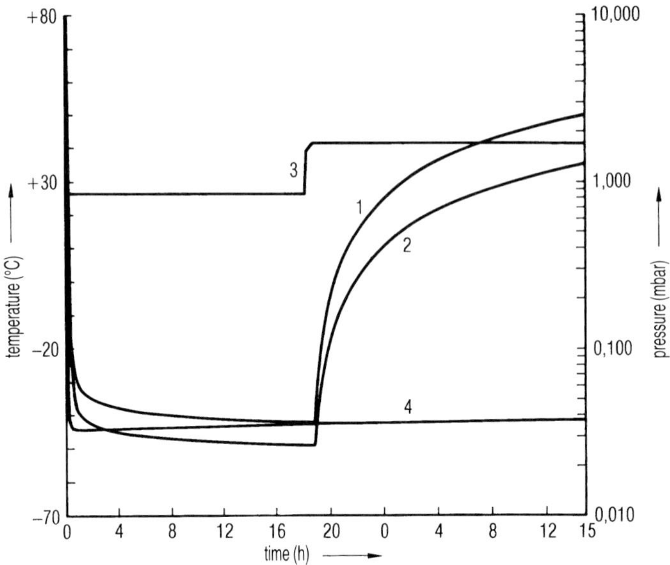

Fig. 2.33.2. Pressure rise measurement as in Fig. 2.33.1 in a plant evacuated for 18 h with heated shelves. T_{st} should not have been changed before measurement of the rise, because the temperature dependent desorption has to be adopted to the new temperature level, as discussed in the text in the remarks about 'history'.
1, p_{ch} measured by TM; 2, p_{ch} measured by CA; 3, T_{sh}; 4, T_{co} .

$7 \cdot 10^{-2}$ mbar/h and during the last 10 h to $4 \cdot 10^{-2}$ mbar/h. This example is to show, that the water desorption during applicable measuring times becomes less and less. If the pressure rise of $8 \cdot 10^{-2}$ mbar/h in this installation is converted into leak rate, LR = $3.6 \cdot 10^{-3}$ mbar L/s, or after 10 h it drops to $1.8 \cdot 10^{-3}$ mbar L/s. In the LR range of 10^{-3} mbar L/s one, has to expect such variations between different measurements, since desorption depends on the history of the plant before measurements start, and variations of this size disturb neither the BTM nor DR measurements.

If the leak rate disturbs the drying process, measurement of the ice temperature or desorption rate, the leak has to be located and closed. Leak 'hunting' at a completely installed freeze drying plant can only be done by a helium leak tester as, shown schematically Fig. 2.34. The installation (1) is evacuated by a vacuum pump (4). With a pistol (5), helium from a pressure bottle (3) is sprayed on the components of the plant most likely to leak: Door seal, valves, windows, lead-through and other flanges. Together with the air, helium diffuses into the plant and is detected by a mass spectrometer (2) specially adjusted to the mass 4 of He and connected to the vacuum pumping line. A freeze drying plant cannot be evacuated by the small pumping system of a mass spectrometer (1.6 m^3/h backing and 33 L/s diffusion vacuum pump); therefore, the mass spectrometer is operated parallel to the pumping system of the freeze dryer. Only a part of the total gas flow passes through the

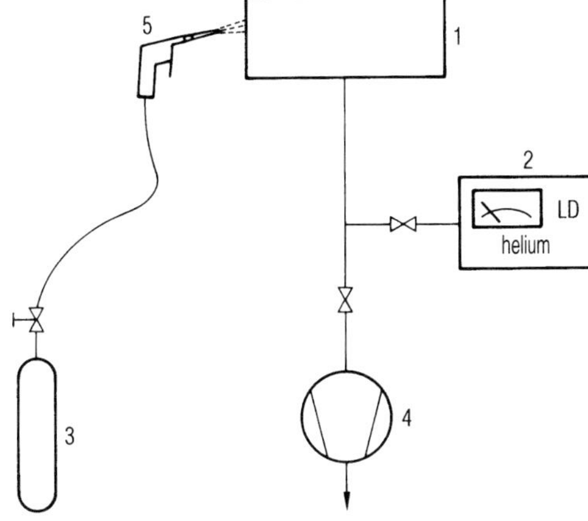

Fig. 2.34. Detection of a leak by a helium leak detector.
1, Chamber with a leak; 2, helium leak detector; 3, pressure bottle with helium; 4, vacuum pumpsystem, evacuating the chamber; 5, helium spray pistol (Fig. 3 from [2.26]).

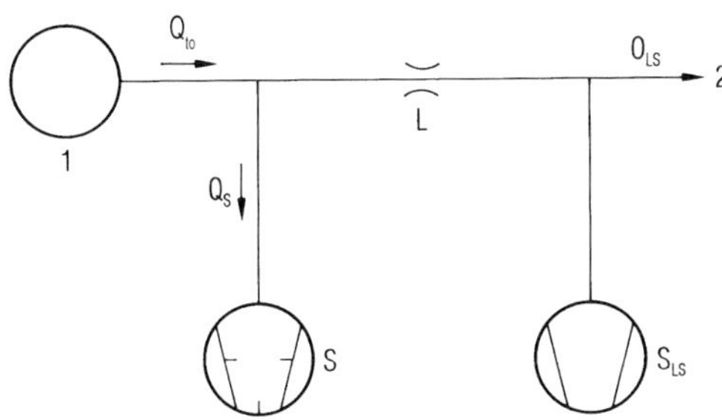

Fig. 2.35. Leak hunting with the part stream method.
1, Object to be tested; 2, helium leak detector; L, slit with a fixed conductivity (Fig. 10 from [2.27]).

mass spectrometer, as shown in Fig. 2.35. If the throttle L has a conductance, which is small compared with the pumping speeds of S and S_{LS} $Q_{LS} = (L/S)\ Q_{to}$. If L and S are known and constant, the part flow measured by the mass spectrometer is only the L/S part of the total He flow. If e. g. S = 360 m^3/h = 100 L/s at 0.1 mbar and L = 0.1 L/s, the leak detector will measure 1/1000 of the total He, which has entered through the leak. Thereby, the smallest leak detectable increases by a factor 1000, but this is not detrimental, as the sensitivity of the mass spectrometer is e. g. $2 \cdot 10^{-10}$ mbar L/s, and if the sensitivity becomes 1000 times smaller, it is still $2 \cdot 10^{-7}$ mbar L/s. For freeze dryer it is sufficient to locate leaks larger than $2 \cdot 10^{-4}$ mbar L/s. L/S could become 1/100 000 and leak detection is still possible at a pressure of 10 mbar. Helium leak testers are available in a transportable box of approx. 50 L with automated measurements, as shown in Fig. 2.36.

Leaks are, besides loss of power, the most unwanted events during freeze drying (see the end of this chapter about defects). Therefore it is recommended, to make a leak test before each freeze drying run, though this routine test does not have to follow the procedure described above. If the leak rate is measured once, the pressure as a function of time during the evacuation period can be recorded and compared with the evacuation plot of the

Fig. 2.36. Helium leak detector in operation at an industrial vacuum plant. The leak detector stands behind the hand rail, the flange is sprayed with He, the mobile indicator is held in the left hand (photograph: Balzers und Leybold Holding AG, D-63450 Hanau).

actual run. If these plots do not deviate from each other, one can conclude that no additional leak has developed. This routine test should only be applied if the cleaning and sterilization of the plant have been the same as before the last run, to keep the desorption qualities of the surfaces as identical as possible.

A freeze drying plant should have as a minimum the following measuring capabilities:

During freezing:

- Inlet and outlet temperatures of the shelves, if there are different blocks of shelves, each block should be measured, if temperature sensors in the product are used at least three should be applied

During main- and secondary drying:

- Shelf temperatures and product temperatures as above
- Condenser temperature(s), in a multiple coil condenser at each separately injected block of coils
- Temperature of the ice at the sublimation front (see Section 1.2.3)
- Total pressure in the chamber by a membrane gauge (CA)
- Total pressure in the condenser by CA, close to the vacuum pump connection (see Section 2.2.4)

- Total pressure by CA between the shut-off valve and the vacuum pump. (The last two CAs can have the same data processing system, since no fast changes have to be measured)
- Software to record DR data and calculate the residual water content (dW) (see Capter 1.2.3)
- Vacuum gauge up to atmospheric pressure during the venting of the plant.

For production plants the following data of machines should be supervised:
- Temperatures of the cooling- and heating medium after the circulation pump(s) and after the heat exchanger(s)
- Temperature of the refrigerant at each injection valve
- Operation temperature of all electric motors
- Open or closed position of all valves (without injection valves)

The control system of a freeze drying plant should allow to program the following data independently from each other during manual operation:

Start of the cooling of the shelves:
- Cooling speed of the shelves *down* to a specified end temperature; the plant manufacturer has to specify the possible maximum and minimum speed for a temperature interval, e. g. 1 °C/min from 0 °C to –40 °C, 2 °C from 0 °C to –20 °C, 0.3 °C from –35 °C to –45 °C.
- Start of evacuation after a preset shelf temperature has been reached
- Selection of shelf temperature as a function of time
- Start of the heating after a preset chamber pressure has been reached, e. g. with a delay of 0.5 or 15 min
- Preset the time between two BTMs, e. g. 15 min or 30 min. If MD is long, e. g. 20 h, the time can be increased to 60 min or decreased to 10 min for a very short MD
- The shelf temperature can be changed during hand operation
- The operation pressure can be preset and changed, e. g. if T_{ce} is too low the p_c can be raised or vice versa

Note for the last two points: A too high or too low ice temperature at the sublimation front can be changed by two operation data: By changing the shelf temperature or by changing the heat transfer from the shelf to the bottom of the vials or trays. As shown in Table 1.10 K_{tot} can vary from ≈ 117 kJ/m^2 h °C at ≈ 0.5 mbar to ≈ 62 kJ/m^2 h° C at 0.36 mbar (all other data constant). The change of T_{ice} with the pressure is rapid as shown in Fig. 2.37.1. Figure 2.37.2 shows, that the range of control is approx. 10 °C (–28 °C to –18 °C) under the conditions of these runs, in which all other data have been constant.

The function of T_{ice} = f (p_c) depends on the design of the plant, quality of the product, type and number of vials, filling height and shelf temperature. With these data, constant, T_{ice} has been found very reproducible during four different runs each with 400 vials filled with 2000 cm^3 10 % saccharose solution, operation pressure (p_c) controlled at 0.15 mbar:

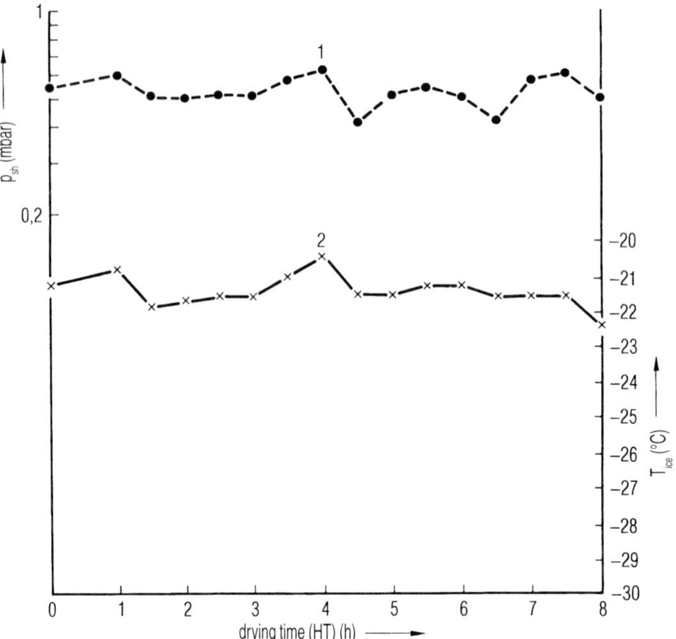

Fig. 2.37.1. Temperature at the sublimation front of the ice, T_{ice}, as a function of the varied operation pressure, p_c.
1, operation pressure; 2, temperature at the sublimation front, T_{ice}. After 5 h drying time the T_{ice} follows the operation pressure more slowly, because the main drying is reaching finalization (measurements by AMSCO Finn-Aqua, D-50354 Hürth).

–32,259 °C (0,332 °C), –32,600 °C(0,423 °C), –32,529 °C (0,375 °C), –32,113 °C (0,304 °C) average of the four runs: –32,375 °C (0,228 °C). Data in () are standard deviation.

The change of the shelf temperature changes T_{ice} much more slowly, since the heat capacity of the shelves and the heating medium is large. It can take 1–2 h before an equilibrium is reached after a shelf temperature change.

For an automatic process control, the first three points can be realized by known technology. The control of T_{ice} should be done by the control of p_c for reasons given above. The automation of the time distances between two BTMs can be done e g. in the following way: At the beginning of the drying 15 min are preset. If three measurements in series differ by less than e. g. ±0.3 °C, the equilibrium is reached and the time interval can be prolonged to 30 or 45 min, depending on the expected time of MD. The end of main drying can be detected automatically by the method described in Fig. 1.78.2: The apparent reduction in T_{ice} indicates the end of MD. $T_{ice/n}$ = average T_{ice} for all measurements 1 to n, maximum $T_{ice/n}$ = maximum of all $T_{ice/n}$ calculated during the run. With these definitions the change from MD to SD can be automated as outlined in the following example: Step 1 if maximum

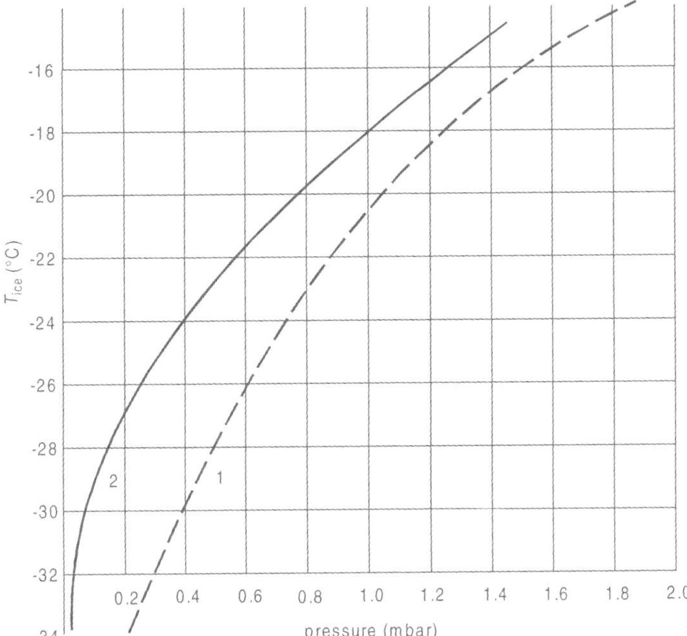

Fig. 2.37.2. T_{ice} as a function of the operation control pressure, p_c. The plot is applicable only for one product in the same type of vials, the same number of vials in the same plant, the same filing height and the same T_{sh} (Fig. 4 from [2.31]).
1, vapor pressure of ice (shown for comparison); 2, T_{ice} as a function of p_c.

$T_{ice/n}$ –1 °C is reached the shelf temperature is raised to the SD level. Step 2 if maximum $T_{ice/n}$ –2 °C is reached the pressure control is switched of. Step 2 could also be delayed until maximum $T_{ice/n}$ –3 °C is reached. The selection, which step to activate at which temperature drop (1, 2 or 3 °C) depends on the slop of the temperature decrease. Step 3 directly after step 2 activates the DR measurements.

The automatic measurement of the 'Desorption Rate' (DR) can be done by the measurement of the pressure rise during 30 s in intervals of e. g. 15 or 30 min. As shown in Fig. 2.38, the DR data after a certain time can be approximated often by a straight line or a curve very close to it in half-logarithmic scale. The data on the left upper part of the plot are characteristic for the change from MD to SD. During this time the slope of the plot changes during the intervals of measurement. After the slop has become stable, the computer can extrapolate the DR plot to a value so small that it does not add substantially to the integrated residual moisture content.

An example is given in Fig. 2.38: plot 1 is extrapolated to DR < 0.1 %/h. At this time, one more hour drying time would contribute to the desired moisture content of e. g. 1 % by only 0.1 %. If the computer starts the integration from this calculated time the result is the residual moisture content as a function of time.

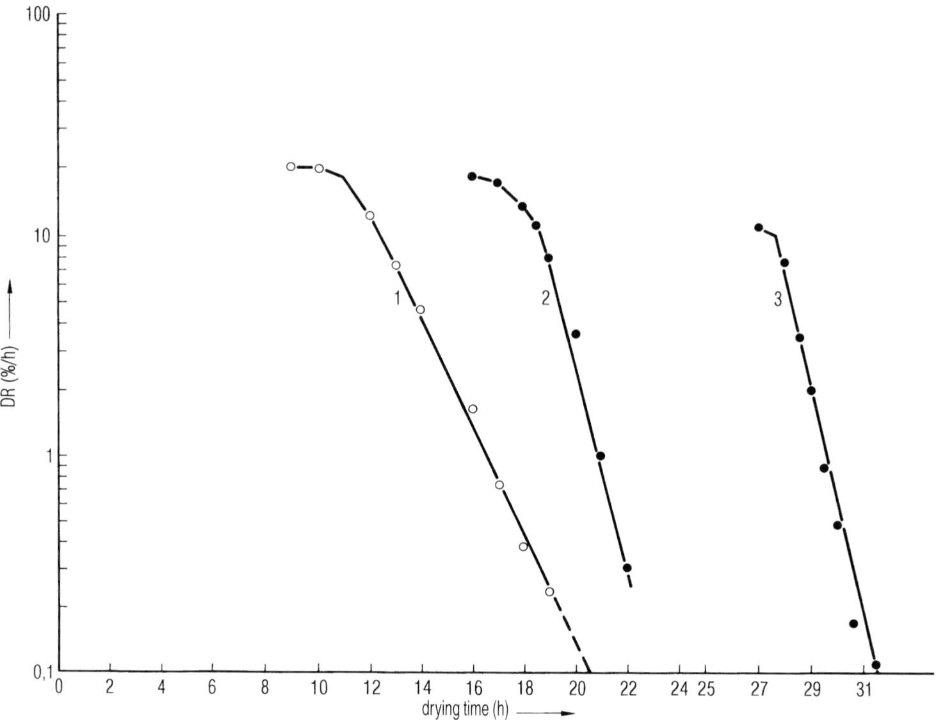

Fig. 2.38. DR measurements of the same product with three different operation pressures and therefore with three different T_{ice} during main drying. The process data were:

	1	2	3
p_c (mbar)	0.55	0.38	0.10
T_{ice} (°C)	–21.5	–24.5	–32.0
Start of SD (h)	11	19	27.5
DR = 0,3 %/h after h	18.5	22	30.5
$T_{sh,\,SD}$ (°C)	+30	+30	+30

The residual moisture content calculated by this method is called desorbable water, d*W*.

$$dW_t = \int_{t=0}^{t=t} \mathrm{DR}\,dt \qquad (18)$$

This residual moisture content d*W* (% of the solids) is calculated as shown in Fig. 2.39. The automatic measurement of DR, which is more accurate than a hand operation of the valve and the visual reading of the pressure gauge, supplies additional information about the main drying as shown in Fig. 2.40.1. The same granulate has been dried at the different p_c 0.27, 0.44 and 1.46 mbar corresponding to three T_{ice} –27, –24.5 and –15 °C during main

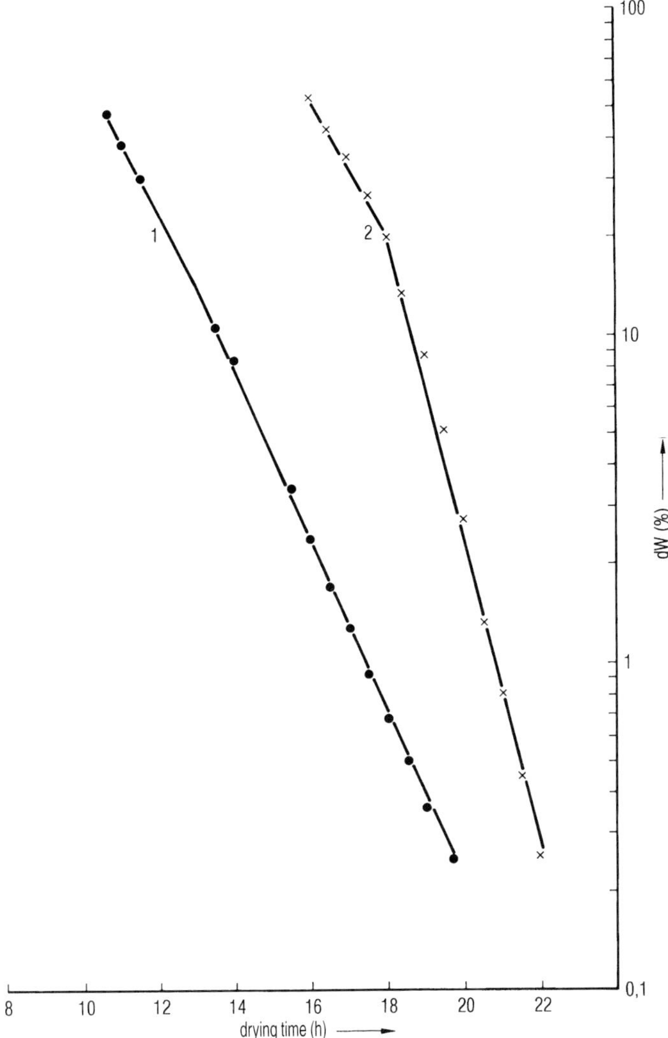

Fig. 2.39. Desorbable water (dW) as a function of the drying time (run 3 not shown) (plots from measurements by AMSCO Finn-Aqua, D-50345 Hürth).

drying. At 1.46 mbar the DR data during secondary drying decrease very slowly and the slop changes during the entire measuring period. The product has been dried at too high a T_{ice} and is collapsed. The water is not only absorbed by the solid, but is partially in a liquid state. SD is not a desorption drying, but a vacuum evaporation of water. The original structure is collapsed and the water is part of a 'syrup' like concentrate, which is difficult or impossible to dry.

The residual moisture content, calculated from DR data, measures the amount of water, which can be desorbed at the product temperature. The procedure corresponds therefore to the thermogravimetric method [1.83] with some practical advantages:

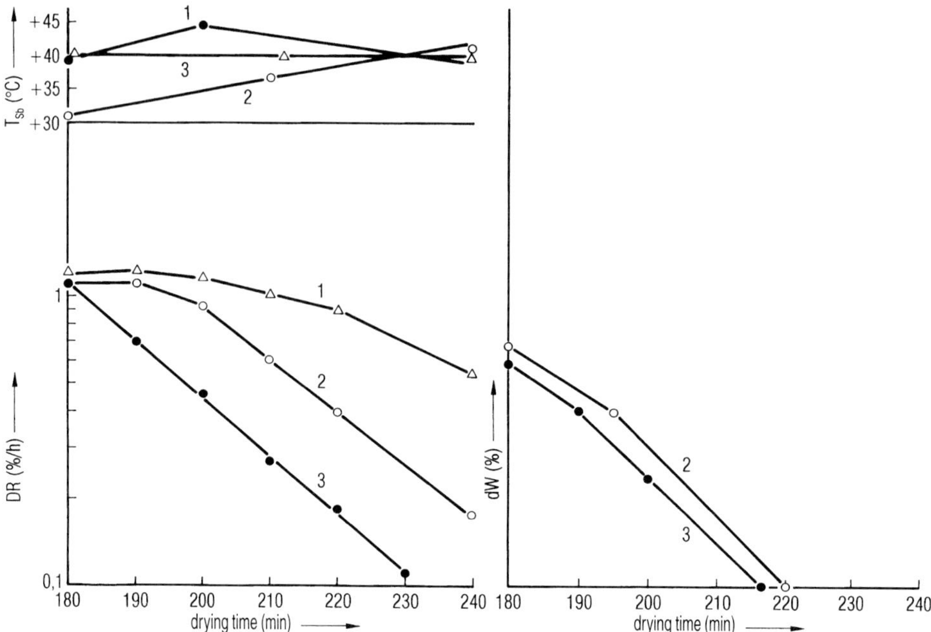

Fig. 2.40.1. DR- and dW-data as a function of drying time for the same granulate with very short drying time. In the upper left part of the figure the plots show the temperature of the tray bottom, which is difficult to control manually during the short drying time (plots from measurements by Dr. Otto Suwelack, D-48723 Billerbeck).

- the value measured is the average of the total charge and not of samples of the charge
- no scale required, the total amount of solids in a charge is known
- the secondary drying can be terminated, when the desired dW is reached (no over-drying) or can be prolonged until dW is reached.

If the production instruction requires an other methods of water determination or an other temperature at which it has to be measured, the data measured can be different, since the water can be bound to the solid by different energy levels as can be seen in [1.83]. In this case a correlation between the two methods or temperatures has to be established (see Section 1.3.1).

In a freeze drying plant automated in this way, the desorption rates and the desorbable water content (in % of solids) can be measured, calculated and documented.

Chase [2.32] presents an alternative method to monitor and control the freeze drying process by measuring the flow of nitrogen to keep the operation control pressure, p_c, constant. The Mass Flow Controller (FMC) consists of a proportional valve, an integral flow meter and a capacitance manometer (CA). The CA measures the total pressure in the plant, the valve opens, if the pressure gets below the preset value and vice versa. The flow of

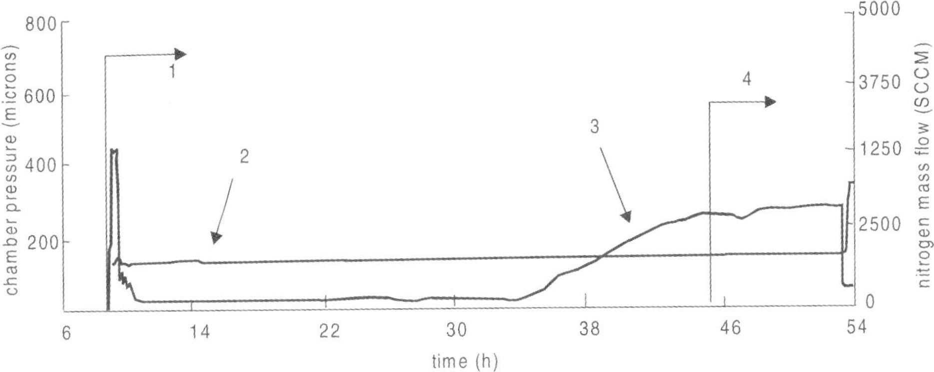

Fig. 2.40.2. Chamber pressure and nitrogen flow rate as a function of drying time. 1 main drying, 2 chamber pressure, 3 nitrogen flow, 4 secondary drying (Fig. 4 from [2.32]).

nitrogen necessary to keep the pressure constant is measured by the flow meter. Figure 2.40.2 shows the high N_2 flow at the start of the drying, because the product is heated up and the water vapor pressure is still low, after approx. 2 h the water vapor pressure rises and the N_2 flow can be reduced to keep the total pressure constant. After 34 h the water vapor flow decreases, and the N_2 flow increases until the water vapor flow becomes very small at approx. 49 h and the N_2 flow almost constant.

Possible problems during the operation of freeze drying plants can mostly be assigned to four categories:

1. Those occuring before the evacuation starts.
2. Those requiring immediate, preferably automatic action.
3. Those to be corrected automatically or by hand within few minutes.
4. Deviations which must be documented.

The analyses of trouble and the required counteractions will depend on the operation principles of the operator, the value of a lyophilizatiion charge, the sensitivity of the product on deviations from the set values, and several other factors. Nevertheless, an analysis of possible problems and a guideline for counteractions should be established for every freeze drying plant as part of the instruction manual, even if it has to be adjusted for new products.

The following summary cannot be a complete list, but reflects only the more likely events of trouble. (In Section 6.1 seven problems are discussed related to product structure, stoppers, traces of volatile components etc.)

Trouble before the evacuation starts:

These events can be avoided in most cases, if the following start-up rules are used: 1. The plant is evacuated down to the lowest operation pressure; if this pressure is reached, possible water or ice from the cleaning or sterilization is evaporated. 2. The condenser is cooled to the operation temperature and the shelf heated to the maximum temperature during SD. 3. If the lowest operation pressure is reached again, the leak test can be done (see Section 2.2.6). 4. The shelves are cooled back to the loading temperature and the plant is vented with a gas as specified (see end of Section 1.2.3 and Fig. 2.33.1, 2.33.2 and 2.34).

With this start-up the most critical machine data are to check: No remaining ice, cooling and heating working, and leak rate acceptable. The procedure takes some time, but minimizes problems occuring during the freeze drying process.

Trouble requiring immediate action:

The most undesirable event is a power failure. In that case, the valve between vacuum pump and the condenser has to close automatically. If the power failure lasts only a few minutes, the condenser has a certain heat capacity to maintain the sublimation of ice. The tolerable time of power failure depends from the plant design.

A power failure exceeding the time limit is most critical during MD. In this case a stand-by power generator with a start up time of e. g. 1 min either for the operation of the whole plant or at least for the critical components is the only answer. The sequence of importance for the components can be:

- operation of a part of the pumping set to maintain the vacuum in the chamber and the condenser
- operation of a part of the refrigerant compressors or the injection valves for LN_2, if the condenser has a separate coil for it (see Section 2.2.5)
- operation of a part of the cooling system for the shelves or the system can also be cooled by LN_2 (see Section 2.2.5)

An other undesirable event, is an air pressure by a leak in the plant which reduces or stops the sublimation. The counteractions could be:

- Reduction of the shelf heating temperature to the minimum possible value
- Using the full pumping capacity or adding an auxiliary pumpset

If the loss of the product can be avoided by these steps, hunting for the leak could become possible (depending on the arrangement of the plant in the building). By experience it is possible to predict, that leaks will not be generated through porous steel, but by seals in flanges or doors. The probability of newly developed leaks after the suggested start up procedure has been used is 10 or 100 times smaller than without this test, but it cannot be fullyexcluded.

Trouble to be corrected automatically or by hand in minutes:

The failure of the condenser compressors can be bearable for some time (see above). For production plants, two smaller compressors are recommended rather than one large unit, as this permits the process to be continued with a prolonged drying time. On failure of the refrigerant compressors for the brine in the shelves the shelf temperature will rise slowly, if the temperature is below room temperature. Stopping the brine pump(s) will slow down this temperature rise during the main drying, and the sublimation of ice will withdraw energy from the shelves. A failure of other components e. g. valves or gauges, could be compensated by replacements or stand-by units. For these types of failures the plant must be completely hand operable and controllable.

Deviations to be documented:

All deviations from the preset data should be documented, even if they are within the given tolerances. For the important process data, not only a comparison of the actual with the preset data is recommended, but also a trend analysis should be installed. This permits the recognition of systematic deviations long before the tolerances are exceeded.

Example: The brine temperature is set at –30 °C ± 1 °C. The brine temperature has been for some time between –31 °C and –30.5 °C, It rises e. g. in 2 h from this range to –30 °C and –29.5 °C. Warning of this trend is helpful Either the pumping speed of the brine pump is reduced or the injection valve of the refrigerant has shifted or the refrigerant compressor has changed its capacity.

Analyze: The inspection system on the motor of the brine pump shows no warning, the temperature of the refrigerant before the injection valve is constant. It follows, that the injection valve has shifted and should be adjusted.

2.3 Installations up to 10 kg Ice Capacity

2.3.1 Universal Laboratory Plants

As indicated by the title of this section, the manufacturers of such plants attempt to make them flexible for different applications by using a modular concept. Often, the basic unit consists of a condenser, a vacuum pump and a vacuum gauge, to which various drying systems can be added: e. g. manifolds for flasks, ampoules or vials, baseplates with belljars or small chambers with temperature-controlled shelves. The following qualities could be important in selecting the most suitable laboratory plant:

- Ice condenser
  - Maximum ice capacity at an ice thickness of 1 to 1.5 cm
  - Can the condenser surface be visually observed?
  - Method of ice defrosting (e. g. electric heating, warm water, warm vapor from the compressor). If the ice must be examined for certain components carried over with the ice (aroma, volatile components), hot air or water may be not applicable.

- Refrigerant compressor:
  - Air- or water-cooling.
  - Lowest operating temperature.
  - Cooling capacity at 5 °C and 10 °C above lowest temperature.
- Vacuum pump:
  - Two stages (recommended) with gasballast (necessary).
  - Valve between the vacuum pump and condenser which closes automatically if the electric power is disrupted.
- Vacuum gauge:
  - If a heat conductivity gauge (TM) is offered, the additional price for a capacitance gauge, which is strongly recommended, should be requested.
- Shelves:
  - Range of controlled temperature.
  - Speed of cooling and heating from xx °C to yy °C for cooling and from zz °C to ww °C for heating.
  - Recommended method of freezing, if the shelves cannot be cooled.
  - Uniformity of temperature for all shelves, e. g. ±1 °C during main drying.
  - Levelness and smoothness of the shelves.
  - Temperature measurement in the product.
- Valve and its free diameter between chamber and condenser:
  - Without this valve no pressure rise measurements can be carried out for BTM (T_{ice}) or DR data for the calculation of the residual moisture.

The process data from manifold installations can hardly be transferred to chamber-type plants. This applies, practically, also to the process transfer from belljar-type installations to chamber plants. Results obtained in laboratory plants of the chamber type must be analyzed carefully, if they are be transferred to another plant. If the product, the layer thickness of the product and the vials or trays are identical, the following conditions should be observed and compared:

- Freezing method and freezing rate must be the same.
- Eveness, smoothness and temperature uniformity of the shelves must be comparable.
- The temperature shielding of the product against wall and door influences have to be comparable.
- The shelf temperature and the controlled operation pressure must be controlled in such a way, that T_{ice} in the laboratory plant is stable and measured with a standard deviation less than ≈ 0.5 °C.
- The conditions of water vapor transport between the chamber and condenser, the condenser surface, the capacity of the refrigerant compressors and the vacuum pumping capacity, must all be the same (pro rata) for the product dried in the laboratory as in the plant to which the process will be transferred.

It is likely asked too much of most laboratory plants, if used as pilot plants for production process development. The best application of laboratory plants is the freeze drying of preparations and products which do not require to be operated within small tolerances, but can be dried under noncritical process data.

2.3.2 Pilot Plants

In the plants described in this chapter the process data can be developed, verified and – in a measurable and reproducible way – modified to achieve the specified quality of the product in the most economical process.

All these plants are of the chamber type (see Fig. 1.88 (c)) with cooled and heated shelves and a condenser which can be separated from the drying chamber by a valve. Refrigeration and vacuum systems should be laid out for temperatures and pressures which can be expected under extreme experimental conditions, even if these extreme data may not be used in the production process. Pilot plants for pharmaceutical or medical products should be laid out differently than those used for food.

The following proposal for a pilot plant specification is given in general terms and will have to be supplemented by specific requirements:

	Pharmaceuticals	Food
Size of chamber	approx. cubic[1], 100–200 L [2]	1 · 0.5 · 0.5 m [3]
Condenser surface	4–6 kg of ice at a layer thickness of 1.5 cm [4]	10–15 kg of ice at a layer thickness of 1 cm [5]
Ccondenser temperature	Minimum –55 °C, preferable –65 °C	Minimum –45 °C, preferable –50 °C
Refrigeration capacity of the compressors at 10 °C higher than the above minimum temperatures	Specification of manufacturer	
Type of valve between chamber and condenser	As shown in Fig. 2.18 (D), preferable no butterfly valve	
Water vapor transport from chamber to condenser [6]	Water vapor g/h at p_{ch} 0.06 and 0.3 mbar, measured by CA	Water vapor g/h at p_{ch} 0.2 and 1.0 mbar measured by CA
Vacuum pumping set	Two stages, with gasballast, in max. 15 min down to 0.02 mbar [7]	Two stages, with gasballast in max. 10 min > 0.1 mbar[7]

leak rate of chamber and condenser[8]	$< 2 \cdot 10^{-3}$ mbar L/s	$< 2 \cdot 10^{-2}$ mbar L/s
Shelves	One shield on top two to four for product	One shield on top two for product
Shelf temperature	Min. –45 °C, preferable –55 °C, max. +60 °C	Min. –10 °C, preferable –25 °C max. +120 °C
Amount of brine circulated through the shelves and specific heat capacity of the brine [9]	Specification of manufacturer	
Method of condenser defrosting, cycle time	Specification of manufacturer	

1 depth of chamber should be limited for cleaning and loading.
2 to measure DR data, the ratio of chamber volume (L): mass of solids (g) should be approx. 1 or (depending on the desired accuracy of DR) maximum 2. If only small amounts of the test product are available, it is recommended, to use a smaller chamber e. g. 50 L, otherwise the free volume of the chamber has to be reduced by glass or aluminum bars.
3 Chamber should be large enough for two production trays on one shelf and two shelves, one on top of the other.
4 With pharmaceuticals the vapor flow (kg/h) during MD is usually smaller than for food; therefore the ice layer produces a relatively smaller temperature difference between coil and ice surfaces, the ice layer can be thicker than for food plants.
5 See [4].
6 Flow of water vapor with pure ice in trays (see Section 1.2.4)
7 Measurements with chamber and condenser at room temperature.
8 Details in Section 2.2.6. For maximum tolerable leak rates see Section 1.2.3 as referred for BTM and DR measurements.
9 With this information the maximum flow of energy can be estimated, which can be transported during freezing and main drying at a desired temperature difference between inlet and outlet temperature of the brine at the shelves, e. g. 2000 kJ/h at a temperature difference of 3 °C. With this amount of energy approx. 0.7 kg ice could be sublimated per hour. (This estimate gives only the maximum possible sublimation-rate, whether it can be achieved or not depends from heat – and mass transfer conditions in the process (see Section 1.2.1 and Eq. (12)).

The requirements on the measure-, control- and safety installations are described in Sections 2.2.8 and 2.4. The degree of automation of pilot plants will depend on the expected operation conditions. However it is recommended to automate the BTM and DR measurements, this being far more accurate and less tedious than hand operation and visual reading.

2.3.3 Manipulators and Stoppering Systems for Vials

In the operation of pilot plants it may be important to close some vials during secondary drying without interruption of the drying process. Vials closed at certain moments can provide various information: The residual moisture content can be determined not only by measuring the DR during SD, but also by other methods (see Section 1.3.1) and compared with data calculated from DR measurements. Furthermore, it can be investigated, whether and how much the chemical structure and or the activity of the product changes at certain times and temperatures during drying. During main drying it is recommended, that closed vials are not removed as this may change the drying conditions for the product in vials close to those removed.

By using manipulators, as shown schematically in Fig. 2.41, certain vials can be closed and left in their position or be removed from the plant by a lock. In a plant without a manipulator it is not possible to close vials after venting the plant for a short time, because during MD the product will collapse or melt and during SD the continuation of the drying will be different, (gas has been absorbed by the solids). Figure 2.42 shows a pilot plant with a manipulator.

An important step in the freeze drying process with vials is the stoppering or closing of the vials either at the end pressure of SD, or at a chosen partial pressure of a specific gas. This avoids handling of open vials, which can lead to contamination and adsorption of water vapor from the atmosphere.

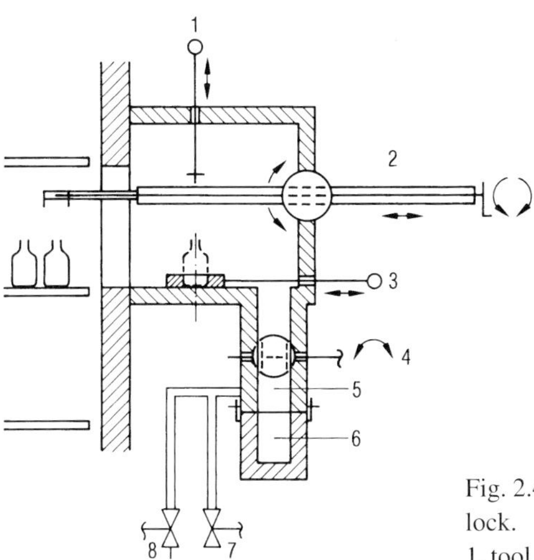

Fig. 2.41. Schema of a manipulator including a vacuum lock.
1, tool to close vials; 2, arm of the manipulator; 3, rod to push vials; 4, ball cock; 5, exit channel; 6, exit container for vials; 7, venting valve; 8, vacuum valve; 9, vacuum pump.

Fig. 2.42. Manipulator as shown in Fig. 2.41, connected to a LYOVAC® GT6 (photograph AMSCO Finn-Aqua, D-50354 Hürth).

This step is so important that some laboratory plants, and usually all pilot and production plants, are equipped with respective mechanisms. The principle is simple: The shelves are connected flexibly with inlet and outlet of the brine. The shelves are pressed together, one after the other, by a plate, which is moved by an external force; in this way the stoppers are pushed into the closed position. If the pressure necessary for this stopper movement is 1 kg per stopper, the resulting total force for 100 vials per shelf is 100 kg, but if 10 000 vials are loaded per shelf, the total force is 10 tonnes, which has to be applied evenly in order to avoid vials breakage.

The technical problems to be solved are two fold: All parts must be steam-sterilized and no abrasion can be tolerated in the chamber. For this the manufacturers offer various solutions: in that the pressure plate is:

- operated by several motor driven spindles; or
- drawn by motor powered ropes; or
- moved by a centralized hydraulic system, as shown in Fig. 2.43.

The shaft (2) is surrounded by a chamber (3), which is evacuated and steam-sterilized. The chamber and the shaft are sealed from the drying chamber by a special sealing system. The advantages are: The slick cylinder is easy to clean and to sterilize, with no possible abrasion. The disadvantage is the greater height, as the cylinder chamber must be as long as the lift of the pressure plate. Another solution to avoid the greater height is to use a bellows as shown in Fig. 2.48.1.

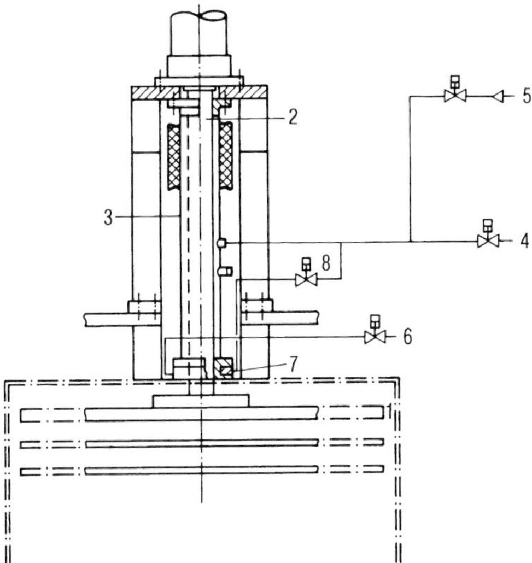

Fig. 2.43. Schema of a steam-sterilizable closing mechanism for vials. The pressure plate (1), by which the stoppers are pushed into the vials is sterilized jointly with the chamber. The shaft (2) to which the pressure plate is connected, moves into the chamber during stoppering. It is not sterilized with the chamber. Therefore the shaft is sterilized in a separate chamber (3). This chamber can also be connected to the vacuum pumping system (4) as to the steam supply (5). Water condensing during the sterilization can be drained by (6). A special seal (7) can (by (8)) also be connected to steam or vacuum, and be sterilized (schematic drawing from information by AMSCO Finn-Aqua, D-50354 Hürth).

2.3.4 Cleaning Installations, Sterilization by Steam and VHP®

Pilot plants are usually cleaned by hand, therefore the depth of chamber and condenser should be limited. Sterilization by steam is possible, as can be seen in Fig. 2.44, but can be replaced by the VHP® (Vaporized Hydrogen Peroxides) process, which works at ambient temperature and without pressure. Nakahira [2.11] describes the development of applicable sterilization cycles, the necessary changes in the freeze drying plant and the sterility test necessary to validate the process. Sterilization by VHP requires certain conditions which result from the nature of the H_2O_2 vapor:

- The plant must be dry, otherwise the H_2O_2 will dissolve in the water.
- The installation must be evacuated to approx. 1 mbar to ensure that the vapor reaches all parts and corners of the plant. H_2O_2 can still be applied at +5 °C, but at higher temperatures, e. g. +25 °C, the time to effect the sterilization is shorter. At higher temperatures, e. g. +50 °C, the time required is still shorter, but H_2O_2 decomposes more readily and higher concentrations of H_2O_2 have to be used. Figure 2.45 shows the results with the resistant *Bacillus stearothermophilus* [2.12].

Fig. 2.44. Steam sterilizable pilot plant LYOVAC® GT 6 D before its installation in the wall of a sterile room (photograph: AMSCO Finn-Aqua, D-50354 Hürth).

Figure 2.45 shows a typical set of D-values (time in which the number of bacteria is reduced by one decade) for *Bacillus Stearothermophilus* for different surface temperatures of the parts to be sterilized and the necessary concentrations of H_2O_2.

Further more some technical prerequisites must be fulfilled:

- Not all materials are resistant to H_2O_2, as shown e. g. in the following list:

resistant	stainless steel
	aluminum
	silicon, viton
	Teflon™

Fig. 2.45. Typical D – values (exposure time in which the number of bacteria is reduced by one decade) for *Bacillus stearothermophilus* for different exposure times and the necessary $H_2 O_2$ concentrations (table from [2.12]).

temperature of the surfaces to be sterilized (°C)	approx. $H_2 O_2$ concentration (mg/L)	typical D-value (min)
+ 4	0,3–0,5	8–12
+25	1–2	1–2
+37	3–4	0,5–1
+55	10–12	0,02

not resistant in all compositions	polyurethane
	Plexiglas™
	EPDM
	not resistant Nylon™

- The pumpset must be laid out to pump water which may be in the plant, but also to be resistant against H_2O_2 . In smaller plants it is possible to place a catalyst to decompose the H_2O_2 in front of the oil-filled vacuum pump. In larger plants it is recommended to use a watering pump combined with an air injector for drying of plant and venting of the H_2O_2.
- The installation does not include long tubes, which are sealed at the end as H_2O_2 will enter these dead ends to only a limited extent. In [2.12] the following example is given: To a drying chamber of 4.5 m^3, a tube of 120 cm length and 1 cm internal diameter is attached. On a stainless steel foil of 120 cm length, 10^6 *Bacillus stearotherophlus* spores are placed at a distance of 10 cm and the foil is placed in the tube. Plant and tube are evacuated to 1 mbar at 25 °C. On four occasions 28–56 g of H_2O_2 are injected into the chamber. All spores, up to a depth of 80 cm in the tube have been killed (22 tests). At a depth of 90 cm 1 of 22 tests was positive, while at 120 cm, 14 of 22 tests were positive and survived spores were found.

The sterilization process consists of three phases, as shown in Fig. 2.46:

- The installation is evacuated for drying to 1 mbar, minimum below 10 mbar. The almost horizontal plot at A is most likely related to some evaporation of water. For the documentation of the process the leak rate of the plant must be measured to prove that no outside perhaps contaminated, air can enter the plant during phase 3. This test (see Section 2.2.8) is best done at the end of phase 1.

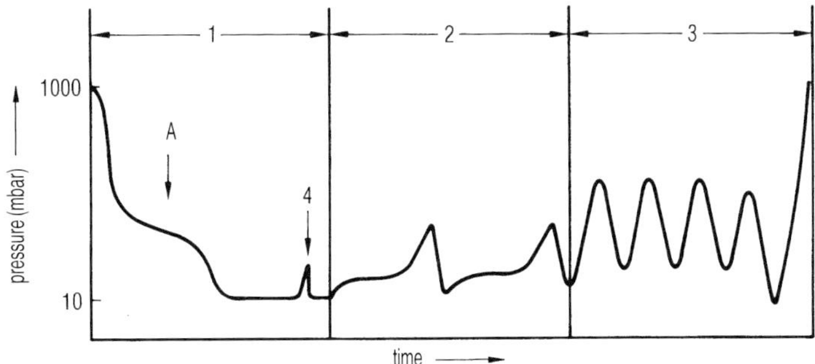

Fig. 2.46. Typical course of pressure during a sterilization by the VHP-process®.
1, Drying of the plant to be sterilized by evacuation down to 1 or 10 mbar; 2, several injections of H_2O_2 each followed by an evacuation; 3, several times venting of the plant with sterile air; 4, leak test of the plant before the start of 2 (VHP® is a process of AMSCO International, Apex, North Carolina 27502 USA; Figure from [2.12]).

- For the sterilization phase (2), a certain amount of H_2O_2 is injected into the plant and pumped off after some time. This procedure is performed once or may be repeated several times.
- Removal of the H_2O_2: The vapor is pumped off by the vacuum pump set, the plant vented with sterile air up to 500 mbar, again evacuated and so on. At the end of this procedure the remaining H_2O_2 concentration is checked, e. g. by the Dräger-H_2O_2 – test tube [2.13], which indicates H_2O_2 concentrations down to 0.1 ppm.

AMSCO International Inc. [2.14] offers complete installations for VHP® sterilization of freeze drying plants, which are programmable and automatic: 4 to 400 g H_2O_2 per injection can be preset, as well as the parameter of the drying-, sterilization- and venting phases. The process is documented. Figure 2.47 shows an AMSCO-VHP® generator.

Nakahari [2.11] describes, on the basis of his experience with VHP®, the advantages of this process: a short sterilization time at room temperature, the possibility to update existing plants, VHP® does not – compared with ethylenoxide and formaldehyde – affect the health of the operators and can be decomposed to water and O_2 without contaminating the atmosphere.

Steam sterilization is at presenzt the most used method [2.15] for freeze drying plants. It requires a temperature of +121 °C for 10 to 30 min, and thus the plant must be pressure-tight up to 2.5 bar.

The rules of steam sterilization are well described [2.15], including some guide-lines for the validation of the sterilization process. The special problems with the steam sterilization of closing systems for vial stoppers has been discussed above. Similar problems

Fig. 2.47. VHP®-generator (left in fore ground) connected to a production plant during factory test.

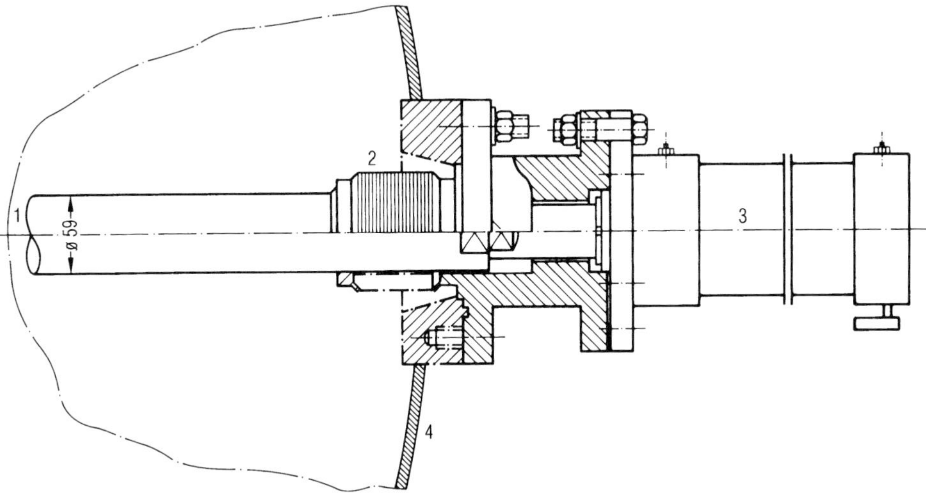

Fig. 2.48.1. Schema of a steam-sterilizable stainless steel bellows, which separates the valve shaft from the condenser chamber.
1, To the valve plate; 2, bellows; 3, valve drive; 4, condenser wall (schema from source material by AMSCO Finn-Aqua, D-50354 Hürth).

arise with central closure systems on doors, which are not closed manually by hand wheels. Central closure systems are normally difficult to clean and sterilize and require mechanical parts, which can lead to abrasion. Therefore, it is preferable to use plain surfaces on the door itself and secure it by eight or more hydraulically operated bolts at the outside. This hydraulic system can at the same time open and close the door and block movement of the door. End position switches can indicate the operation situation of the door.

The third critical part in a steam-sterilized plant is the valve between the chamber and condenser. Its design should influence the flow of water vapor from the chamber to the condenser as little as possible, and its component must be easily cleaned and sterilized. Several technical solutions are offered for this problem. In Fig. 2.48.1 a design is shown in which the valve plate is moved back-and-forth by a hydraulic cylinder. This fulfils the two requirements:

(i) The vapor flow is conducted to the outer condenser coils by the special form of the plate (the design shown is only schematic in Fig. 2.48.2).
(ii) The hydraulically operated valve shaft is sealed by a sterilizable membrane bellows from the chamber. There are no other movable parts.

Freeze drying plants must be easily cleaned, a task which for pilot installations may be done by hand. This is difficult to document and one has to rely on the experience and conscientiousness of the operator. Gaster [2.16] presents a survey of advantages and disadvantages of cleaning methods from hand operation to spray systems with a sequence of

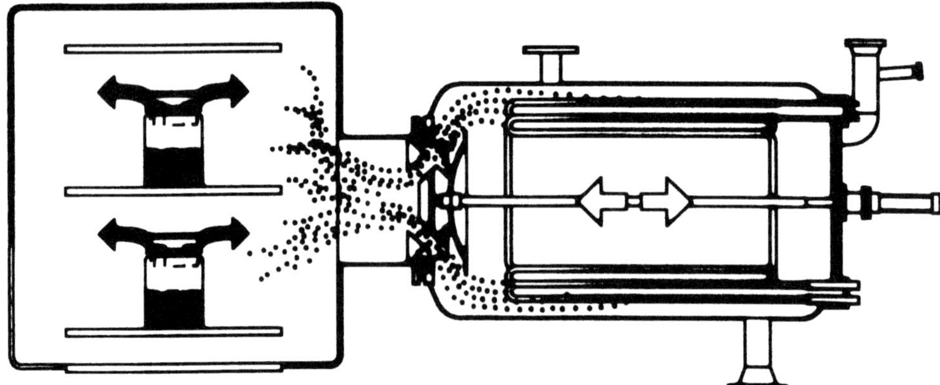

Fig. 2.48.2. Water vapor flow from the drying chamber (left) to the condenser.

cleaning agents and water (see Fig. 2.49.2) and to automatic systems with pressurized steam. It is difficult to outline general criteria for such systems, since one has to find an optimum for a given plant between efficiency, technical complexity, moving parts, types of seals in the plant, and a program which can be documented. Also the steam cleaning proposed by Gaster has technical problems which can only be assessed from case to case.

2.4 Chamber Production Plants

For a production freeze drying plant, no general guidelines for a specification can be given as for pilot plants. The specification must follow the intended production process., but one design criterion should always be paid attention to: The shortest possible connection between chamber and condenser and its internal diameter. As shown in Figs. 1.89 and 1.90, the specific water vapor flow (g/cm^2 h) drops at 0.1 by a factor of almost 2 if *l*/*d* (length/diameter) for the connection goes from 1 to 5 and at 0.04 mbar the factor is approx. 3.

Fig. 2.49.1 shows a freeze drying plant in which each chamber and the related condenser are flanged together. The plant drawn schematically in Fig. 2.49.2 goes one step further, the chamber and condenser being in one housing, separated internally by a plate. The plant is cooled by LN_2. The directly cooled condenser consists of plates, as can be seen in Fig. 2.49.3. Figure 2.49.4 is a view in another production plant from the chamber into the condenser, in which the seal of the valve can be exchanged without dismantling the condenser.

Both installations have full-size doors which are opened during loading and unloading. The loading of 50 000 or 100 000 vials takes some time, and the shelves should therefore be at room temperature to avoid condensation of ice from the humidity of the atmosphere.

Fig. 2.49.1. Steam sterilizable-production plant with two LYOVAC® GT 500-D.
The condensers are directly flanged to the chambers and have an ice capacity of 500 kg each (photograph: AMSCO Finn-Aqua, D-50354 Hürth).

If the loading of vials has to be done on cold shelves, a smaller loading door as shown in Fig. 2.50 and 2.51.1 should be built in to reduce the amount of air diffusing into the chamber. In addition, a small overpressure of sterile air or N_2 in the chamber reduces the condensation of ice. If N_2 is used, the O_2 content near the loading door should be monitored.

The process control, and the monitoring and documentation of all relevant data and a hierarchic warning and alarm system can be planned and installed as suggested in Section 2.2.8, and 6.2.4, and 6.2.5. Some of the suggestions may appear overdone, e. g. inlet and outlet temperatures of each shelf or shelf package. However these data can prove that the freezing of the product on all shelves has been uniform, or a deviation can be seen on shelf 'XX'. Similar analysis can be made for MD. An other example: Temperature at all injection valves for the refrigerant. If one (of several) valve malfunctions, the time of MD may be prolonged (no condensation of ice on one coil) or the final end pressure during SD is reached much later than usual (ice sublimates from the coil warmer than the others). Production freeze drying plants are like airplanes: They carry a very precious load. The earlier a deviation from the normal or preset data can be analyzed, the more likely the load is safely brought home.

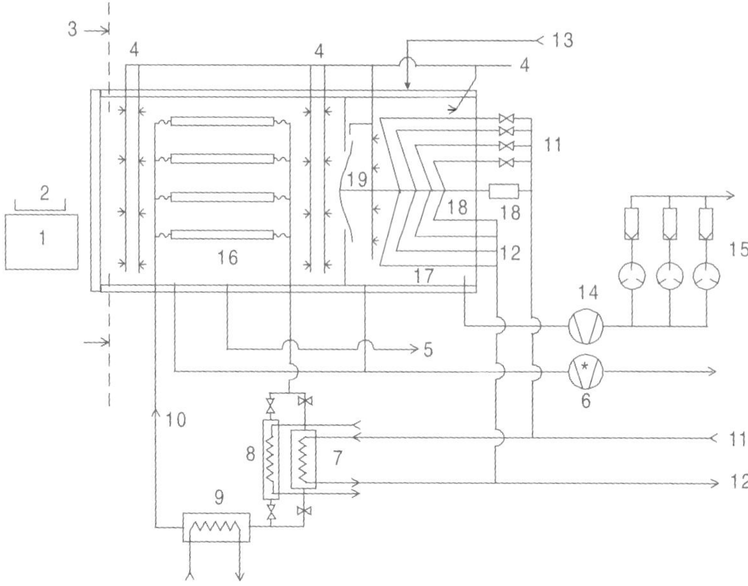

Fig. 2.49.2. Schema of a freeze drying production plant with approx. 20 m^2 shelf area. The chamber and condenser are in the same vacuum chamber, separated by a wall in which the valve is built, providing the shortest possible path for the water vapor. The condenser and the brine heat exchanger are cooled by LN_2. The condenser surface is made from plates (Fig. 2.49.3), its temperature can be controlled between –110 °C and –60 °C. The shelves can be controlled by the circulated brine between –70 °C and +50 °C. The trays with product can be automatically loaded and unloaded from a trolley. The shelves can be pressed together in one block and the trays are loaded to the shelves by pushing one shelf after another in front of the trolley.

1, Trolley for loading and unloading; 2, product in trays; 3, sterile room; 4, CIP system; 5, water drain; 6, water ring pump; 7, heat exchanger for the brine cooled by LN_2; 8, heat exchanger for the brine cooled by water (+20 °C); 9, heat exchanger for the brine, electrically heated; 10, brine (heat transfer medium) circuit; 11, LN_2 inlet; 12, N_2 gas outlet; 13, water for defrosting; 14, roots vacuum pump; 15, three two-stage pumpsets; 16, chamber; 17, condenser; 18, condenser plates; 19, hydraulically operated valve between chamber and condenser.

Fig. 2.49.3. Condenser plates (18) in Fig. 2.49.2.

Fig. 2.49.4. View of a valve inside the condenser (all shelves in their lowest position). The condenser coils are seen in the background. In front to the right is a part of the CIP system. The seal of the valve can be replaced in the position shown.
(Fig. 2.49.1–2.49.4, AMSCO Finn- Aqua, D-50354 Hürth).

Fig. 2.50. Automatic loading system in front of a freeze drying plant with partially opened loading door (photograph: AMSCO Finn-Aqua, D-50354 Hürth).

2.4.1 Loading and Unloading Systems

The loading and unloading of thousands or tens of thousands of vials cannot be done by hand. The equipment manufacturers offer installation for loading and unloading from carrier assisted to fully automated systems. In Fig. 2.50 a part of a loading procedure can be seen in front of a freeze drier. Figure 2.51.1 shows schematically a loading and unloading system for two plants from the loading of the filled vials, the closing of the vials in the chamber, and unloading of the vials to the crimping machine, which fixes the stopper to the vial. The technical solutions are not discussed in detail, they depend on the local situation, the required loading speed and the partial or total absence of personnel. An example of an enclosed loading and unloading system which is completely in isolators is shown in Fig. 2.51.2, the isolators being seen from the operation room, while in Fig. 2.51.3 the loading system can be seen in front of the freeze drying plant. Even with loading speeds of e. g. 10 000 vials per hour, the operation takes several hours.

Therefore, it is decisive for the freeze drying process which follows that the filled vials during this time pass through a known and reproducible course of temperature. The prod-

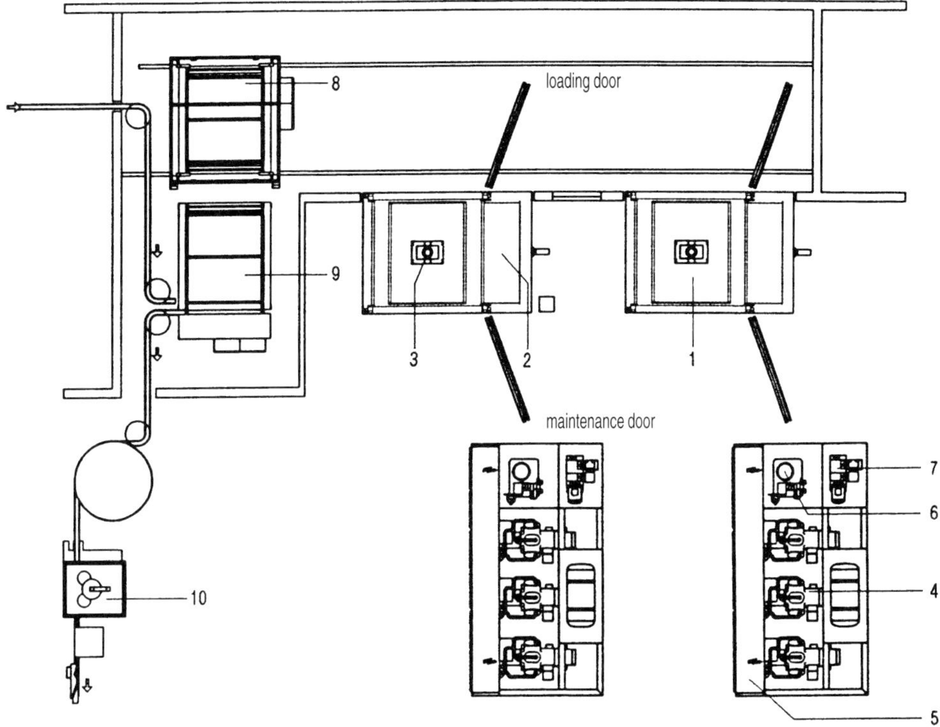

Fig. 2.51.1. Schema of a loading and unloading system for two freeze drying plants.
1, Drying chamber; 2, ice condenser; 3, stoppering device for closing the vials; 4, refrigerant compressors; 5, power supply; 6, hydraulic system; 7, vacuum-system; 8, loading- and unloading carrier, which is loaded in the position shown from the formatting table and moved to one of the chambers. During unloading the closed vials are loaded from each chamber on to the same carrier. It will be moved to the position shown and the vials are transported via the formatting- unloading table to the cramping machine; 9, formatting- unloading table; 10, cramping machine. Both chambers have a small loading and unloading door (similar to Fig. 2.50) and a large door for maintenance (schematic drawing AMSCO Finn-Aqua, D-50354 Hürth).

uct e. g. of +10 °C is filled in vials of +10 °C, the stoppers are set into position and transported to the formatting space of the loading system. During this time the temperature of the vial and the product should be kept constant e. g. within ±1 °C. When the vials are pushed onto the cold shelves, e. g. –40 °C, the product will start freezing, e. g. with a freezing rate of 1 °C/min between +10 °C and –35 °C. The product in the first vials will be frozen down to –35 °C in 45 min and approach –38 °C in the time thereafter. If the loading time is e. g. 5 h. the freezing of the last vials to –35 °C will take 5h 45 min. The first vials have been kept for 5 h at approx. –35 °C. It must be tested, whether the structure of frozen product in the vials is uniform enough to obtain the specified quality of the dried product within the specified drying time. The ice crystals will grow in 5 h at approx. –40 °C. Theo-

Fig. 2.51.2. The completely in isolators enclosed transport- and loading and unloading system.

Fig. 2.51.3. The loading and unloading system inside the isolators in front of the freeze drying plant (Fig. 2.50.1–2.50.3, AMSCO Finn-Aqua, D-50354 Hürth).

retically, the ice crystals in the first vials will be larger than in the last ones and therefore will – in general – dry faster, which could lead to an over-drying of product in the first vials before the last ones are dried. The opposite is also possible, if the growing ice crystals push well-distributed small enclosures of concentrated solids into larger areas with unfrozen water. These areas may take longer to dry or require a lower T_{ice} during MD. Whether the crystal growth has a measurable influence on the product or the process can be estimated by tests with methods described in Sections 1.1.5 and 1.1.6, e. g. by ER measurements and observation of the drying process in a cryomicroscope.

There will be some differences and it is a quantitative question, whether or not they can be tolerated or not. For a final decision, test runs in a pilot plant should be carried out with freshly frozen product and such which has been resting for 5 h before drying. These tests are recommended because the methods mentioned above use different sized samples in different configuration than are used in the production. The amount of product and its geometrical dimension will also influence the structure as well as the number of crystallization nuclei in the product, which can be very different in a normal laboratory and in a clean production area.

2.5 Production Plant for Food

2.5.1 Discontinuous Plants

Basically the chamber plants described in Section 2.4 can be used for foodstuffs and other products, as described in Sections 5.1 and 5.2. Freeze drying plants for food and similar products have to handle large quantities of product. The cleaning requirements remain, but no sterilization is necessary. The product can be transported in trays as described in Section 2.2.2 and dried in cylindrical tunnels. Figure 2.52 shows the two systems most commonly used to day. Their characteristic features are:

	A	B
Drying chamber	Tunnel	Tunnel
Condenser	Flanged to the tunnel	In the lower part of the tunnel
Connection between chamber and condenser	Vacuum tight valve	Barrier plate
Trays	Ribbed trays[1]	Flat trays[1]
Shelves	Steam- or brine heated	Brine heated
Tray transport	In carriers on an overhead rail	Pushed over rails
Heat input	Trays lowered on the shelves during drying	Mostly radiation
Defrosting of the condenser	Steam	Warm water
Loading and unloading	Carrier with trays is moved in and out	Trays are moved to each shelf level by a lift
BTM/DR measurement	Possible, vacuum-tight valve	Not possible, barrier plate not tight enough

[1] see Section 2.2.2

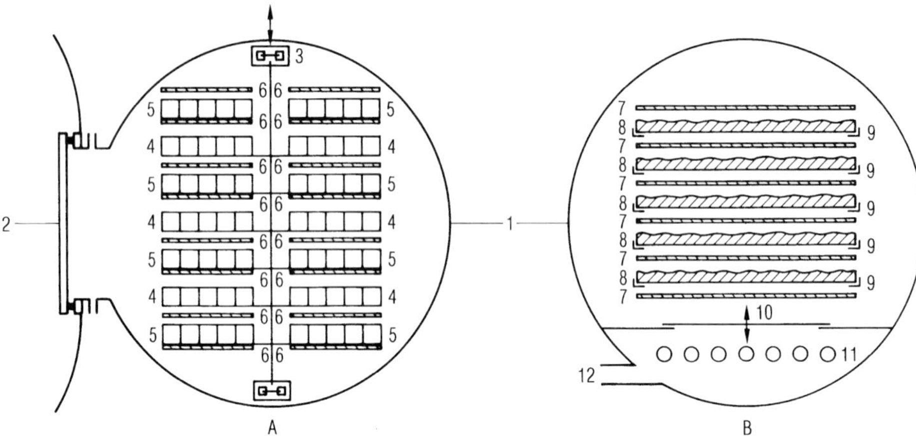

Fig. 2.52. Schematic comparison of two commonly used tunnel freeze drying systems for the freeze drying of food and luxury food.
1, Drying tunnel; in A: 2, valve before the condenser; 3, lift- and transport device; 4, trays in transport position; 5, trays on the shelves during drying; 6, heated shelves; in B: 7, radiation plates; 8, trays; 9, guide rails for the tray transport; 10, separation plate between tunnel and condenser; 11, condenser; 12, vacuum connection.

2.5.2 Continuos Plants with Tray Transport

Both systems A and B can be used for continuous operation with vacuum locks. In system A (Fig. 2.53), one or two carriers are be moved into a lock (4) in front of several connected tunnels (8). The lock is evacuated and the carrier(s) are moved into the tunnels. At the same time, the equal number of carriers is moved from the tunnels into an exit lock (10). Both locks can be separated from the tunnels by two large slide valves (7). In system B (Fig. 2.54), each tray (1) passes through an entrance lock (2) into a paternoster lift (3), which moves the tray to a certain level and pushes it into the drying zone. The last tray is pushed by the entering tray into the exit paternoster lift (6), which moves the tray into the exit lock (7). These plants are illustrated in Fig. 2.55 and 2.56.

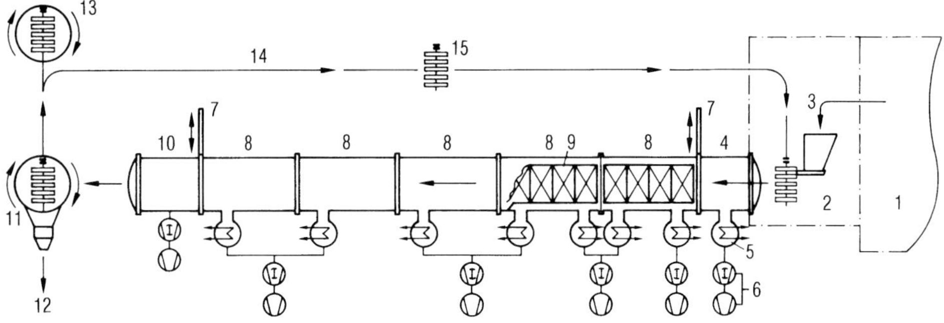

Fig. 2.53. Schema of a CQC-freeze drying plant as shown in Fig. 2.55.
1, Preparation room, e. g. freezing, grinding, sieving; 2, loading area; 3, loading of trays and carriers; 4, entrance lock; 5, ice condenser; 6, 2-stage vacuum pump set; 7, sliding gate valve between tunnel and lock; 8, four drying tunnels; 9, transport carrier with trays; 10, exit lock; 11, unloading of trays; 12, exit of dry product; 13, washing of carriers and trays; 14, return of trays; 15, carrier with empty trays.

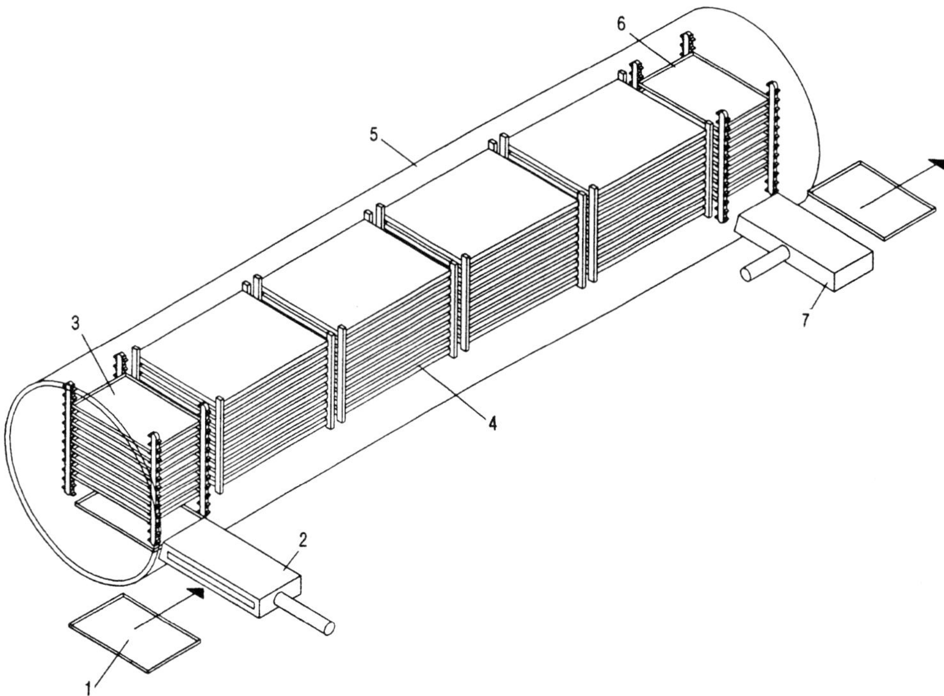

Fig. 2.54. Schema of a CONRAD® freeze drying plant as shown in Fig. 2.56.
1, tray; 2, entrance lock; 3, lift for the trays; 4, heating zones; 5, tunnel wall; 6, lift for the trays; 7, exit lock.

Fig. 2.55. Freeze drying plant, type CQC, as shown in Fig. 2.53. In this plant, two carriers are moved in parallel through the tunnel. The photograph shows the two heated shelves systems. The sliding gate valve is located in the large extension on the left side. In this plant approx. 5000 tonnes of freeze dried coffee granulate can be produced annually (photograph: ALD Vacuum Technologies GmbH, D-63526 Erlensee).

Fig. 2.56. Freeze drying plant, type CONRAD® 800, as shown in Fig. 2.54. In this plant, two trays are pushed in parallel through the tunnel. The two heating systems can be seen, and in front of these are the two lifts, which hoist the tray to required level. The two locks for the trays are on the left and right side, in the lower part of the tunnel. This plant produces approx. 3000 tonnes of freeze dried coffee granulate per year (photograph: Atlas Industries A/S, DK-2750 Ballerup).

2.5.3 Continuous Plants with Product Transport by Wipers or by Vibration

Oetjen and Eilenberg [2.17] have shown, that granulated products (which do not stick together) can be freeze dried with a 5 to 10 times higher ice sublimation rate, if the product is rolled over on the heating surface, compared with a static layer. Figure 2.57 shows the drying time of granulated product for different layer thicknesses as a function of the mixing frequency. In the disc dryer Fig. 2.58, the product passes through a vacuum lock onto the first disc. Wiper blades distribute the granulates on the disc and push it over the edge of the first disc to the second one and so on. The product is brought back to atmosphere through an exit lock. This form of transport works with a mechanically stable product, but with foamed granulated coffee extract a substantial amount of fines is produced, because the wiper blades not only push the product but partially abrade or mill the granulate. The particles produced with a size smaller than 0.5 mm can be 20 % of the throughput or more. Certain products e. g. those which roll or have too soft a surface, cannot be dried in such a plant. As shown by Oetjen [2.18], the abration can be kept as low as by the filling of trays (1.5 % of the throughput), if the product is transported on a vibrating bed. Figure 2.59 shows the schema of a vibration freeze dryer. Figure 2.60 illustrates this type of plant: The vacuum chamber and the entrance lock is shown in the upper picture and assembly of the vibrated and heated shelves in the lower picture.

The output of such high-speed dryers is limited by the increasing density of the water vapor flow. The grains of the product are floating in the vapor stream as in a fluidized bed, and the smallest particles are carried along with the vapor to the condenser. Even if only 1 % of the dried product is carried away, it sumps up to 10 kg per day if the throughput is 1000 kg per day. In 4 weeks, this totals to 280 kg or 1 m^3 of coffee powder. To remove this out from the vapor stream very large filters have to be used in order to minimize the pressure drop in the filters.

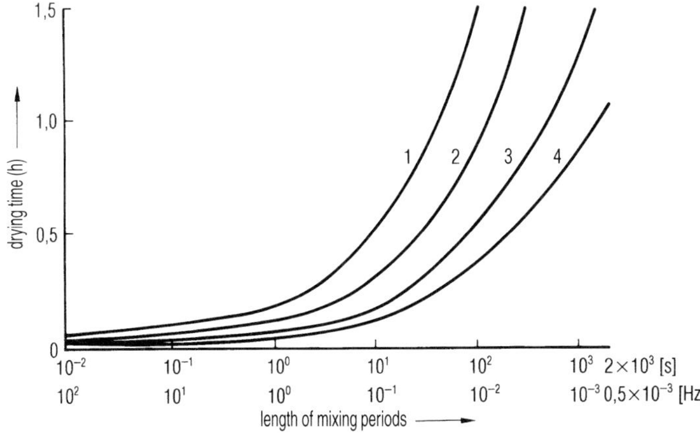

Fig. 2.57. Drying time as a function of a periodic, thorough mixing of a granulated product. Layer thickness: 1, 20 mm; 2, 12 mm; 3, 8 mm; 4, 5 mm (Fig. 12 from [2.17]).

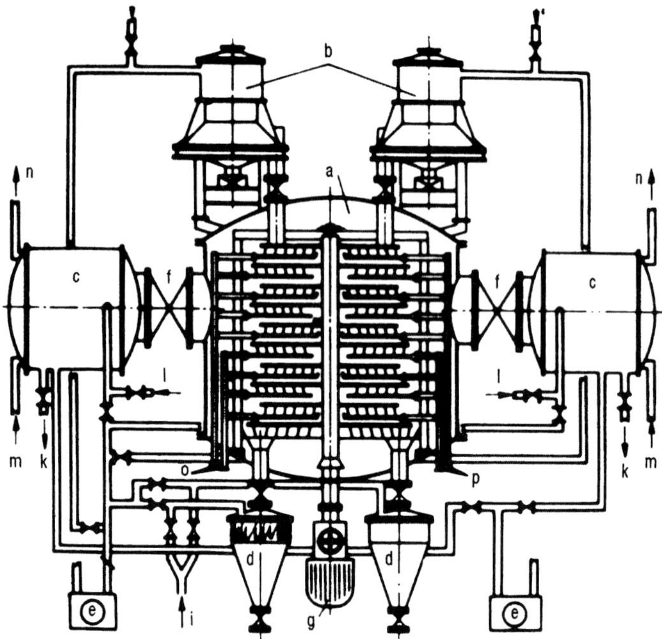

Fig. 2.58. Disc dryer, heated disc surface 95 m^2.
a, dryer housing; b, alternating locks with product storage; c, ice condenser; d, alternating locks for product removal; e, vacuum pumps; f, shut-off valve for the condenser; g, drive of product wipers; i, venting of product locks (N_2); k, drain of water after defrosting; l, evacuation of the condensers; m, refrigerant to the condenser; n, refrigerant outlet; o, heating medium to the discs; p, heating medium outlet (Fig. 13 from [2.28]).

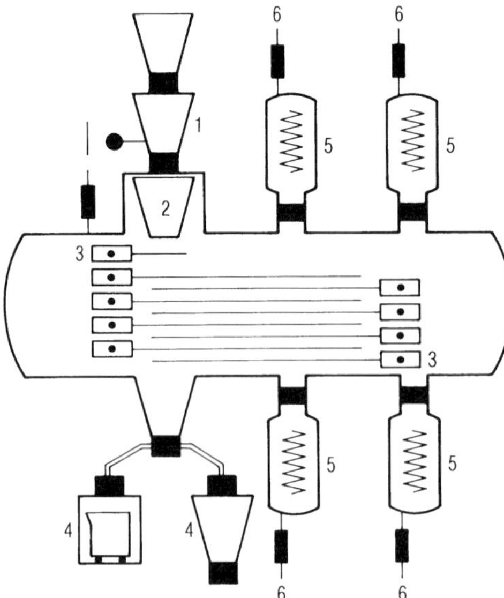

Fig. 2.59. Schema of a horizontal vibration dryer with 10 m^2 drying surface.
1, entrance lock for the product; 2, storage- and dosage unit; 3, heated shelves vibrated at 50 Hz; 4, alternating locks for product removal; 5, condensers; 6, connection to the vacuum system (Fig. 7 from [2.18]).

Fig. 2.60. Photograph of a vibration dryer with 10 m^2 surface vibrated with a frequency of 50 Hz. Top: Freeze dryer with the product feeding system in front. Bottom: Heated vibration shelves removed from the chamber (slides 21 and 22 from [2.29]).

References for Chapter 2

[2.1] Yokota, T.: A continuous method for freezing droplets by a wetted- wall column in freeze-drying. Kagaku Kogaku Ronbashi 15, p. 877–880, 1989

[2.2] Rolfgaard, J.: Industrial freeze-drying for the food and coffee industry. Atlas Industries A/S, DK- 2750 Ballerup

[2.3] Unpublished data from Dr. Otto Suwelack D-48723 Billerbeck.

[2 4] Haseley, P.: New cooling and sterilization technology in freeze-dryers and a comparisation of operating costs. PDA Aisan Symposium, p. 305–317, Tokyo 1994

[2.5] Wiilemer, H.: Freeze-drying plants with modern refrigerants including liquid nitrogen. 1st. World meeting of APGI/APV Arbeitsgemeinschaft für pharmazeutische Verfahrenstechnik, Budapest, May 1995

[2.6] Snowman, J. W.: Replacement of conventional refrigeration systems in freeze-drying by liquid nitrogen cooled systems. PDA Asian Symposium, p. 329–345, Tokyo, 1994 Copyright © 1994 PDA, Inc. Bethesda, Maryland, USA

[2.7] Cully, R.: Refrigerants, the environment and the liquid nitrogen option. International Society of Pharmaceutical Engineering (ISPE), Antwerp, 1994

[2. 8] AMSCO-Finn-Aqua publication FA 00.20.02./10.91

[2. 9] Neumann, K. H.: Grundriß der Gefriertrocknung. 2. edition, Musterschmidt Wissenschaftlicher Verlag, Göttingen 1952

[2.10] Leybold AG, Katalog HV 300, Teil A 11, p. 29

[2.11] Nakahira, K.: Validation of deep vacuum vapor phase hydrogen peroxide sterilizer retrofit to a production lyophilizer. PDA Asian Symposium, p. 1/6–6/6, Tokyo 1994

[2.12] Steiner, R.: VHP®-Sterilisation of freeze-dryers. ISPE-Seminar: Lyophilisation, Antwerp, Nov. 1994. Internnational Society of Pharmaceutical Engineering (ISPE).

[2.13] Dräger – Prüfröhrchen for H_2O_2, Drägerwerk AG, D-23542 Lübeck

[2.14] AMSCO International Inc. 1002 Lufkin Road, P. O. Box 747, Apex/North Carolina 27502

[2.15] Sterilization of freeze-dryers. Technical Monograph No. 5. The Parenteral Society, 6 Frankton Gardens, Stratton St. Magaret, Swindon, Wiltshire, 1994

[2.16] Gaster, A. J.: In place cleaning systems for freeze dryers. PDA Asian Symposium, p. 289–293, Tokyo, 1994

[2.17] Oetjen, G. W., Eilenberg, H. J: Heat transfer during freeze-drying with moved particles, p. 19–35, International Institute of Refrigeration <IIR> (Com. X, Lausanne 1969)

[2.18] Oetjen, G. W.: Continuos freeze-drying of granulates with drying time in the 5–10 minutes range, p. 697–706, International Institute of Refrigeration <IIR> (XIII. Congress, Washington 1971)

[2.19] Wutz, M., Adam, H., Walcher, W.: Theorie und Praxis der Vakuumtechnik, 2. Auflage 1982 Copyright © 1982 Friedr. Vieweg & Sohn, Verlagsgesellschaft mbH, D-65048 Wiesbaden

[2.20] Catalog, Martin Christ Gefriertrocknungsanlagen GmbH, D 37507 Osterode, p. 7

[2.21] Leybold catalog HV 300, Teil A 7, edition 03/91

[2.22] Leybold catalog HV 300, Teil A 2, p. 29

[2.23] Leybold catalog HV 300, Teil A 6, p. 4

[2.24] Leybold catalog HV 300, Teil A 4, p. 4

[2.25] Leybold catalog HV 300, Teil A 4, p. 8

[2.26] Leybold catalog HV 300, Teil B 10, p. 4

[2.27] Leybold catalog HV 300, Teil B 10, p. 7

[2.28] Festschrift, 25. anniversary Mr. Wolfgang Suwelack

[2.29] Oetjen, G. W.: Freeze-Drying Processes and Equipment, XVIII. Congresso Nationale del Freddo, Padua, June 1969

[2.30] Haseley, P., Spreckelmeyer, J., Steinkamp, H.: Betriebsbedingungen in Gefriertrocknungsanlagen bei Einsatz von Schrauben- und Hubkolbenverdichtern. DKV, 1997, p. 155–170

[2.31] Haseley, P., Oetjen, G. W.: Equipment data, thermodynamic measurements, and in-process control quality control during freeze-drying, Fig. 4. PDA International Congress, p. 139–150, Basel 1998

[2.32] Chase, D., R.: Monitoring and control of the lyophilization process using a mass flow controller. Pharmaceutical engineering, p. 92–98, *Jan.*/Feb. 1998

3 Pharmaceutical, Biological and Medical Products

De Luca et al. [3.1] nominate, under the heading of 'sensitive biologicals', three product groups for therapy and diagnostics, which are mostly or at least partially freeze-dried:

	Number of formulations studied	Number of formulations freeze dried	%
antibiotics	46	14	30
macromolecules	26	24	92
others (organic electrolytes)	31	16	52

The freeze drying of these products has been performed by one or more of the following reasons: The ingredients of the formulation are not stabile in the liquid state; the amount of the active ingredient is very small; the dosing of liquids is safer to control than that of a powder; the sterilization of the vials before filling, and filling into the final container minimizes handling and reduces the possible contamination.

3.1 Proteins and Hormones

Carpenter et al. [3.2] show that the protection of the native structure of proteins requires two different mechanisms during freezing and freeze drying. Phosphofructokinase (PFK) has been chosen as model substance, because PFK is irreversible denatured during freezing and thawing. During freezing the best substances to minimize denaturation have higher repellent forces between the protein and excipient than attractive forces. A 3 M NaCl solution destabilizes the activity of PFK by 80 %, while a 1 M polyethylenglycol (molecular weight 600) solution protects the activity completely. During freeze drying and in the dry state, only such carbohydrates prevent a loss of activity, which can be bound to the protein molecule by replacing water molecules and thereby forming hydrogen bonding with the protein. By adding 0.6 mM zinc sulfate solution to the CPA, e. g. 50 mM trehalose solution, the activity of PFK is completely stabilized. Some thought is given to the possible mechanism of this effect.

Remmele et al. [3.55] studied with infrared spectroscopy the structure-hydration behavior of a 49.4 mg/mL lysozyme D_2O solution with and without 10 % sucrose. The sample was cooled n in the measuring chamber to –100 °C and then connected to a freeze drying installation, after which the temperature of the sample was raised to +40 °C:

Time (h)	Duration (h)	Temperature (°C)	Observation
0–2.4		–100 to –45	
2.4–3.9	1.5	–45	D_2O crystal growth
3.9–4.9	1.0	–45 to –24	Sublimation of ice
4.9–5.2	0.3	–24 to –13	Decreasing sublimation of ice, remaining structure: amorphous
5.2–10.5	5.3	–13 to –10	Substantial loss of water
10.5–17.5	7.0	–10 to +18	Ice completely removed
17.5–18.4	0.9	+18 to +20	Noticeable loss of water
18.4–26.6	8.2	+20 to +27	Noticeable loss of water
26.6–46.0	9.4	+27 to +40	End of drying

From the form and the peak location of the O-D stretch band of the spectra is concluded, that the main drying is completed at –10 °C; between –13 °C and –10 °C a combination of main and secondary drying has occurred. $T_{g'}$ for sucrose is reported as –40 °C [3.56] and as –32 °C [3.6]. Both temperatures have been exceeded long before the main drying was completed, and the product collapsed, as confirmed by the appearance of the product and the very slow secondary drying. During a second test, the temperature was kept below $T_{g'}$. The residual moisture of the second run was 4 % instead of 7 % in the first test. From the changes in the protein band it was concluded that sucrose had substituted for water by hydrogen bonding.

Pikal [3.3] suggests also, that the mechanism of protection of proteins is different during freezing and drying. For the selection of CPAs, Pikal recommends to observe three points, which can be summarized:

- CPAs should, at least partially, solidify in the amorphous state. However amorphous state alone does not assure protection. Izutsu et al. [3.4] showed, for beta-galactosidase by X-ray diffraction, that only such additives avoid denaturation, which do not crystallize. A dilution of protein in the solidified protective agent reduces the chance of reactions between the protein molecules. Amorphous substances dry more slowly, which makes it easier to avoid overdrying.
- $T_{g'}$ should be chosen as high as possible, since chemical substances in general are more instable above $T_{g'}$ and RM should be small, since high RM reduces $T_{g'}$ data.
- Buffer substances or other salts should be used in the smallest possible concentration, since they may partially crystallize (changes of pH-value) and they will mostly reduce $T_{g'}$.

De Luca [3.5] recommends furthermore the addition of e. g. tertiary butylalcohol (TBA), to increase the transport of water vapor out of the product and to avoid collapse in sucrose-, lactose- and sorbitol solutions. Thereby, higher temperatures during drying (e. g. for hemoglobin in sucrose solution) can be applied.

Skrabanja et al. [3.6.] do not accept a combination of 12 different excipients e. g. for erythropoetin as the efficacy of each component cannot be proven. The mostly frequently used excipients are listed in five groups:

- protein: human serum albumin, gelatin
- amino acids: glycine, arginine, alanine
- alcohols: mannitol, PEG (polyethyleneglycol)
- carbohydrates:
  - monosaccharides: glucose, fructose
  - disaccharids: lactose, maltose, sucrose, trehalose
  - polysaccharides: dextran, HP-beta CD
- other:
  - metals
  - surfactants
  - polymers
  - buffer salts

Quick cooling is often advantageous, e. g. for recombinant DNA-proteins, in order to avoid crystallization of salts and to obtain the best possible, homogeneous cake.

Figure 3.1 shows the influence of different freezing methods on the activity of a recombinant DNA protein. However, in the case of other macromolecules, for example a monoclonal antibody, there may be exceptions from the rule and different factors (e. g. pH-value) play an important role.

As shown in the following table (from [3.6]) the selection of the excipient also defines T_g, and the amount of unfreezable water (UFW) in the glass phase.

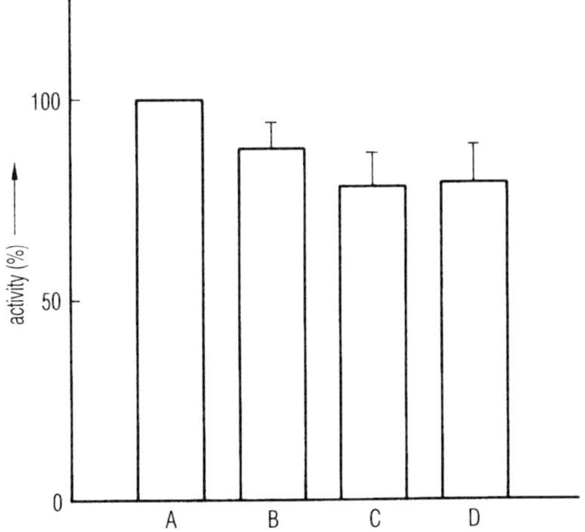

Fig. 3.1. Influence of different freezing processes on the activity of recombinant DNA protein.
A, Solution before freezing; B, freeze dried after quick freezing; C, freeze dried after freezing on precooled shelves; D, freeze dried after cooling shelves and product simultaneously (Fig. 1 from [3.6]).

Excipient	$T_{g'}$ (°C)	UFW (%)
Sucrose	–32	35,9
Maltose	–30	20
Lactose	–28	40,8
Trehalose	–30	16.7
Fructose	–42	49.0
Glucose	–43	29.1
Glycerol	–65	45.9
Sorbitol	–43	18.7
Dextran	–9	
P beta CD	–8	
Albumin	–10	
PVP	–19.5	
PEG	–13	
Sodium citrate/citric acid	–40	
Na_2PO_4 / KH_2PO_4 (1 : 1)	–80	

The data of the mixture of two ore more excipients cannot be calculated from the individual data, as shown in Fig. 3.2 for a sucrose-citrate solution. This can also be seen in Fig. 3.3.1: The pure solution (without factor VIII) (right) would have to be dried – according to the plot d(log R)/dt – below –50 °C, while the solution with factor VIII (left) (following the same rule) could be dried at –43 °C (the minimum of the plot d(log R)/dt shifts by almost 10 °C). Figure 3.3.2 shows three factor VIII solutions with different excipients: Identical data of resistance are found at –55 °C, –51 °C and –35 °C.

During secondary drying, a small RM should be reached, since T_g of the dry product increases with decreasing water content. $T_{g'}$ of amorphous, freeze dried sucrose increases from 16 °C with 8.5 % RM to 63 °C or 64 °C between 1.0 and 0.7 % RM. It should be taken into account, that RM cannot be taken only at the end of drying, but a possible increase during storage by water desorbed from the stopper has to be considered (see Section 1.3.2 and Pikal et al. [3.7]).

Srabanje also shows, that storage below T_g alone is not a sufficient criteria to protect the activity of the protein, since O_2 can have a important influence (Pikal et al. [3.8]).

The viscosity in the vicinity of T_g follows the Williams-Landel-Ferry equation [3.9], as well as the stability of KS 1/4-DAVBL (deacetylvinblastin hydrazide conjugate), as shown in Fig. 3.4 [3.10].

Jensen [3.11] as well as Teeter [3.12] studied by X-ray diffraction the structure of water molecules in the vicinity, at the surface and inside of protein crystals. Jensen used rubredoxin (CEB) crystals to deduce the structure of water from the density distribution of electrons, calculated from diffraction pictures. Jensen found that water molecules which are placed within approx. 60 nm of the protein surface form a net, which is most dense in the distance of a hydrogen bond at the donor- or acceptor- molecules of a protein. In distances larger than 60 nm, the structure of water becomes increasingly blurred, ending in a structureless phase. Water molecules are also in the inside of proteins, but are more strongly bound than

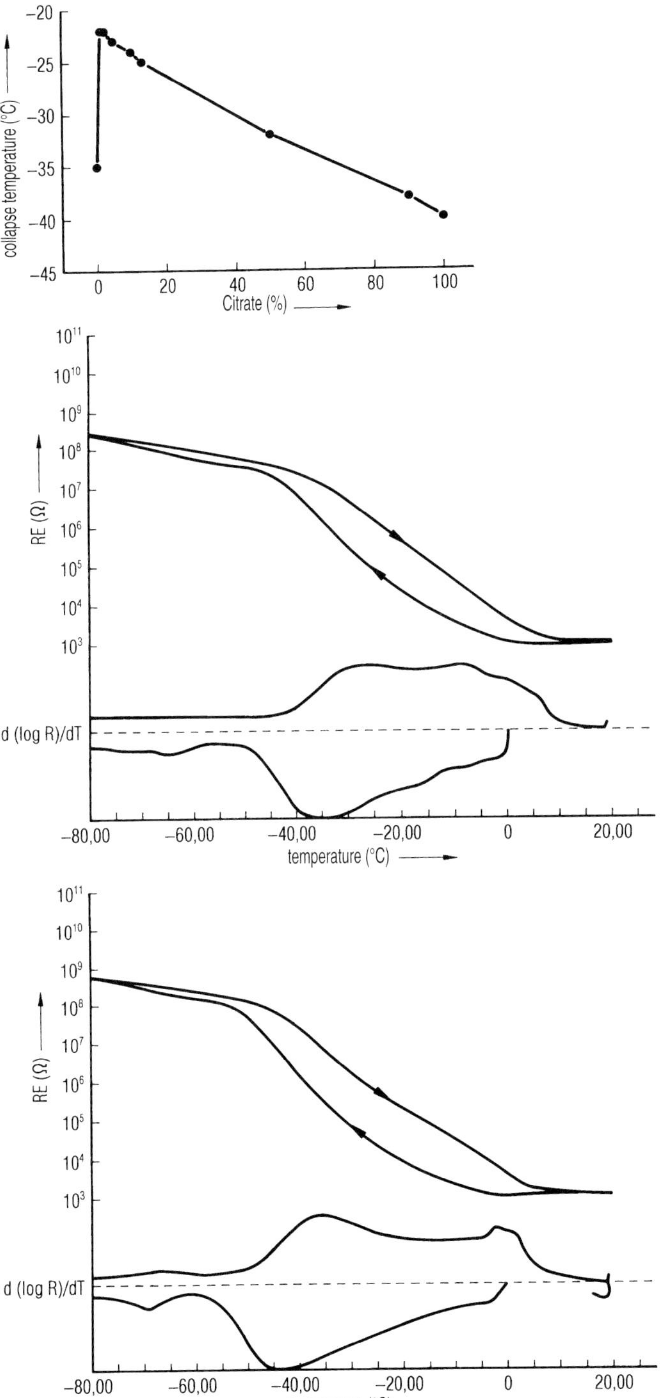

Fig. 3.2. Collapse temperature of a sucrose solution as a function of the added citrate solution (%) (Fig. 3 from [3.6]).

Fig. 3.3.1. Electrical resistance as a function of temperature. Upper: excipient solution; lower: solution with factor VIII. Cooling rate 15 °C/min. Measurements for [3.23], not published.

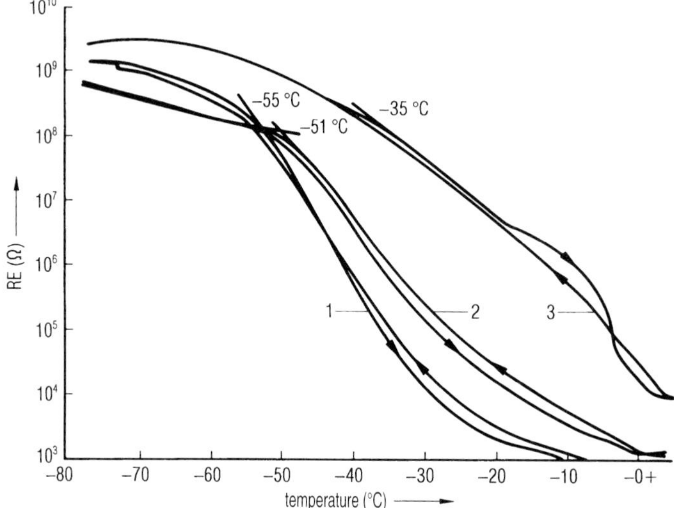

Fig. 3.3.2. Electrical resistance as a function of temperature of three factor VIII solutions with different excipients (Fig. 4 from [3.23]).

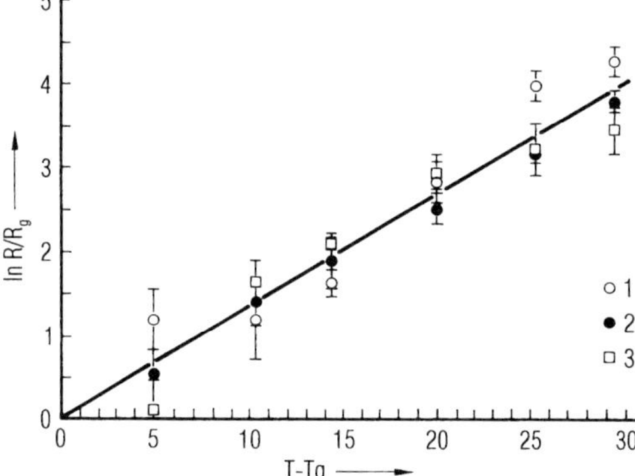

Fig. 3.4. Analysis of the degradation of KS1/4 hydrazide conjugate by the Williams-Landel-Ferry glass transition theory (Fig. 9 from [3.10]).
1, Formation of dimers; 2, free vinca generation; 3, decomposition of vinca; R, rate of degradation in %/month at a given temperature and water content; R_g, rate of degradation at $T_g \cdot R_g = 0.10$ (dimer formation), 0.92 (free vinca generation, 3.9 (vinca decomposition).

those on the surface. Teeter [3.12] used crambin, a hydrophobic protein (MW 4700) to show that at 130 K, two different nets of water exist, one which forms rings of a pentagon shape and the other in chain-like configurations which are strongly influenced by the surface of the molecule and affect the stability of the molecule. Teeter has not studied the water molecules inside a protein.

Hagemann et al. [3.13] calculated the absorption isotherms for recombinant *bovine Somatotropin* (rbSt) and found 5–8 g water in 100 g protein, which was not only on the surface but also inside the protein molecule.

Townsend and De Luca have studied the influence of lyoprotectans (LP) on ribonuclease (Ri) [3.14–3.17] as a protein model. Lyoprotection is defined as stabilization and prevention of degeneration of macromolecules during freeze drying as well as during storage. With phosphate buffer at pH 3 to pH 10, Ri in the dry stage loses its activity at 45 °C over time by forming aggregates with covalent bonds. Ficoll 70 is the most effective LP in the range of pH 3.0 to pH 10.0. All three Ri-LP products were amorphous and not crystalline, though for optimum protection the mass ratio of LP to Ri had to be 6:1. The negative influence of the phosphate buffer is attributed to possible heavy metal ions in the buffer. Increasing RM in the dry Ri increases the loss of activity and the aggregation. The same applies with increasing buffer salt concentration, as air in the closed vials was increased, though Ar or N_2 decreased the denaturation.

Carpenter et al. [3.18] showed by IR-spectroscopy, that besides the H-bonds between protein and carbohydrate carbohydrate bonds are also necessary to stabilize proteins during drying and reconstitution.

A review by Dong et al. [3.57] provides an overview of how Fourier transform IR spectroscopy can be used to study protein stabilization and to prevent lyophilization- induced protein aggregation. An introduction to the study of protein secondary structures and the processing and interpretation of protein IR spectra is given.

Hora et al. [3.19] described the complexity of protein stabilization by the example of recombinant, human Interleukin-2 (rhIL-2). Formulations with amino acids and mannitol/ sucrose are sensitive to mechanical stress e. g. by pumping. Hydroxypropyl-beta-cyclodextrin (HPcD) provides stability, but increases the sensitivity to oxygen. Polysorbate 80 forms a mechanically stable product, but results in oxidation. In both cases contamination in the HPcD or traces of H_2O_2 in the Polysorbate may have been the starter for the oxidation. Brewster [3.20] reports, that HPcD stabilizes interleukin without forming aggregations and this results in 100 % biopotency.

Prestrelski et al. [3.58] studied the pH conditions and different stabilizers to provide optimal storage stability for IL-2 by Fourier-transform IR spectroscopy. Different pH conditions in the absence of excipients change the dry state confirmation of IL-2 dramatically. At pH 7, IL-2 unfolds extensively, while below pH 5 it remains essentially native. IL-2 at pH 5 is approx. one order of magnitude more stable than at pH 7, taking into account the amount of soluble and insoluble aggregates. A similar pH profile was observed in the presence of excipient, although excipients change the overall stability profile. Excipients with the capacity to substitute for water during drying preserve the native structure best. Those with a high glass transition temperature provide the highest level of storage stability, but do not prevent unfolding during drying.

Carpenter et al. [3.21] studied the stabilization of proteins by non ionic surfactants. It is known, that the critical micelle concentration of the surfactant protects proteins best in aqueous solutions. With certain proteins, e. g. human growth hormone, the maximum protection depends upon the surfactant binding stoichimetry. The binding of the surfactant to the protein sterically hinders aggregate formation between protein molecules. During freezing, damage might be caused by the ice-water interface, and this could be inhibited by the surfactant.

Contrary to the quoted experiences, Vermuri et al. [3.22] reported for recombinant alpha-antitrypsin (rAAT) in a phosphate-citrate buffer of pH 7.0, that there was no need of CPAs during freezing, thawing and freeze drying. Comparisons of rAAT in lactose, sucrose, and polyvinylpyrrolidone showed generally no significantly better protection. Freezing

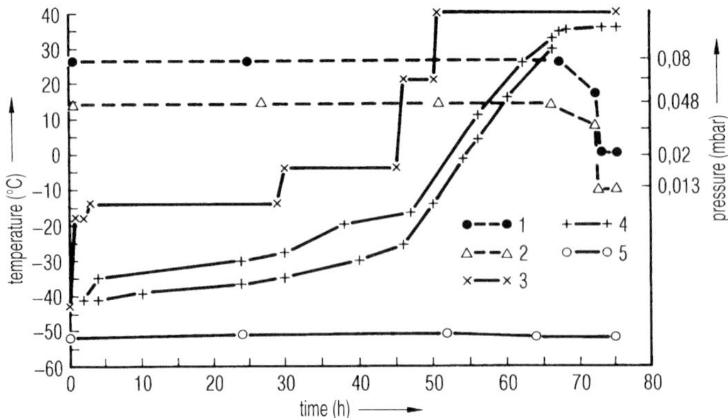

Fig. 3.5. Freeze drying course of factor VIII solution after freezing in LN_2.
1, p_{ch} (TM); 2, p_{ch} (CA); 3, T_{sh}; 4, T_{pr} (two sensors); 5, T_{co} (Fig. 7 from [3.23]).

in LN_2 and an increase in concentration from 10 mg/ml to 50 mg/ml rAAT did not alter the criteria of stabilization.

Figure 3.5 is a diagram of the freeze drying of a factor VIII-solution (the same solution as used in Fig. 3.3.1)

1 The following facts can taken from the diagram:

1.1 The condenser temperature has been constant at –52 °C.

1.2 The temperature at the sublimation front, T_{ice}, (Fig. 3.6.1) measured by BTM has been –41 °C after 6 h and –40 °C after 30 h, the controlled operation pressure (p_c) has been 0.048 mbar (CA), respectively 0.078 mbar (TM), T_{sh} –13 °C.

1.3 The product temperature, T_{pr}, measured by two PT 100 is an average of –38 °C in the beginning and –31 °C after 30 h.

1.4 The increase of T_{sh} between 28.5 and 30 h to –4 °C changes T_{ice} in 2 h to –38 °C.

1.5 After 46 h pressure rise measurements permitted the raising of T_{sh} to +40 °C, while T_{pr} rose to approx. +20 °C.

1.6 The pressure rises increase with increasing T_{sh}, but fall after 52 h and reach 1 %/h at 68 h.

1.7 After 66 h, the pressure control could no longer maintain the selected pressure of 0.048 mbar; thus, the pressure control is cut off, the pressure drops to 0.012 mbar (CA). The RM measurement by Karl Fischer showed 0.8 ± 0.06 %.

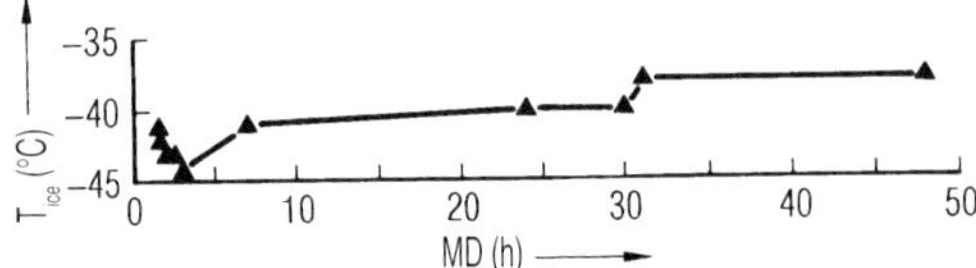

Fig. 3.6.1. Plot of the temperature at the sublimation front (T_{ice}) of the run shown in Fig. 3.5 (Fig. 10 b from [3.23]).

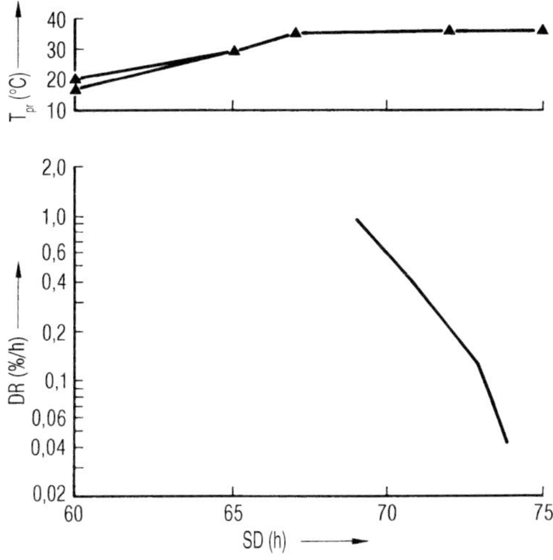

Fig. 3.6.2. Plots of the product temperature and DR values of the run shown in Fig. 3.5. (Fig. 10 c from [3.23]).

2 The following conditions applied to the process:

2.1 Each vial was filled with 30 ml factor VIII-solution, filling height 27 mm.

2.2 The product in the vials has been frozen in LN_2 in approx. 4 min from +22 °C to –50 °C (approx. 18 °C/min).

2.3 After freezing, the vials rested for 7 h on the shelves at –42 °C (Fig. 3.7), to simulate the loading of a production plant (growing of crystals in that time).

2.4 From photos (Fig. 3.8) taken by a cryomicroscope, in which freeze drying is possible and from ER-measurements (Fig. 3.3.1), the maximum possible T_{ice} was determined as –38 °C, and –40 °C is chosen as T_{ice} for the beginning of MD.

3 The following deductions for a shorter drying time can be made from Fig. 3.5, Fig. 3.6.1, Fig. 3.6.2 and the RM data:

3.1 T_{sh} can be raised in one step from –42 °C to e. g., –6 °C.

3.2 The pressure difference between the sublimation surface (–40 °C ≈ 0.128 mbar) and the chamber pressure (0.048 mbar) is almost a factor of 3, which is unnecessary large.

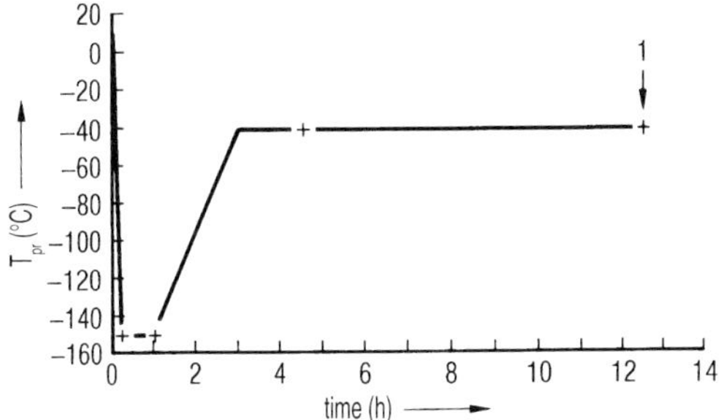

Fig. 3.7. Plot of the product temperature during freezing before the course shown in Fig. 3.5. 1, start of MD (Fig. 10a from [3.23]).

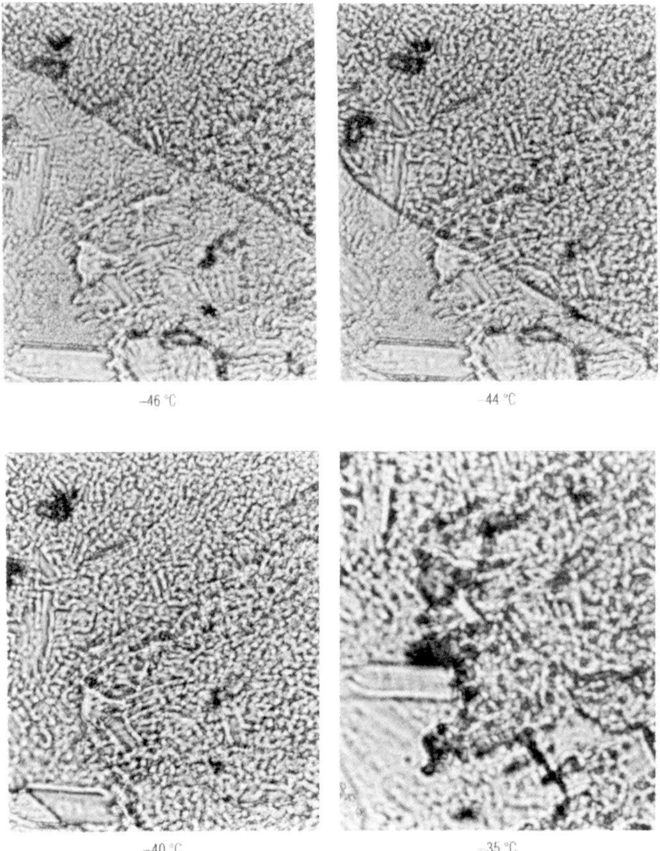

Fig. 3.8. Photographs taken with a cryomicroscope of factor VIII solutions at four temperatures. At –40 °C the structure is still visible, but is more coarse compared with –44 °C. At –35 °C the structure is collapsed (Fig. 5 from [3.23]).

In choosing a factor of 2 (which is conservative), p_c could be raised to 0.06 mbar, reducing the transported water vapor volume by 25 %. Since the heat transfer coefficient in this pressure range depends very little on the pressure, the higher pressure is only advantageous if the vapor transport from the chamber to the condenser is the bottle neck of the process (see Section 1.2.4). At a higher pressure, one could then load more vials in the chamber. In the experiment described, the vapor transport is not the bottle neck, therefore one cannot expect a reduction in drying time as a result of higher pressure.

3.3 Based upon the earlier increase of T_{sh} (3.1) and the higher p_c (resulting in a little higher T_{ice}) the temperature increase to +20 °C, respectively +40 °C could be done earlier, probably by a few hours.

3.4 The temperature in the product rises only slowly, even after the increase of T_{sh}, (in 10 h from –20 °C to 0 °C and in additional 10 h from 0 °C to 30 °C). The temperature rise could be accelerated: The shelf temperature is increased, e. g. to 50 °C until the product has reached +25 °C. At that time T_{sh} is reduced to +40 °C. This temperature increase has taken approx. 16 h in the diagram, with a temperature difference $T_{sh} - T_{pr}$ ≈ 30 to 35 °C. The higher difference of ≈ 40 to 45 °C should save again several hours.

3.5 The pressure control could be cut off at 66 h instead of 72 h, shortening the drying time by an other 3–6 h.

3.6 In total, 10–15 h can be saved, resulting in a drying cycle of 64 to 59 h instead of 72 h, which could mean the completion of a cycle in 3 rather than 4 days.
On the other hand, it does not seem likely to reduce the drying time under the given conditions (filling height 27 mm and T_{ice} –40 °C) to 40 h, or one cycle in 48 h.

Fig. 3.9. Photograsphs of two vials with freeze dried factor VIII. Left: frozen in LN_2; right: frozen on the shelves of the freeze drying plant (Fig. 8 from [3.23]).

Fig. 3.10. Photographs of the freeze dried factor VIII cake. Left: and right as in Fig. 3.9 (Fig. 9 from [3.23]).

This example is only valuable with the conditions given. Different types of vials, different filling heights, different excipients with different maximum T_{ice} and different temperatures during SD will change the data discussed here. Figure 3.9 shows the end product in vials, frozen in LN_2 (left) and on the shelves at –42 °C (right) and Fig. 3.10 shows the stable cake of the factor VIII frozen in LN_2 (left) and the powder (right) after freezing on the shelves.

3.2 Viruses, Vaccines, Bacteria, and Yeasts

All substances of this chapter can only be dried in the present of CPAs, if their natural qualities are to be protected.

Greiff [3.24] studied the stability of purified influenza virus of the strain PR 8 in physiological NaCl solution with calcium lactobionate and human serum albumin (each 1 % in the solution). The freezing rate was approx. 1 °C/min down to –30 °C. During the freeze drying, the product temperature was raised in 12 to 16 h from –30 °C to 0 °C and the product was dried at this temperature. After 24 h, the first 145 vials were removed, and additional vials after intervals of 24 h each. The residual moisture content was 3.0, 2.0, 1.5, 1.0 and 0.5 %. The stability of the freeze dried virus (expressed in days during which the titer of the infectivity decreased by a factor of 10 was most unfavorable at 0.4 % and 3.2 % RM, (4 and 7 days respectively at +10 °C) and best at 1.7 % RM: 145 days or more than 1000 days at –10 °C.

Overdrying (0.4 % RM) and to high RM (3,2 %) results in unstable dry products. Overdrying removes bond water, which is essential to keep the protein structure; furthermore the hydrophilic locations of the protein are exposed to gases, e. g. O_2. At too high an RM, free water remains in the dry product and induces reactions which change the protein molecule.

Greiff [3.25] classified the virus into five categories: 1, Nucleic acid type (either DNA-core or RNA-core); 2, sensitivity against lipid solvents; 3, envelope about the nucleocapsid or not (naked); 4, pH sensitivity, exposure to pH 3 for 30 min differentiates between those viruses, which lose more than a decade in titer and those which lose no titer or less than one decade; 5, heat-sensitive virus cannot be exposed to +50 °C for 30 min.

During the freeze drying tests the virus suspensions are either basic salt medium (BSM) or BSM plus calcium lactobionate (CL) plus serum albumin (SA) frozen at –76 °C and dried either at 0 °C or at –40 °C. The activity has been evaluated after 30 days storage at –4 °C or –65 °C. Rehydratation was done with distilled water at 0 °C. The results with the freeze dried viruses indicate:

- Al RNA-viruses in BSM showed a marked decrease in titer. With the addition of CL and SA, no or only small decreases in titer are found.
- Al DNA-virus suspensions in BSM changed only slightly.

- DNA-viruses with envelopes, which are solvent-sensitive are less affected by freeze drying than solvent-resistant, naked DNA viruses.
- pH-sensitive DNA-viruses are less affected by freeze drying than the pH stable DNA viruses.
- Changes in titer of lyophylized DNA-viruses were independent of temperature sensitivity.

Doner et al. [3.26] studied bovine corona-virus (BCV) and respiratory syncytial virus (RSV). Both can be frozen without CPAs at 0.2 to 0.3 °C/min with no loss of titer. Faster freezing (0,4 °C/min to 30 °C/min) results in an increasing loss of titer of 1 to 3 decades. The freeze drying experiments have been started therefore by freezing at 0.25 °C/min and with the addition of various CPAs. No loss in titer was observed only with 3.6 % dextran + 10 % sucrose in the suspension for RSV and BCV. RSV could also be dried in 10 % sucrose + 1,5 % gelatin suspension, without loss of titer. Both viruses belong to the RNA-virus group; thus it should be possible to dry them without loss of titer [3.25] with CPAs. The results of [3.26] show, that the conclusions of Greiff cannot be applied to other CPAs without further studies. Bennett [3.27] studied the freeze drying of varicella zoster viruses (VZV), – a DNA-virus with an envelope – which are very labile in cell free suspensions. Freezing of 0.7 ml in 3 cm^3 vials is performed in LN_2, after which the vials are placed on precooled shelves at –45 °C, and the chamber is evacuated to p_{ch} < 0.07 mbar for 1 h. The freeze drying is carried out with three different process data:

	t_{MD} (h)	p_{ch} (mbar)	$T_{sh,MD}$ (°C)	t_{SD} (h) p_{ch} (mbar)	$T_{sh,SD}$ (°C)	RM (%)
I	3.5	0.31*	–45/+30[1]	4.5 0.035**	+30 7.1	±0.6
II	9	0.091***	–45/+30[1]	5 0.035**	+30 4.1	±1.5
III	40	0.035**	–26 [2]	8 0.035**	+30 0.9	±0.4

[1] $T_{sh,MD}$ is raised from –45 °C to +30 °C during t_{MD}

[2] $T_{sh,MD}$ is constant during t_{MD}

* (0.47 mbar TM · 0.65 = 0.31 mbar CA)

** (0.07 mbar TM · 0.5 = 0.035 mbar CA)

*** (0.14 mbar TM · 0.65 = 0.09 mbar CA)

For pressure conversion, see Section 1.2.3. In the table, CA-data are converted from TM data as shown.

Following Eq. (12), the $t_{MD,}$ if all data except p_c and T_{sh} are constant, should have a ratio between the three tests as follows:

$$t_{MD\ I} : t_{MD\ II} : t_{MD\ III} = 1/T_{tot\ I}\,(1/K_{tot\ I}) : 1/T_{tot\ II}\,(1/K_{tot\ II}) : 1/T_{tot\ III}\,(1/K_{tot\ III})$$

K_{tot} for II and III are practically independent of the pressure and therefor identical. $K_{tot\ I}$ is approximately twice as large as $K_{tot\ II}$ or $K_{tot\ III}$ as shown in Fig. 1.58. With these data one can conclude:

$2\ T_{tot\ I} : T_{to\ II} : T_{to\ III} = 1 : 0.388 : 0.088$

Assuming (Fig. 1.75, column 3) that at a pressure of 0.3 mbar T_{ice} will be approx. –27.5 °C, one can draw the following picture of the three tests:

	I	II	III
	°C	°C	°C
T_{tot} $(T_{sh} - T_{ice})$	–20	–15.5	–3.5
T_{ice}	–27.5	–23	–29

If the assumed T_{ice} = –27.5 °C is replaced by –22 °C (Table 1.9, column 3), the data are as follows:

	I	II	III
	°C	°C	°C
T_{tot} $(T_{sh} - T_{ice})$	–14.5	–11.2	–2.6
T_{ice}	–22	–18.7	–28.6

The factor 2 as the difference of the two K_{to} is often confirmed under different conditions. To complete the picture, K_{tot} shall be assumed as 1.5, resulting in the following data set with T_{ice} = –27,5 °C:

	I	II	III
	°C	°C	°C
T_{tot} $(T_{sh} - T_{ice})$	–20	–11.6	–2.6
T_{ice}	–27.5	–19.1	–28.6

These estimates are simplified. They are made only to show the following:

1. The measured t_{MD}, can only be achieved, if T_{ice} in test II is 3–8 °C higher than in tests I or III. The short t_{MD} in test I and the long one in III result from the respective differences between T_{sh} and T_{ice}. The stability and the better yield in test I are not necessarily the result of the residual moisture content, but could also be related to the different T_{ice}. This may also be indicated by a question during the discussion of the paper [3.27]: "If it is not the water, who else might it be, that is leading to higher potency and stability in the short term drying cycles?"
2. The unfavorable results in III could also be related to a too small RM.
3. The importance to measure also T_{ice}. From the published data, the behavior of the product during desorption cannot be estimated. One could conclude, that Dr-measurements during SD would have given additional informations, e. g. if test III would have been terminated after 7 % RM had been reached.

Terentier and Kadeter [3.28] described the freeze drying of the vaccine *Yersinia pestis* EV 76 in a solution containing 10 % sucrose, 1 % gelatin, and 0,5 % thiourea. The product has been frozen on the shelves of a freeze drying plant with approx. 8 °C/min to –40 °C. From ER- measurements it was concluded, that below –24.4 °C a glass phase starts and the eutectic temperature, T_e, was –17.1 °C. The drying time was determined as 9 h. If T_{sh} was controlled in such a way that T_e is exceeded after 4.5 h, the survival rate fell to approx. 50 %, if T_e was reached after 6 h, the survival rate is approx. 80 %. One can assume, that MD was only terminated after 5 h or more, at which time the temperature could be raised.

Morichi et al. [3.29] showed with 54 different bacteria and bacteria-strains, that the α-COOH-, the α-NH_2- and the guaidino-groups played an important role in the protective behavior of arginine. In the opinion of the authors, the common quality of the three groups was their ability to form hydrogen bonds.

Gehrke et al. [3.30] studied the course of freeze drying on *Eschirichia coli* (E) and *Lactobacillus plantarum* (L) with a specially developed plant (Fig. 3.11), which permitted the weighing of samples during freeze drying, the locking of samples from the chamber into an isolator (glove box), and the measurement of RM in the isolator using the Karl Fischer method. The schema of the plant is shown, since its design is almost ideal for freeze drying studies to follow its course quantitatively. For quick freezing of the cultures (1.7 to 2.2 °C/min) the containers, with the samples were placed on the precooled shelves at –45 °C. Figure 3.12.1 shows the process data of a drying of E, and Fig. 3.12.2 the plot of the weight loss of during this test. Figure 3.13 presents the decrease of viable organisms (CFU) per g of mass during freezing and drying. 10 %(w/w) skimmed milk and 10 %(w/w) of glycerol were added to the suspensions of cells in all tests. Figure 3.14 shows the drying time for E as a function of T_{sh} and the Fig. 3.15 the sublimation rate in g of water per open surface area of the container as a function of the water content in the product. As

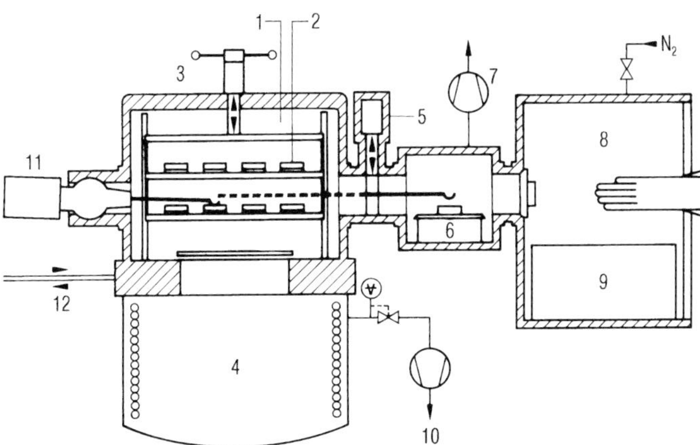

Fig. 3.11. Schema of a laboratory freeze drying plant.
1, Vacuum chamber with tempered shelves; 2, container with probe; 3, lift for shelves; 4, condenser; 5, lockgate; 6, balance in the lock; 7, vacuum pump for the lock; 8, glove box; 9, Karl-Fischer measuring system: 10, pressure controlled vacuum pump; 11, manipulator; 12, tempered medium (Fig. 1 from [3.30]).

shown, the sublimation rate is independent of the layer thickness of the product. This is only possible during MD, if the heat transfer to the sublimation front (Eq. (15)) is decisive under the test conditions. The drying time remains dependent of *d*. The influence of the operation pressure on the drying time is shown as approx. 25 % shorter at 0.18 mbar than at 0.05 mbar.

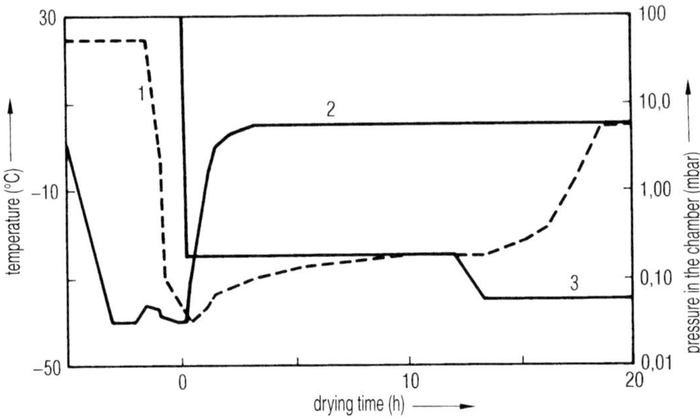

Fig. 3.12.1. Freeze drying run with *Escherichia coli* (E), $d = 20$ mm. 1, T_{pr}; 2, T_{sh}; 3, p_{ch} (Fig. 2 from [3.30]).

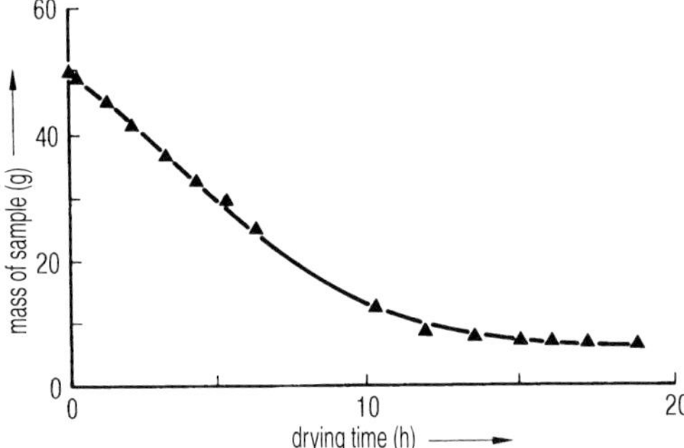

Fig. 3.12.2. Mass of sample as a function of time for (E) at $p_{ch} = 0.18$ mbar, $d = 20$ mm (Fig. 3 from [3.30]).

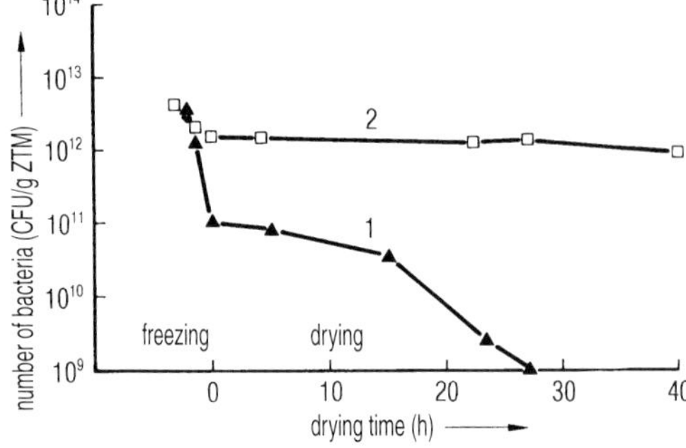

Fig. 3.13. Number of bacteria as a function of the drying time for (E) and (L). 1, (E); 2, (L); CFU = measure of number of viable bacteria; ZTM, solids of cells (Fig. 6 from [3.30]).

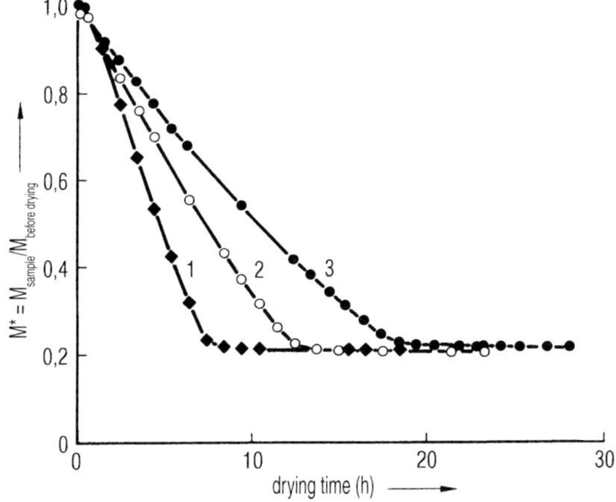

Fig. 3.14. Influence of the shelf temperature T_{sh} on the drying time for (E). 1, T_{sh} +13 °C; 2, T_{sh} −5 °C; 3, T_{sh} −13 °C (Fig. 11 from [3.30]).

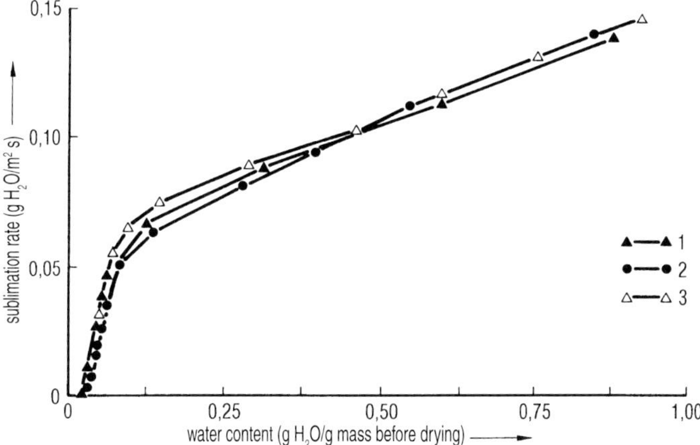

Fig. 3.15. Sublimation rate as function of the water content for different product layers. 1, 10 mm; 2, 15 mm; 3, 20 mm (Fig. 10 from [3.30]).
(Authors note: The measured independence of the sublimation rate from the layer thickness shows, that the sublimation rate under the conditions of the experiment depends only from the heat transfer to the product and not from the water vapor transportation through the dried product.)

Israeli et al. [3.31] found, that trehalose is a very good stabilizer for E, even if the freeze dried suspension of E was stored at 21 °C and 60 % relative humidity, and/or was exposed to visible light. In 3 h, the survival rate decreased without trehalose to 0.01 % under the influence of light and air, with trehalose, 35 % survived. The optimum trehalose concentration was found to be 100 mM. This corresponded with the number of trehalose molecules necessary to replace the water molecules in the outer membrane of the phospholipid molecules.

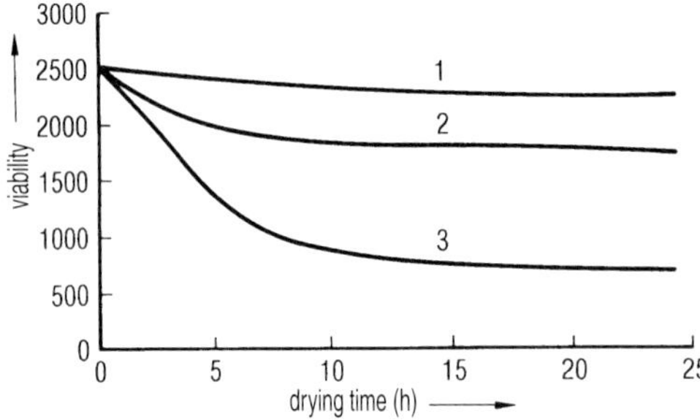

Fig. 3.16. Viability of *Saccharomyces cerevisiae* as a function of drying time, frozen with three different freezing rates. 1, 5 °C/min; 2, 1.5 °C/min; 3, 0.5 °C/min (Fig. 2 from [3.33]).

To increase the activity and capability of reproduction of *Saccharomyces cerevisiae* (SC) Kabatov et al. [3.32] proposed the addition of 10 % skimmed milk, which has been saturated with Ar or N_2. The freezing down to –25 °C was done under pressure and continued down to –55 °C. The freeze dried suspension did not change its quality during storage at +4 °C.

Pitombo et al. [3.33] found that 0.010 M succinate buffer at pH 4.6 was the best stabilizer for SC. The influence of three different freezing rates (0.5, 1.5 and 5 °C/min) on the capability of reproduction is shown in Fig. 3.16. During 235 days storage at +25 °C, no measurable decrease in invertase activity was observed, if the RM was below 4 %. With RM approx. 14 %, the invertase activity decreased in 20 days to half and was immeasurable after 57 days, since an insoluble cluster had been formed. A 4 % RM correspond at +25 °C with a monomolecular layer of water.

3.3 Antibiotics, Cytostatics and Ibuprofen

The freeze drying of antibiotics and blood serum have largely represented the beginning of industrial lyophilization. Neumann [3.34] wrote in 1952 "The (freeze drying) temperature for the older, not well purified Penicillin preparations had to be kept surprisingly low. It could not exceed –25 °C or –40 °C" and later on "Today Peniciilin is manufactured as crystals without the need of freeze drying".

Other antibiotics still require freeze drying, e. g. Na-Cephalotin (Na-CET). Takeda [1.32] showed, that thermal treatment of Na-CET was not sufficient to produce pure crystalline Na-CET, as the amorphous fraction discolors during storage and must be avoided. Takeda described the production of pure crystalline Na-CET by adding microcrystals of Na-CET to a saturated solution of Na-CET. If this mixture was frozen and freeze dried, then no amorphous or quasi-crystalline were found. Koyama et al. [3.35] described, that after thermal treatment for 24 h some parts remained incompletely crystallized. After adding 5 % (w/w) isopropylalcohol, a thermal treatment of 1 h was sufficient. Furthermore, the product could be dried at a higher pressure. Thus the drying time could be reduced and 100 % of the product could be used.

Ikeda [3.36] presented a two stage freezing process for an antibiotic (Panipenem), which reacts with another component of the drug (Betamipron) and has therefore to be separated until its use. The first substance is filled into the vials and frozen. The precooled second substance is filled then into cooled vials and frozen. By this process, the amount of undesirable reaction product could limited to 0.5 % during a 6-months storage at 40 °C. If the two products were frozen simultaneously the amount of reaction product was 1.2 %.

Vries et al. [3.59] described the development of a stable parenteral dosage form of the cytotoxic drug E 09. E 09 dissolves poorly in water and its solution is unstable. With the addition of 200 mg of lactose per vial containing 8 mg of E 09, an optimum formulation was developed with respect to solubility, dosage of E 09 and length of the freeze drying cycle. DSC studies have been used to select the most effective parameters. The freeze dried product remains stable for 1 year when stored at 4 °C in a dark environment.

Kagkadis et al. [3.60] have developed an injectable form of ibuprofen {(±)-2-(*p*-isobutylphenyl) propionic acid}, which is very slightly soluble in water and has a poor wettability. *b*-hydroxypropylcyclodextrin (*b*-HPCD) is used to form a better soluble complex with ibuprofen. This solution has been successfully freeze dried. The freezing and freeze drying process have been kept uniform in all experiments, though the freezing and freeze drying cycle itself cannot be discussed from the data presented as an optimum, as the product data as a function of concentration and freezing speed are not given.

3.4 Liposomes and Nanoparticles

Phospholipids are capable of forming vesicles under certain conditions of excess water, and this can be described schematically (Fig. 3.17). Liposomes can have a variety of structures (Fig. 3.18), as described e. g. by Talsma [1.34].

Liposomes can, generally speaking, only be frozen without damage if the suspension is frozen in a glass phase of water. This requires the addition of CPAs e. g. mannitol, dextran or trehalose, and quick freezing (e. g. 10 °C/min by LN_2) [3.37], (page 363).

Talsma [1.34] showed with phospholipon 100 H, a hydrated phosphatidyl-choline of soya beans, (Nattermann, Cologne) and dicetylphosphate (DCP) (molar ratio 10:1) from which bilayer liposomes have been produced, the influence of one CPA (I), of several CPAs (II), of the vesicle size (III), and the cooling rate (IV). In all of the following tests Tris buffer of pH 7.4 is used.

(I). In this example, mannitol is applied as CPA. Size of the vesicles was 0.27 to 0.32 µm, and vesicle concentration was 0.4 per μm^3. The energy flows during cooling (10 °C/min) and rewarming (10 °C/min) were measured by DSC. Figure 3.19 (a–d) show, that peak 2 does not exist, if the mannitol is within the liposomes. (Pure Tris buffer and pure liposome suspension do not show this peak either.) Note: During the quick freezing of mannitol solutions not all water crystallizes but forms an amorphous glass

Fig. 3.17. Schematic construction of small unilamellar vesicles, 25–100 nm.
1, water; 2, phospholipid (Fig. 1 from [3.38]).

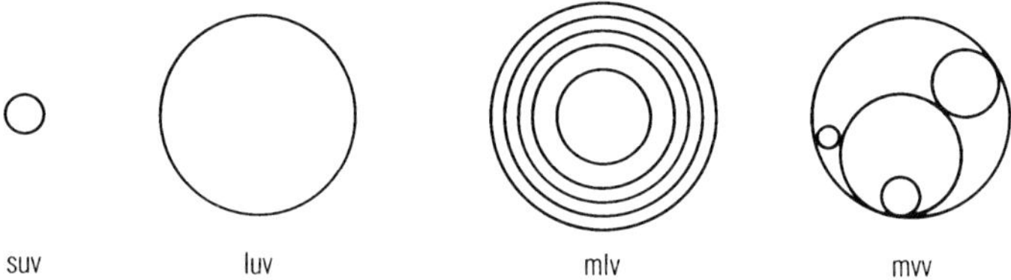

Fig. 3.18. Morphology of different liposome structure.
Suv, small unilamellare vesicles; luv, large unilamellare vesicles; mlv, multilamellare vesicles; mvv, multivesiculare vesicles (Fig. 4 from [1.34]).

phase. During rewarming, the viscosity and specific heat change. This happens e. g. in Fig. 3.19 (b) at -33.6 ± 0.6 °C and changes at –22.6 ± 0.3 °C into an exothermic process. At this temperature the ice is so much softened that water can crystallize, the water clusters can now migrate. Only in approx. 11 % mannitol solution can water solidify amorphously under the conditions of the experiment.

(II). In the following table which is different from that in Fig. 3.19, the temperatures during freezing at which the homogeneous crystallization of ice starts are listed. This is shown by the temperature of pure ice (–41.9 °C):
Start of homogeneous ice crystallization in different CPAs (lipoid as in Fig. 3.19) at a concentration of 30–50 µmol/ml, lipoid size 0.3 µm; 10 mM Tris buffer, pH 7.4; CPA and Tris buffer within or outside the liposomes [1.34, page 68]

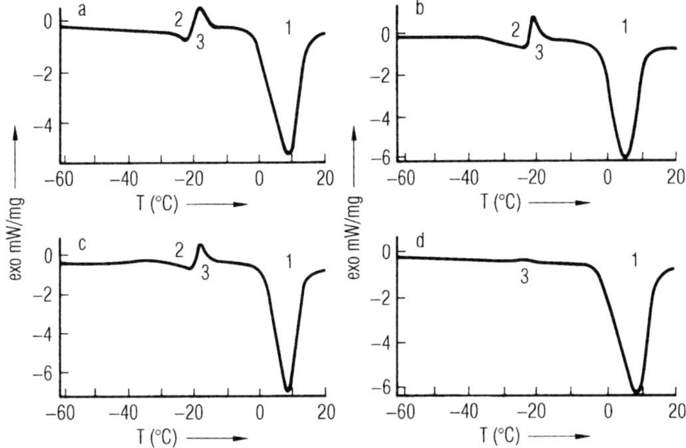

Fig. 3.19. Heat flow as a function of time during rewarming of the samples measured by DSC. (a) 11 % mannitol in 10 mM Tris buffer, pH = 7.4. (b) 35 µmol lipid/ml; 11.2 % mannitol in 10 mM Tris buffer, pH = 7.4, inside the vesicles as well as in the surrounding medium, particle size 0.32 µm. (c) 33 mmol lipid/ml, 11.2 % mannitol in 10 mM Tris buffer, pH = 7.4 in the surrounding medium and 10 mM Tris buffer pH = 7.4 inside the vesicles, particle size 0.27 µm. (d) 35 µmol lipid/ml, 11.2 % mannitol in 10 mM Tris buffer pH = 7.4 inside the vesicles surrounding medium 10 mM Tris buffer pH = 7.4, particle size 0.32 µm (Fig. 1B from [1.34]).

CPA %	T_{start}	T_c
None	–41.9	
Mannitol 11,2	–44.8	–33.6
Glycerol-mannitol 10/10	–48.7	
Glucose 30	–527	

T_{start} = extrapolated temperature of start of crystallization, T_c = collapse temperature.

The temperature of the homogeneous crystallization can be changed by changing the CPAs or their mixture.

(III). Talsma shows that peak 2 changes only from –39.8 °C to –40.4 °C (liposomes, liposomes concentration, buffer and cooling speed as in (I), but no mannitol) if the liposomes size is decreased from 0.87 µm to 0.14 µm. With small liposomes, the start of the homogeneous crystallization is delayed. This can also be deduced from the weakly performed crystallization (Fig. 3.19 (d), peak 3), if mannitol is only within the liposome.

(IV). If the liposomes (as in (I), mannitol within and outside) are cooled at 5 °C/min instead of 10 °C/min, the peak 2 starts at approx. 10 °C higher temperatures and shows a saddle-like form.

Under the conditions described, a quick freezing is desirable to produce a maximum of amorphous ice, which can be proven by a crystallization at peak 3. Until now, the freezing criteria are judged by the changes which they produce during cooling/rewarming. However the most important goal in the studies of liposomes is to find ways that water-soluble substances can be encapsulated in liposomes in such a way that they do not leak from the liposomes during transportation and storage, and are released in a controlled manner during application.

To test and measure the retention rate of liposomes, carboxyfluorescin (CF) can be used. Ausborn and Nuhn [3.38] studied different lipid vesicles, e. g. egg lecithin (EPC), hydrated egg lecithin (HEPC), cholesterol (CHOL) and mixtures thereof. For centrifuged EPC a retention rate of 67.5 %, and for centrifuged HEPC liposomes 75 %, has been found in 0.4 mol/l saccharose with 0.15 mol/L phosphate buffer. Furthermore, the results with different mixtures are reported: HEPC/CHOL with 1 mol/L sucrose has a retention rate of almost 100 %, while HEPC-liposomes in a 0.4 mol/L saccharose solution reach approx. 85 %. Talsma [1.34] established some quantitative connections between retention rates and particle size and storage temperatures. The retention rate of PL 100 H/DCP (10 : 1) (30 µmol/ml) increased from 51 % to 98 % if the liposome size decreased from 0.2 µm to 0.12 µm (liposomes in 10 % (w/w) saccharose with 10 mM Tris buffer). The retention rate of the same liposomes suspension also depends on the lipome size as well as the storage temperature after freezing in an acetone-dry ice mixture [1.34, page 92]:

Liposomes size (µm)	Retention rate for CF (%) after 65 h Storage at –25 °C	in LN_2
0.20	22	47
0.18	41	84
0.12	55	97

To freeze dry liposomes requires stabilization not only during freezing but also during drying and storage of the dry product. Talsma [1.34, page 106] showed that the retention rate of small liposomes (0.13 µm) was 24.1 % and greater than of large liposomes (0.28 µm) which had only a 7.4 % retention rate.

Crowe and Crowe [3.39] proved that it is sufficient for certain liposomes, e. g. egg phosphatidyl-choline (DPPC), to be vitrified by trehalose or dextran during freezing and freeze drying. In trehalose the retention rate was almost 100 %, and in dextran more than 80 %. This did not apply to egg PC-liposomes: Dextran as CPA alone led to an almost total loss of the CF-indicator, but addition of dextran into a trehalose solution (Fig. 3.20) also reduced the retention rate of CF substantially, e. g. from 90 % in a pure trehalose to approx. 45 % if trehalose and dextran were in equal amounts in the solution. Since $T_{g'}$ of dextran is approx. –10 °C and $T_{g'}$ of trehalose is –30 to –32 °C, dextran should form a glass phase at much higher temperatures than trehalose. Therefore the stabilization of egg- PC with trehalose cannot be related with the vitrification. Crowe showd with IR spectroscopy that egg-PC freeze dried with 2 g trehalose/g lipid had almost the identical spectrographic characteristics as the hydrous lipid: Trehalose molecules replaced the water molecules, and hydrogen

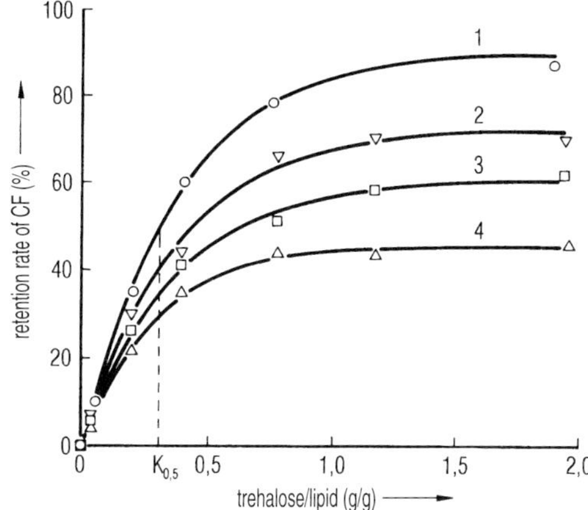

Fig. 3.20. Retention of CF as function of the trehalose/lipid concentration with varied dextran additions. 1, no dextran; 2, 0.08g dextran/g lipid;. 3, 1.2 g dextran/g lipid, 4, 2.0 g dextran/g lipid (Fig. 9 from [3.37]).

bonds were formed between the lipid- and trehalose molecules. Thereby, the stability of the lipids was preserved even if the water was removed. Crowe and Crowe compared this process with the survival of plants at low temperatures by producing trehalose.

Hauser and Strauss [3.40] assumed the hydrogen bond between sucrose and phospholipid as the cause of the integrity of the unilamellar vesicles and showed, that enclosed ions cannot migrate to the surroundings.

Ausborn et al. [3.61] confirmed by IR spectroscopy the strong hydrogen bonds between sucrose and SPS monoester (sucrose-palmitate/stearate) with the phosphate head groups, which supports the replacement theory of water molecules.

Suzuki et al. [3.62] concluded from their measurements that glucose and maltose completely prevent the aggregation or fusion of liposomes during freeze drying, but other maltodrextins support the aggregation due to their weak hydrophobic behavior.

Jizomoto and Hirano [3.41] tried to increase the amount of drug inclusions in liposomes by inserting Ca^{2+} ions in dipalmitoylphosphatidylcholine (DPPC) liposomes. The included volume (mL) per g of liposomes is called V_{cap}, and this can be increased as a function of the Ca^{2+} concentration up to ten fold of the minimum V_{cap} The increase in V_{cap} is attributed to the electrostatic repulsion between the Ca ions, which reduces the number of lamella and increases the diameter of the liposomes to a certain extent, but increases V_{cap} substantially. A calculated simulation of this thesis is in reasonable agreement with the measurements

The inclusion of drugs in liposomes is discussed in four examples:

Gu and Gao [3.42] reported that freeze dried cyclophosphamide in liposomes (CPL) reconstitutes well and has a larger antitumor activity and a smaller toxitity than CPL in aqueous solution.

Rudolf and Cliff [3.43] described the inclusion of hemoglobin in liposomes (LEH), to produce a stable blood substitute. The liposomes were formed from a solution of soya bean – phosphatidylcholine (soy PC), cholesterol, dimyristoyl-phosphatidyl, DL-glycerol (DMPG), and alpha-tocopherol with a ratio of 10:9:0.9:0.1. The product was dried and

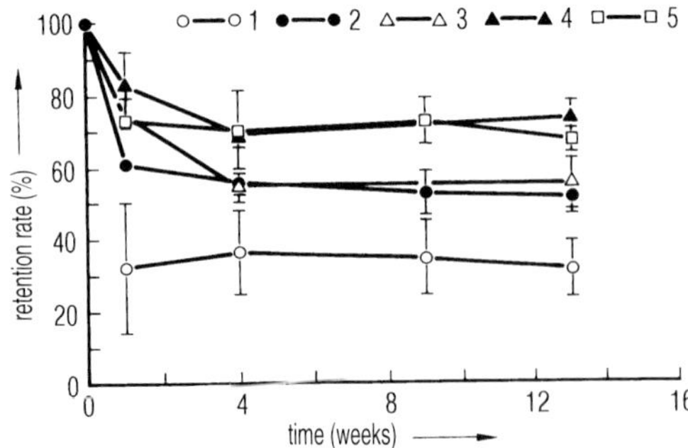

Fig. 3.21. Retention rate of liposome-encapsulated hemoglobin as a function of the time elapsed after the reconstitution of the freeze dried LEH, with different trehalose concentrations as CPA. 1, no trehalose; 2, 10 mM; 3, 50 mM; 4, 150 mM; 5, 300 mM trehalose (Fig. 2 from [3.43]).

rehydrated in a solution of 30 mM trehalose with phosphate buffer, pH 7.4. The evolved multilamellar vesicles were transform into large unilamellar vesicles (LUV), frozen in LN_2 and freeze dried. The LEH had an average diameter of 0.4 μm. The retention rate of hemogoblin in freeze dried LEH is shown in Fig. 3.21: After freeze drying in 150 mM trehalose solution, approx. 87 % of the hemoglobin remained within the liposomes after storage for 13 weeks. In the dry product, stored under vacuum, the level of methemoglobin rose to approx. 15 % after four weeks and remained constant at that level up to 12 weeks. That corresponds approx. to the data achieved with liquid LEH stored at +4 °C. The authors expect to develop a storable blood substitute with liposomes.

Foradada and Estelrich [3.63] studied the encapsulation of thioguanine (TG) in three types of liposomes produced by: extrusion, ethanol injection and dehydration-rehydration vesicles. The entrapment has been examined at three different concentrations (1, 0.1 and 0.01 mM) and at three different pH values (4.7, 7.4 and 9.2). The dehydration-rehydration vesicles were found to be the optimum method to encapsulate TG, independent of the pH value. At pH 4.7, 12 mmol/mol of lipid were entrapped, while with the other methods a maximum of 3 mmol/mol of lipid has been achieved. The authors related this behavior to the formation of hydrogen bridges between the TG and the liposomes.

Kim and Jeong [3.64] developed freeze dried liposomes containing recombinant hepatitis B surface antigen (HBsAg) to enhance the immunogenicity of HBsAg and to produce a stable product during storage. Dehydration-rehydratation vesicles with HBsAg were filtered through a 400 nm polycarbonate filter, and freeze dried in a 4 g trehalose/g lipid solution. After 1 year of storage at 4 °C the vesicles showed a similar size distribution as before freeze drying, and an approx. 70 % immunogenicity of HBsAg. Dried liposomes with HBsAg included showed an earlier sero conversion and a higher titer than free HBsAg or a mixture of aluminum phosphate and HBsAg.

van Winden and Crommelin [3.65] summarized the freeze-drying of liposomes as follows:

The requirements of a liposome drug formulation are

- chemically stable
- drug remains encapsulated in the liposomes
- liposome size unchanged during storage

The freeze drying of such a formulation is done for two main reasons: 1. The hydrolysis of the phospholipids without most of the water is substantially delayed or avoided. 2. Other degradation processes are delayed, as the mobility of the molecules is much smaller in the solid state than in the liquid phase. However, damages occur during the freezing and the freeze drying process itself. To avoid these damages, in most cases lyoprotectants must be used, thogh It might be possible to avoid them if certain interactions between the drug and the vesicle can be identified.

The lyoprotectant e. g. disaccharides, forms an amorphous matrix between the liposomes, and thus prevents aggregation and fusion during freeze-drying. If the protected liposomes are loaded with drugs which interact with the vesicles neither separation between drug and liposome nor damage to the liposomes is expected. However, if the drug is water-soluble leakage may even occur at a high ratio of sugar to liposomes (2 g sugar/g liposomes). The retention of the water-soluble carboxyfluorescein (CF) after freeze drying and reconstitution depends on the lipid composition, vesicle size and freezing rate.

These influences are described under I to IV at the beginning of this Section. The authors conclude from comparisons with freezing and thawing experiments, that the leakage may occur during rehydration of the liposomes, and not during the freezing process. From the FTIR analysis of the freeze dried cakes it is further concluded that the influence of size and lipid composition cannot be explained by different levels of bonds between the lyoprotectant and the bilayer component of the liposomes. The CF retention by liposomes based on saturated phospholipid DPPC increases when slowly frozen (0.5 °C/min) from 42 % to 80 % when frozen in LN_2. DPPC liposomes with cholesterol increase the retention rate under the same conditions from 75 % to 90 %. On the other hand, no effect of freezing rate has been found with liposomes based on unsaturated phospholipid egg phosphatidylcholine (EPC). From these and other experiments the authors conclude that: During freezing, the leakage of liposomes may not be induced, but conditions are created which can induce damage during rehydration. Despite the presence of lyoprotectants a 'repacking' of the bilayers can occur during and after rehydratation.

Studies with the freeze dried DPPC liposomes in trehalose solution showed, that not T_g of the amorphous sugar is the critical temperature during storage, but the bilayer transition temperature T_m, for the lyposomes determines the short term stability of the formulation. With trehalose as lyoprotectant and a low residual water content, T_m proved to be 10 to 30 °C below the onset of T_g. 30 min heating above T_m but well below T_g decreased the retention of CF after rehydration. T_m, after the heating was reduced from 40 to 80 °C to below 25 °C.

The exact mechanism of leaking induced by heating the dry product above T_m is unclear, but the authors exclude a bilayer phase transition during rehydratation or a fusion between the liposomes as a cause of the leakage.

Freeze dried liposomes loaded with doxorubicin (DXR) have been stored for 6 months at temperatures between –20 and +50 °C. Up to 30 °C, no sign of degradation was found, but at 40 to 50 °C – well below T_g of the dried cake – the total DXR content and the retention of the drug after dehydration decreased, while the size of the liposomes increased to a certain extent. The stability with RM below 1 % has been better than with 2.5–3.5 %.

Lactose, trehalose and maltose have equally lyoprotectant properties, while liposomes with sucrose showed an increase in size.

In summary [3.65], with an optimized formulation and freeze drying protocol, liposomes loaded with water-soluble CF or DXR can be freeze dried with a 90 % retention upon rehydration. The cake is stable for at least 6 months at temperatures up to 30 °C.

Auvillain et al. [3.44] studied the possibilities of drying nanospheres and nanocapsules without changing their diameter. Besides a suitable CPA, two conditions during freezing and freeze drying were decisive: The freezing rate and the melting temperature of the encapsulated oil. The CPA was a 30 % trehalose solution (10 % was not sufficient). Rapid freezing in an alcohol bath (approx. 4 °C/min) or in LN_2 (approx. 100 °C/min) was necessary to protect the capsule diameter. The included oils did not affect the diameter during freezing and freeze drying as long as the solidification temperature of the included oil was lower than the essential freezing temperature of the suspension. Oil having a solidification temperature of +4 °C was less suited than one with a temperature of –25 °C: moreover and a solidification temperature of –65 °C was generally best, as it increased the diameters only marginally or not at all. The authors presumed, that the nanocapsules withstood the freezing and drying more easily, if the surrounding of the capsules solidifies while the oil in the capsule remained soft.

For nanospheres, a slow freezing was requested by the authors, though in the opinion of the writer this was a misleading definition of slow freezing. To freeze the nanospheres quickly the containers with the suspension was placed directly on the precooled shelves of –40 °C, producing a 'small number of large crystals'. During the 'slow' freezing, an isolating layer was placed between the shelves and the containers. This may have led to a substantial subcooling followed by an abrupt crystallization, which produced a large number of small crystals. However such a process should not be denominated as 'slow'. The freezing may be very fast after a deep subcooling. If this course of events is accepted, the results for nanospheres were comparable with those for nanocapsules. For both nanoparticles it seems important to produce small crystals quickly. This is possible by freezing in LN_2 and in an alcohol bath with solid CO_2, but subcooling and abrupt freezing can lead to a similar result.

Nemati et al. [3.66] described the freeze drying of nanoparticles produced from monomer iso-hexylcyanoacrylate (IHCA) in which doxorubicin was encapsulated. The suspension contained: 1 % dextran 70, 5 % glucose, 10 mg doxorubicin chlorate and 50 mg lactose and was adjusted to pH 2.3. The product (1.3 ml/vial) was frozen on shelves of –50 °C for 3 h and thermal treated for 24 h at –35 °C. After freeze drying, the vials with the product were placed for 48 h in a dryer containing P_2O_5. Only after this additional drying the nanoparticles with doxorubicin have the same size after rehydration (∅ 351 ± 52 nm) as before freeze drying (∅ 334 ± 55 nm). Authors note: This second drying could also be done in the freeze drying plant, if the condenser temperature is low enough.

Fouarge and Dewulf [3.45] reported about the freeze drying of poly (isohexylcyanocrylat) nanoparticles, which were loaded with dehydroemetine (DHE). The load of absorbed DHE was uniform and reproducible. The stability remained good during 24 months, and the acute toxicity of DHE was reduced by combination with nanoparticles, as was the radical concentration.

Fattale et al. [3.46] compared negatively charged liposomes with nanoparticles from poly- (isohexyl-cyanoacrylate), which of both were loaded with ampicillin. Both carriers were of approximately the same size: 200 nm; but the nanoparticles could be loaded with approx. twenty times more ampicilin. After freeze drying and storage at –4 °C, no ampicillin leaked from the nanoparticles, while it migrated quickly from the liposomes.

3.5 Transplants

The freezing and conservation of viable cells or organs is not discussed here, but only the freezing and freeze drying of transplants, in which preservation of the structure and its chemical composition is the goal.

Hyatt [3.47] showed in Fig. 3.22 a typical temperature- and residual moisture course during the freeze drying of spongiosa and bone corticalis. The long drying times are due to three reasons:

- The heat transfer from the shelves to the sublimation front is much smaller compared with the transfer to vials with frozen liquids, as the material has an irregular form and the heat must be transferred through the already dried material (see Section 1.2.1).

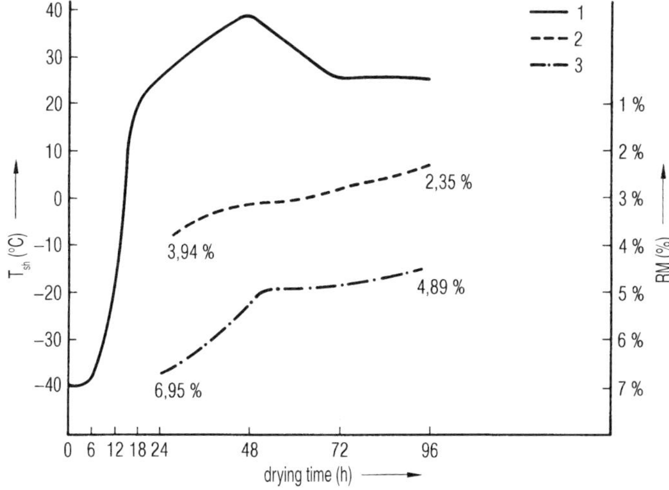

Fig. 3.22. Shelf temperature (T_{sh}) as a function of drying time during the freeze drying of bone corticalis and spongiosa and the related residual moisture content (RM).
1, T_{sh}; 2, RM in spongiosa; 3, RM in bone corticalis (Fig. 14 from [3.47]).

- Furthermore, the transplants are often packed in aluminum boxes, which are sealed by sterile filters permeable to water vapor. Even if the box can be designed with a negligible resistance to water vapor flow, the heat transfer is substantially reduced.
- In addition, the transplants may have a larger layer thickness to be dried than in other freeze drying processes. It is recommended in one charge to freeze dry transplants of similar layer thickness (the size can vary).

Bassett [3.48] requested that the transplants be cooled quickly to –78 °C, or lower. The transplants should be stored at this temperature for maximum of one year before freeze drying. The drying process should be terminated only after 1–3 % RM are reached. Krietsch et al. [3.49] list 15 different freeze dried preserves, which have been produced over a period of ten years, in which bone chips are the largest group with 34 %, followed by spongiosa with 22 %, dura with 19 %, and sinew with 11 %.

Marx et al. [3.50] described the application of transplants in the jaw area. All transplants were frozen at –70 to –80 °C or in LN_2, and freeze dried at a pressure of 0.01 mbar for 21 days to RM smaller than 5 %.

Merika [3.51] emphasized from his 17 years of experience with the quality control of freeze dried transplants the importance of sterility and residual moisture control as the decisive characteristics. Furthermore, the leak tightness of the storage containers was constantly controlled. Merika did not measure the product temperature during drying, but controled the process by measuring water vapor pressure and temperatures of the shelves and the condenser. The residual moisture content after 2 years of storage must be below 5 %. All products were sterilized by gamma radiation.

Malinin et al. [3.52] discussed the measurement of residual moisture content (RM) in freeze dried bones and compared three methods: The gravimetric, the Karl-Fischer titration, and NMR. The three methods are discussed in Section 1.3.1. All transplants in this comparison were frozen in LN_2 and remained for several weeks at this temperature. The temperature of the condenser during freeze drying was –60 to –70 °C. The shelves were kept at –30 to –35 °C for the first 3 days. During the last days of the drying the shelf temperature was raised to +25 or to +35 °C. The chamber pressure (p_{ch}) has been 0.1 mbar. During the initial phase of the process the amount of water vapor transported to the condenser was so large that the ice surface on the condenser was 20 °C warmer than the refrigerant.

After 3 days cooling of the shelves was terminated, and the transplant temperature rose to –15 °C.

The RM of the dried product was measured at 50 °C over P_2O_5 or in an oven with circulating air at 50 °C, or in the same oven at 90 °C over silica gel. Identical measurements were made with fresh bones. For NMR measurements, a known amount of D_2O was added to the bone in a glass container. After equilibrium between D_2O and H_2O was reached, a known amount of the product was taken from the solution and studied in a Perkin Elmer NMR-spectrometer. In Fig. 3.23 the water contents of fresh and freeze dried bones are listed measured by NMR and the gravimetric methods at 90 °C. The data show that only a certain amount of the total water can be removed at 90 °C, while another amount is so

Fig. 3.23. water content of fresh and freeze dried bone corticalis.
1, water content measured by NMR; 2, water removed in an oven at 90 °C over P_2O_5; 3, water content measured by NMR after the drying according to 2 has been completed; 4, sum calculated from 2 and 3 (follows on to Table 16–3 from [3.52]).

Type of bone	1 Total water by NMR (%)	2 Water removed at 90 °C (%)	3 Water content by NMR after 2 (%)	4 Sum of water 2 + 3 (%)
Fresh, ground Corticalis (tibia)	17.34	12.56	4.43	16.99
Fresh, ground Corticalis (femur)	20.5	14.62	5.28	19.90
Freeze dried, ground corticalis	7.55	5.73	4.20	9.93

strongly bound that it cannot be removed by heating. The data show also that the sum of the water measurable by gravimetry and by NMR agrees with the total amount of water.

The merit of Malinin's work is the comparative study of water content of bones by reproducible methods. The measurement of water vapor pressure during the drying cannot be used dirctly to determine the RM, as Malinin correctly states. Measurement of the description rates (DR) provide a means to follow quantitatively the course of desorption drying. The method is described in Section 1.2.2, but cannot be applied in an installation used by Malinin because the condenser cannot be separated from the chamber by a valve. By using the data given in the paper of Malinin it is possible to estimate the freeze drying process of bone transplants as follows:

In the first 3 days T_{sh} has been –30 to –35 °C and T_{co} between –60 and –75 °C. In the early stages of the process so much water was transported to the condenser that a thick ice layer condensed on the surface, producing a surface temperature of –50 °C. This amount of water cannot sublime from the bones. At T_{sh}· –35 °C and T_{pr} at –35 °C only such ice was sublimated, which was condensed during the loading on the cold shelves and transport boxes. This was also confirmed by the temperature rise in the bones from –70/–65 °C by 5–10 °C after commencing evacuation and the further increase in bone temperature to –35 °C. If the water were to sublimate in large amounts from the bones, their temperature should be lowered by the energy of sublimation. The temperature increase showed that the water did not come from the bones. This ice load of the condenser and the time taken may be avoided. The transportation boxes should be isolated during transport as much as is practicable, and the door of the installation only opened for a short time. During this short period a small stream of dry (dew point –40 °C) and sterile air should flow through the chamber, producing a small positive pressure (a few mm water column). If the condenser can be separated from the chamber by a valve, as assumed in the further discussion, it cannot ice over.

After 3 days, cooling of the shelves was terminated and the temperature of the bones rose to –15 °C. The operation pressure in the plant was always 0.1 mbar or less. The first 3 days can be saved, since at T_{pr} –35 °C p_s is approx. 0.2 mbar. The amount of water sublimated during these 3 days was small compared with the amount sublimated thereafter at T_{pr} –15 °C, since $p_s \approx$ 1,5 mbar. From the third day on, seven times more water/unit time can be sublimated than in the previous days. Therefore it is recommended to rise T_{sh} from the beginning to –15 °C. The ice temperature at the sublimation front T_{ice}, established under these conditions, can be measured by BTM, as shown in Section 1.2.1.

The easiest way to control T_{ice} is by the operation pressure. T_{ice} depends only on two processes:

- heat transfer from the shelves to the sublimation front; and
- mass transfer from the sublimation front to the condenser.

Heat transfer from the shelves to the sublimation front depends on the pressure and the distance between shelf and product (Fig. 1.58). Mass transfer (g/s) increases with the pressure, but also depends on the flow resistance of the already dry product and of the packing of the bones. If the maximum tolerable T_{ice} is defined, the drying time depends only on the two processes mentioned above. It cannot be shortened under a given geometric situation and the chosen T_{ice}. This method of T_{ice} control does not require thermocouples, and does not contaminate the product.

Selection of the maximum tolerable T_{ice} can not be done by the methods described in Section 1.1.5. as no uniform mass exists. Willemer [3.53] concluded from measurements of homogenized tissue a T_{ice} maximum of –25 °C, while Malinin [3.52] used –15 °C after 3 days. Authors note: –15 °C may not represent T_{ice}, but the surface temperature of a bone.

During MD the T_{ice} can be closely controlled and the change from MD to SD documented and, if required, automatically executed. During SD the pressure control can be switched of, the pressure will drop and the progress of SD can be followed by DR measurements. DRs give the amount of water desorbed/h in % of solids. By the integration over time (Eq. (20)) it can be decided, when e. g. 5 % desorbable water are reached and which result can be expected in another 48 h of drying.

Figures 3.24.1 and 3.24.2 [3.54] present the data of the freeze drying of 622 g bone corticalis from swine, which were cut into pieces of 6 · 11 cm and 2 cm thickness or 2.5 · 4.5 cm and 3 cm thickness. The pieces were placed in aluminum trays . Three thermocouples were fixed into bores in the bones. Cooling from +20 °C to –45 °C took approx. 2.5 h. After the evacuation, T_{sh} was raised to +22 °C and the operation pressure controlled at 1.2 mbar TM, corresponding to 0.8 mbar CA. After 2 h an equilibriu was reaches between shelf-, product surface- and T_{ice} at approx. ca –35 °C. This remained for a little less than 5 h between –32 °C and –35 °C, while the temperatures on the bones increased to –5 °C. After 19 h, the controlled operation pressure could no longer be kept constant and DR-measurements showed after 21 h at a constant T_{pr} of approx. +20 °C a falling tendency. The increase of T_{sh} to +40 °C can be seen in the DR data (Fig. 3.24.2). From the DR data it is possible to estimate the amount of desorbable water (dW) after 19 h (in % solids) as shown by the straight line in Fig. 3.24.2 . At 19 h, dW was approx. 0.9 %, at 28 h approx.

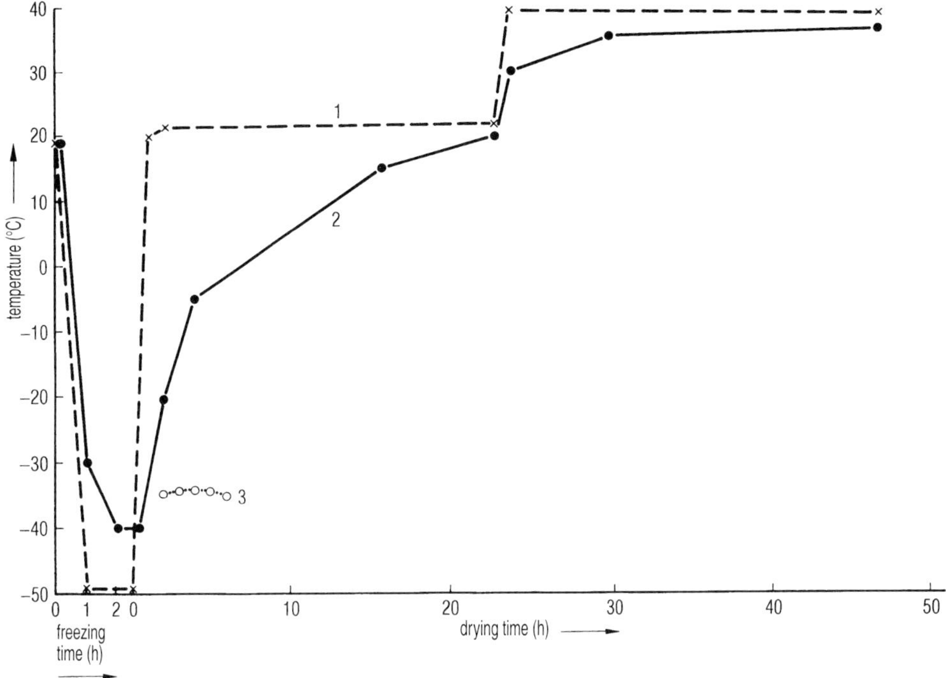

Fig. 3.24.1. Course of the freeze drying of 622 g corticalis of pork.
1, temperature of the shelves; 2, average temperature of the bones (three sensors); 3, temperature of the sublimation front (T_{ice}), measured by BTM (measurements by AMSCO Finn-Aqua, D-50345 Hürth).

0.24 % and at 38 h approx. 0.05 %. Any further drying would not reduce the dW data measurably. The dW data, in this example at 37 °C, provided the information at which time the drying could be terminated. At a different temperature, e. g. +90 °C. a different equilibrium will be aimed at. dW data and the RM measured by Malinin at +90 °C in a vacuum oven and by NMR [3.52] are different numerical quantities, but can be combined:

Residual moisture content in % of solids	
1. dW at +37 °C	0.06
2. removable water at 90 °C in a vacuum oven over silica gel	4.60
3. water content by NMR after operation 2. is concluded	2.49
total water content after freeze drying	7.15

At the end of the freeze drying process shown in Fig. 3.24.1, four pieces of bone were weighed and had lost 24.5, 24.2, 20.9 and 26.5 % (average 24.4 %) of their original weight. Drying was then continued for additional 40 hs, at which time the average weight loss was 24,6 %. The desorbable water dW could not be reduced any further after 46 h of the first drying.

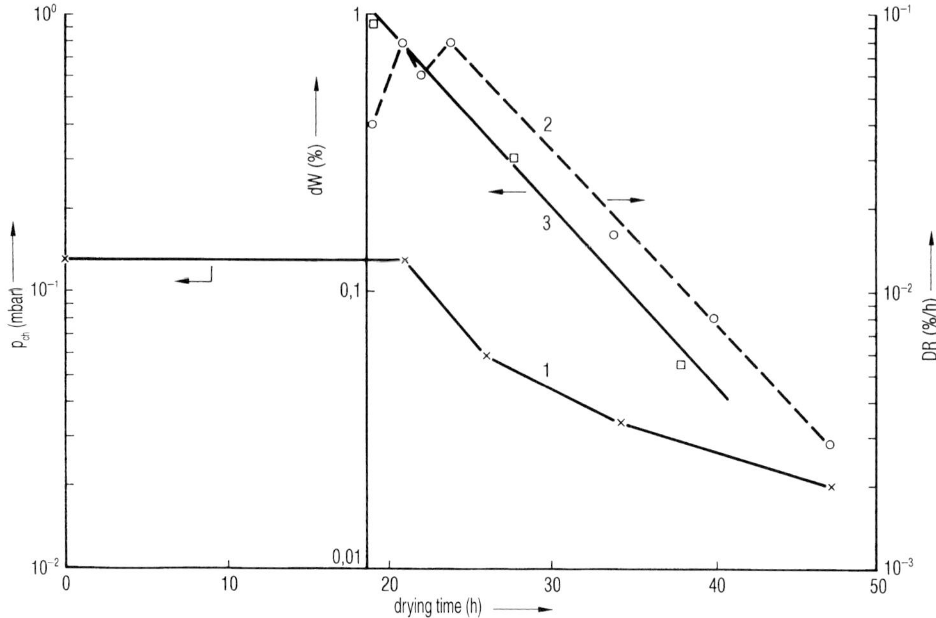

Fig. 3.24.2. Plots of pressure, DR and dW during the drying shown in Fig. 3.24.1.
1, chamber pressure p_{ch}, ordinate on the left side; 2, desorption rate, DR = water desorbed in % of solids/h, ordinate on the right side; 3, desorbable water, dW = water in % of solids (residual moisture at the given T_{pr}). DR and dW calculated from pressure rise measurements by AMSCO Finn-Aqua, D-50345 Hürth.

Schoof et al. [3.67] described the development of a collagen product with homogeneous pores whose sizes can be determined by the freezing conditions. The authors quoted several publications in which the homogeneous pore size distribution was shown as essential to optimize the population of the implanted collagen by living cells. The desirable sizes varieed between 20 µm and 125 µm. The collagen suspension, supplied by Dr. Otto Suwelack Nachf. GmbH & Co., Billerbeck, contained 1.8 % collagen, water and HCl at pH 3.2. Furthermore 3.8 % acetic acid was added (pH 2.5). The samples were frozen by the Power-Down process, in which a temperature gradient is applied between the two ends of the sample. Subsequently the end temperatures are lowered with a constant cooling rate while the temperature gradient is kept constant. In comparison the same collagen suspension containing 3.8 % acetic acid was conventionally frozen in a cold bath of –25 °C. The frozen samples were freeze dried. Figure 3.25.1 shows the structure of freeze dried collagen sponges that were conventionally frozen and freeze dried. In Fig. 3.25.2, the collagen was frozen by the Power-Down method and freeze dried; Fig. 3.25.3 shows a part of Fig. 3.25.2 enlarged approx. 10 times. The pore size can be influenced by the acetic acid content: From 3.8 % (w/w) to 1.5 % the size was reduced from approx. 40 µm to 20 µm, and the temperature gradient was 50 K/cm in this experiment.

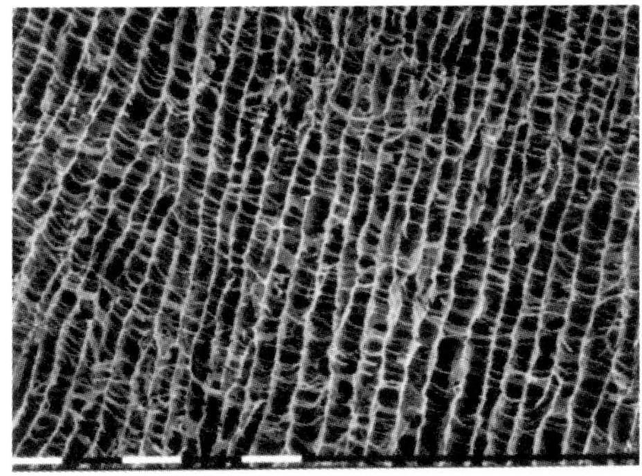

Fig. 3.25.1. Porous structure of collagen sponge produced by freezing in a cryogenic bath at –25 °C and subsequently freeze dried (scanning electron microscope, white bar = 1 mm) (from [3.67]).

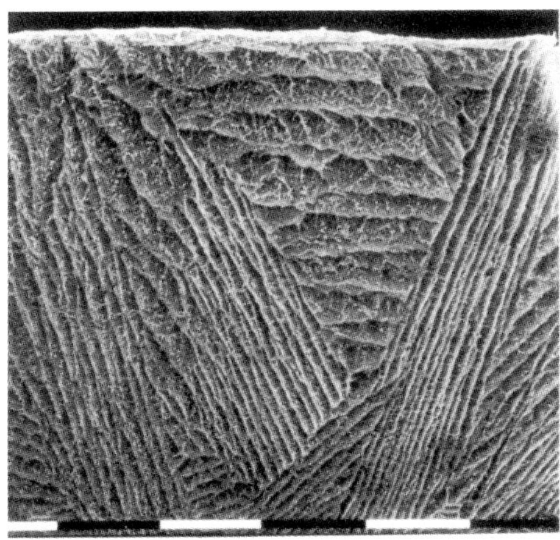

Fig. 3.25.2. Porous structure of collagen sponge produced by directional solidification according to the Power-Down method and subsequently freeze dried (scanning electron microscope, white bar = 1 mm) (from [3.67]).

Fig. 3.25.3. Magnification of the defined and homogeneous porous structure of a collage sponge produced by directional solidification according to the Power-Down method and subsequent freeze drying (scanning electron microscope, white bar = 0.1 mm) (from [3.67]).

References for Chapter 3

[3.1] De Luca, P. P. and Trappler, Ed.: Design and operational requirements of freeze-drying facilities for pharmaceuticals. The American type culture election and the Merieux Foundation. A compact freeze-drying course on the fundamental aspects on the preservation of sensitive biologicals . Washington DC. April 2–4, 1990

[3.2] Carpenter, J. F., Crowe, J. H., Arakawa, T.: Comparison of solute – induced protein stabilization in aqueous solution and in the frozen and dried states. Developments in biological Standardization Vol. 74, p. 225–239. Acting Editors: Joan C. May – F. Brown. S. Karger AG, CH-4009 Basel (Switzerland), 1992

[3.3] Pikal, M. J.: Freeze-drying of proteins, part II, formulation selection. BioPharm. 3, p. 26–30, 1990

[3.4] Izutsu, K., Yoshioka, S., Takeda, K.: The effect of additives on the stability of freeze-dried 'Beta-Galactosidase' stored at elevated temperature. Int. J. Pharm. 71, p. 137–146, 1991

[3.5] De Luca, P. P.: Development of lyophilization formulations. International Colloquium on Ind.-Pharm. Lyophilization, Gent, 1992

[3.6] Skrabanja, A. T. P., de Meere, A. L. J., de Ruiter Rien, A., van der Oetelaar, P. J. M.: Lyophilization of biotechnology products. PDA Journal of Pharmaceutical Science and Technology, Vol. 48, p. 311–317, 1994. Copyright © 1994 PDA, Inc., Bethesda, Maryland, USA

[3.7] Pikal, M. J., Shah, S.: Moisture transfer from stopper to product and resulting stability applications. Developments in Biological Standardization, Vol. 74, p. 165–179. Acting Editors: Joan C. May – F. Brown. S.K arger AG, CH-4009 Basel (Switzerland), 1992

[3.8] Pikal, M. J., Dellermann, K., Roy, M. L: Formulation and stability of freeze-dried proteins: Effects of moisture and oxygen on the freeze-dried formulation of human growth hormones. Developments in Biological Standardization, Vol. 74, p. 21–38. Acting Editors: Joan C. May – F. Brown. S. Karger AG, CH-4009 Basel (Switzerland), 1992

[3.9] Williams, M. L., Landel, R. F., Ferry, J. D.: The temperature dependence of relaxation mechanisms in amorphous polymers and other glass forming liquids. J. Am. Chem. Soc. 77, p. 3701–3707, 1955

[3.10] Roy, M. L., Pikal, M. J., Rickard, R. C., Maloney, A. M.: The effects of formulation and moisture on the stability of freeze-dried monoclonal antibody – vinca conjugate: A test of the WLF glass transition theory. Developments in Biological Standardization Vol. 74, p. 323–340. Acting Editors: Joan C. May – F. Brown. S. Karger AG, CH-4009 Basel (Switzerland), 1992

[3.11] Jensen, L. H.: The structure of water in protein crystals. Developments in Biological Standardization, Vol. 74, p. 53–61. Acting Editors: Joan C. May – F. Brown. S. Karger AG, CH-4009 Basel (Switzerland), 1992

[3.12] Teeter, M. M.: Order and disorder in water structure of crystalline proteins. Developments in Biological Standardization, Vol. 74, p. 63–72. Acting Editors: Joan C. May – F. Brown. S. Karger AG, CH-4009 Basel (Switzerland), 1992

[3.13] Hageman, M. J., Possert, P., Bauer, J. M.: Prediction and characterization of the water sorption isotherm for bovine somatropin recombinant. Journal of Agricultural Food Chemistry, 40 (2), p. 342–347, 1992.

[3.14] Townsend, M. W., de Luca, P. P.: Use of lyoprotectants in the freeze-drying of a model protein ribonuclease. PDA Journal of Pharmaceutical Science and Technology, Vol. 42 (6), p. 190–196, 1988. Copyright © 1988 PDA, Inc., Bethesda, Maryland, USA

[3.15] Townsend, M. W., Byron, P. R., de Luca, P. P.: The effects of formulation additives on the degradation of freeze-drying ribonuclease A. Pharm. Res. 7 (10), p. 1086–1091, 1990

[3.16] Townsend, M. W., de Luca, P. P.: Stability of ribonuclease A in a solution and the freeze-dried state. J. Pharm. Sci., 79 (12), p. 1083–1086, 1990

[3.17] De Luca, P. P., Townsend M. W.: Stability of ribonuclease-A in solution and the freeze-dried state. Congr. Int. Technol. Pharm. 5th. Vol. 1, p. 457–465, 1989

[3.18] Carpenter, John F., Crowe, John H.: An infrared spectroscopy study of the interactions of carbohydrates with dried proteins. Biochemistry 28 (9), p. 3916–3927, 1989

[3.19] Hora, M. S., Rana, R. K., Wilcox, Cynthia L., Katre, N. V., Hirtser, Pamela, Wolfe, S. V., Thomson, J. W.: Development of lyophilized formulation of interleukin-2 (rhIL-2). Developments in Biological Standardization Vol. 74, p. 295–306. Acting Editors: Joan C. May – F. Brown. S. Karger AG, CH-4009 Basel (Switzerland), 1992

[3.20] Brewster, M. E.: Use of 2-Hydroxypropyl-beta-cyclodextrin (HPCD) as a solulizing and stabilizing excipient for protein drugs. Pharm. Res., 8 (6), p. 792–795, 1991

[3.21] Carpenter, J. F., Kreilgaard, L., Jones, L. S., Webb, S., Randolph, T. W.: Mechanisms of protein stabilization by nonionic surfactants. Freeze-Drying of Pharmaceuticals and Biologicals, presented by National Science Foundation, Industry/University Cooperative Research Center for Pharmaceutical Processing, CPPR, Brownsville, Vermont USA, 1998

[3.22] Vermuri, S., Yu, Chang, Roosdorp, N.: Effect of cryoprotectants on freezing, lyophilization and storage of lyophilized recombinant Alpha- Antitrypsin formulations. PDA Journal of Pharmaceutical Science and Technology, Vol. 48 (5), p. 214–246, 1994

[3.23] Willemer, H.: Freeze-drying process data determination for human blood derivates with factor VIII as example. PDA Fourth International Congress, p. 142–151, Vienna, 1996. Copyright © 1996, PDA Inc. Bethesda, Maryland, USA

[3.24] Greiff, D.: Important variables in the long term stability of viruses dried by sublimation of ice in vacuo. International Institute of Refrigeration <IIR> (XIII th International Congress of Refrigeration, p. 657–667, Washington DC, 1971

[3.25] Greiff, D.: The cryobiology of viruses classified according to their chemical, physical and structural characteristics. International Institute of Refrigeration <IIR> Comm. C 1, p. 8–11, 1982

[3.26] Doner, T., Dundrarova, D., Teparicharova, I., Orozeva, M., Bustandzieva, R., Mitor, B.: Choice of cryoprotective media and freeze-drying parameters of bovine corona-virus and respiratory syncytial virus. IV. International School Cryobiology and Freeze-Drying p. 31,32, Sofia, 1989

[3.27] Bennett, P. S., Maigetter, R. Z., Olson, Margit, G., Provost, P. J., Scattergood, E. M., Schofield, T. L.: The effects on freeze-drying on the potency and stability of live varicella virus vaccine. Developments in Biological Standardization, Vol. 74, p. 215–221. Acting Editors: Joan C. May – F. Brown. S. Karger AG, CH-4009 Basel (Switzerland), 1992

[3.28] Terentier, A. N., Kadeter, V. V.: Freeze-drying of the vaccine strain *Yersinia pestis* EV 76. IV. International School Cryobiology and Freeze-Drying, p. 29,30, Sofia, Bulgarien, 1989

[3.29] Morichi, T., Irie, R., Yano, N., Kembo, H.: Protective effect of organic and its related compounds on bacterial calls during freeze-drying. Agr. Biol. Chem., Vol. 29, No 1, p. 61–65, 1965

[3.30] Gehrke, H.-H., Krützfeld, R.; Deckwer, W. D.: Gefriertrocknen von Mikroorganismen. I. Experimentelle Methoden und typische Ergebnisse. Chem. Ing. Tech. MS 1832/1990

[3.31] Israeli, E., Shaffer, B. T., Lighthart, B.: Protection of freeze-dried *Escherichia coli* by trehalose upon exposure to environmental conditions. Cryobiology 30, p. 510–523, 1993

[3.32] Kabatov, A. I., Nikonov, B. A., Sventitskii, E. N., Afanasi, Eva, S.: Working out the means of recreating the biological activity of *Saccharomyces cerevisiae* yeast at sublimation drying. Biotekhnologiya (1), p. 45–46, 1991

[3.33] Pitombo, R. N. M., Spring, C., Passos, R. F., Tonato, M., Vitalo, M.: Effect of moisture content on the interface activity of freeze-dried *S. cerevisiae*. Cryobiology 31, p. 383–392, 1994. Copyright © 1994 by Academic Press, Inc., New York

[3.34] Neumann, K. H.: Grundriß der Gefriertrocknng . 2. Auflage, p. 102–103, Musterschmidt Wissenschaftlicher Verlag, Göttingen, 1952

[3.35] Koyama, Y., De Angelis, R. J., De Luca, P. P.: Effect of solvent addition and thermal treatment on freeze-drying of cefazolin. PDA Journal of Pharmaceutical Science and Technology, Vol. 42 (2), p. 47–52, 1988

[3.36] Ikeda, M.: Development of a multilayer lyophilization technique for parenteral dosage forms. PDA Asian Symposium, p. 261–266, Tokyo, 1994

[3.37] Crowe, J., H., Leslie, S. B., Crowe, L. M.: Is vitrification sufficient to preserve liposomes during freeze-drying? Cryobiology, 31, p. 355–366, 1994. Copyright © 1994 by Academic Press Inc., New York

[3.38] Ausborn, M., Nuhn, P.: Möglichkeiten und Probleme der Stabilisierung von Liposomen durch Frier- und Lyophilisationsverfahren. 2. Mitteilung: Einfluß von Saccharose und Saccharose-Fettsäureestern auf das Verhalten von Lecithin-Cholosterol-Liposomen. Pharm. Ztg. Wiss., Nr. 1–4./136. Jahrgang, p. 17–24, 1991. Govi-Verlag GmbH, D-6760 Eschborn

[3.39] Crowe, L. M.; Crowe, J. H.: Stabilization of dry liposomes by carbohydrates. Developments in Biological Standardization, Vol. 74, p. 285–294. Acting Editors: Joan C. May – F. Brown. S. Karger AG, CH-4009 Basel (Switzerland), 1992

[3.40] Hauser, H., Strauss, G.: Stabilization of small unilamellar phospholipid vesicles by sucrose during freezing and dehydration. Adv. Exp. Med. Biol., p. 71–80, 1988

[3.41] Jizomoto, H; Hirano, K.: Encapsulating of drugs by lyophilized empty dipalmitoyl-choline (DPPC) liposomes: Effects of calcium ion. Chem. Pharm. Bull., 37 (11), p. 3066–3069, 1989

[3.42] Gu, X. Q., Gao, X. Y.: A novel procedure for preparing liposome entrapment of cyclophosphamine (CPL) in its reconstituted form, the properties and antitumor activities of the reconstituted CPL. Congr. Int. Technol. Pharm. 5th Vol. 3, p. 60–65, 1989

[3.43] Rudolph, A. S., Cliff, R. O.: Dry storage of liposome-encapsulated hemoglobin: A blood substitution. Cryobiology, 27, p. 585–590, 1990. Copyright © 1990 by Academic Press, Inc., New York

[3.44] Auvillain, M., Caré, G., Fessi, H., Devissaguet, J. P.: Lyophilisation de vecteurs colloidaux submicromiques. S. T. P. Pharma, 5, p. 738–747, 1989

[3.45] Fouarge, M., Dewulf, D.: Development of dehydroematine (DHE) nanoparticles for the treatment of visceral leishmaniasis. J. Microencapsulation 6 (1), p. 29–34, 1989

[3.46] Fattale, E.; Rojas, J.; Roblot-Treupal, L.; Andremont, A., Couveur, P.: Ampicillin-loaded liposomes and nanoparticles: Comparison of drug loading, drug release and in vitro antimicrobial activity. J. Microcapsulation 8 (1), p. 29–36, 1991

[3.47] Hyatt, G. W.: Procedes employes pour obtenir des tissus humain a usage chirurgical et, en particulier methode de conversation par lyophilisation, p. 279–301
Rey, L. et collaborateurs: Traité de Lyophilisation. Hermann, Paris VI, 1960

[3.48] Bassett, C. A. L.: A survey of the current status of tissue procurement, processing and use. Aspect theoriques et industriels de la lyophilisation, p. 332–339, Hermann, Paris VI, 1964

[3.49] Krietsch, P., Hackensellner H. A., Näther, J.: 10-jährige Erfahrung bei der Herstellung und Anwendung von Gewebekonserven in der DDR. 6. Gefriertrocknungstagung Leybold- Hochvakuum-Anlagen GmbH Köln 1965

[3.50] Marx, R. E., Kline, S. N., Johnson, R. P., Malinin, T. I., Matthews II., J. G., Gambil, V. The use of freeze dried allogeneic bone in oral and maxillofacial surgery. J. Oral Surgery Vol. 39, p. 264–274, 1981

[3.51] Merika, P.: Quality control of freeze dried tissue grafts. International Institute of Refrigeration <IIR>, Com. C1, p. 102–105, Paris, 1983

[3.52] Malinin, T. I., Wu, N. M., Flores, A.: Freeze-drying of bone for allotransplantation, p. 183–192, Osteochondral allografts, Editors: Friedlaender, G. E., Mankin, H. J., Sell, K. W., publisher Little, Brown & Co., Boston/Toronto, 1983

[3.53] Willemer, H.: Data to be considered for freeze dryers to be used in freeze drying of transplants (especially bones). 1th European congress of tissue banking and clinical application. Berlin, Oct. 1991

[3.54] Measurements by AMSCO Finn-Aqua, D-50354 Hürth, Germany

[3.55] Remmele, R. L., Stushoff, Carpenter, J. F.: Real-time spectroscopy analysis of lysozyme during Lyophilization: structure-hydration behavior and influence of sucrose. American Chemical Society Symposium, Ser. 567 (Formulation and delivery of proteins and peptides) 1994. © 1994 American Chemical Society

[3.56] Franks, F., Hatley, R. H. M., Mathias, S. F. Pharm. Technol. Int 3, p. 24–34, 1991

[3.57] Dong, A., Prestrelski, S. J., Allison, S. D., Carpenter, J. F.: Infrared spectroscopy studies of lyophilization- and temperature-induced protein aggregation. J. Pharm. Sci. 84 (4), p. 415–424, 1995

[3.58] Prestrelski, S. J., Pikal, K., Arakawa, T.: Optimization of lyophilization conditions for recombinant interleukin-2 by dried state conformational analysis using Fourier-transform infrared spectroscopy. Pharm. Res. 12 (9), p. 1250–1259, 1995

[3.59] Vries, J. D., Talsma, H., Henrar, R. E. C., Bosch, J. J., Bult, A., Beijnen, J. H.: Pharmaceutical development of a parenteral lyophilized formulation of the novel indoloquinone antitumor agent E 09. Cancer Chemother. Pharmacol. 34 (5), p. 416–422, 1994

[3.60] Kagkadis, K. A., Rekkas, D. M., Dallas, P. P., Choulis, N. H.: A freeze-dried injectable form of ibuprofen: Development and optimization using respond surface methodology. PDA Journal of Pharmaceitical Science & Technology, 50 (5), p. 317–323, 1996

[3.61] Ausborn, M., Schreier, H., Brezesinnski, G., Fabian, H., Meyer, H. W., Nuhn, P.: The protective effect of free and membrane-bound cryoprotctants during freezing and freeze drying of liposomes. J. Controlled Release, 30 (2), p. 105–116, 1994

[3.62] Suzuki, T., Komtatse, H., Miyajima, K.: Effects of glucose and its oligomers on the stability of freeze dried liposomes. Biochim. Biophys. Acta 1278 (2), p. 176–182, 1996

[3.63] Foradada, M., Estelrich, J.: Encapsulation of thioguanine in liposomes. Int. J. Pharm. 124 (2), p. 261–269, 1995

[3.64] Kim, Ch. K., Jeong, E. J.: Development of dried liposomes as effective immuno-adjuvant for hepatitis B surface antigen. Int. J. Pharm. 115 (2), p. 193–199, 1995

[3.65] van Winden, E. C. A., Crommelin, D. J. A.: Freeze drying of liposomes. National Science Foundation, Industry/University Cooperative Research Center for Pharmaceutical Processing, CPPR, International Conference on Freeze Drying, Brownsville, Verm., USA, 1998

[3.66] Nemati, F., Cave, G. N, Couvreur, P.: Lyophilization of substances with low water permeability by a modification of crystallized structures during freezing. Assc. Pharm. Galenique Ind., Chatenay Malabry, 3, p. 487–493, 1992

[3.67] Schoof, H., Apel, J., Heschel, I., Rau, G.: Influence of the freezing process on the porous structure of freeze dried collagen sponges (unpublished results, Helmholtz-Institute, Aachen, Germany

4 Food and Luxury Food

In the early 1960s the freeze drying of food was first welcomed world-wide as a new method of food preservation. Three typical quotations on the subject were: from the *Wallstreet Journal* [4.1]: 'Steaks and other items by freeze-drying become very light weight, keep for years, retain their original taste to an extent rarely strained through older drying methods', while Tschigeov [4.2] said: 'The lyophilization method is becoming used of present in the soviet food industry mainly for drying meat and fish in the production of concentrated foods', and in *The Times* [4.3] 'Freeze-drying has been found to be fully effective. It is the acceptable technique for preserving ... pharmaceutical products and can be applied with equal success to the drying of food'.

In the following ten years, the scientific and technical presumptions have been the subject of many studies, to produce freeze dried food economically for long-term storage. However, there have always been four questions to be answered:

1. Which factors determine the optimum freezing and freeze drying time of a given product.
2. Which conditions must be fulfilled by the product so that it remains unchanged during storage.
3. How a freeze drying plant must be designed, in order to produce food according to questions 1 and 2.
4. How the products have to be packed during storage.

The freezing of a product in the vacuum chamber by the evaporation of 15–25 % of its water would reduce the process time, the investment- and the operation cost, since the evaporation of this water could be done in a very short time, and no extra freezing installations are required. Oetjen [4.4, 4.5] points out, that this process has substantial disadvantages for the quality of the product and is only applicable for a limited number of products. Therefore this process is no longer used (see Section 2.1.5). The freezing processes for food are discussed in the Sections 2.1.1 and 2.1.3, and the possible freezing rates in Section 1.1.1. The decisive quantities for the freezing rastes are:

- the heat transfer coefficient from the cooling medium to the product
- the heat conductivity in the product to be frozen
- the amount of unfreezable water (Section 1.1.1)

The general rule for all freeze drying processes applies also to food: The method of freezing, the freezing rate and the final temperature of freezing largely determine the quality of

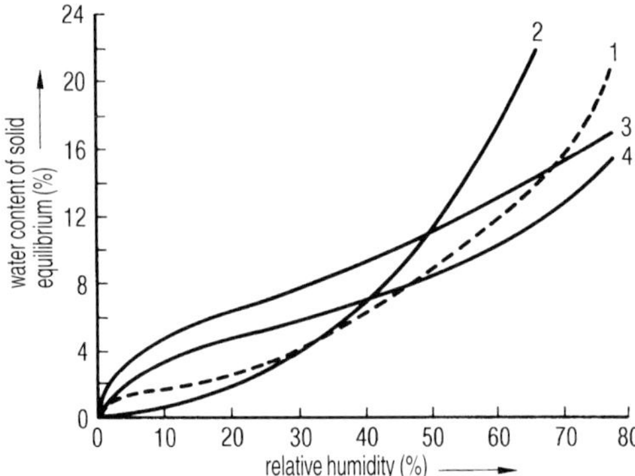

Fig. 4.1. Absorption isotherms: water content in % of solids as a function of the relative humidity in air at 22 °C.
1, green pepper; 2, peaches; 3, potatoes; 4, beef meat (Fig. 7b from [4.5]).

the end product. Slow freezing has the advantage that it results in large crystals, which sublime more rapidly. However, there are two disadvantages: A noticeable freeze concentration of the still unfrozen solution leads to highly viscous, glass-like solidified inclusions between the ice crystals; and a destruction of cell structure by the large crystals. In general, freezing as rapidly as possible is recommended, e. g. 0.5 to 3 cm/h [4.6] or in other terms 0.5–1 °C/min.

The possible drying times for the main drying are estimated in Section 1.2.1 and complemented by examples. The decisive qualities are the heat transfer coefficient from the shelf to the sublimation front of the ice (K_{tot}). The heat conductivity in the product does normally not play an important part (see Fig. 1.67), except that a granulated product is dried from the surface to the center (see Fig. 1.68). The shortest possible main drying time can be estimated with 5 or 10 % error, if the dimensions of the product and the maximum tolerable T_{ice} (e. g. –10 °C) are given (Eq. (12), (12 a–c) in Section 1.2.1).

The storability of the dried product depends to a large extend on the selected type, e. g. strawberries, carrots and green beans [4.7]. For meat, the fat content can be important. Karel [4.8] studied the influence of the water content in stored dried food, and found that not only was the amount of water of influence, but also the kind of bond to the solids. This link can be described by adsorption isotherms, as shown in Fig. 4.1. In food technology, the bond of water is often given by the term water activity, a_w:

$$a_w = p/p_s$$

p = partial pressure of water of the food
p_s = water vapor pressure of pure water at the given temperature

Figures 4.2.1 and 4.2.2 [4.9] show sorption isotherms: 1, for such products which become less hygroscopic with increased temperature, or 2, for glucose, fats and oils which become more hygroscopic with increased temperature. Since freeze drying lowers a_w values, the growth of bacteria, fungus and yeast below $a_w = 0.8$ is reduced, or impossible. On the other hand, the Maillard reaction increases with decreasing a_w up to a maximum at a_w 0,6–0,7,

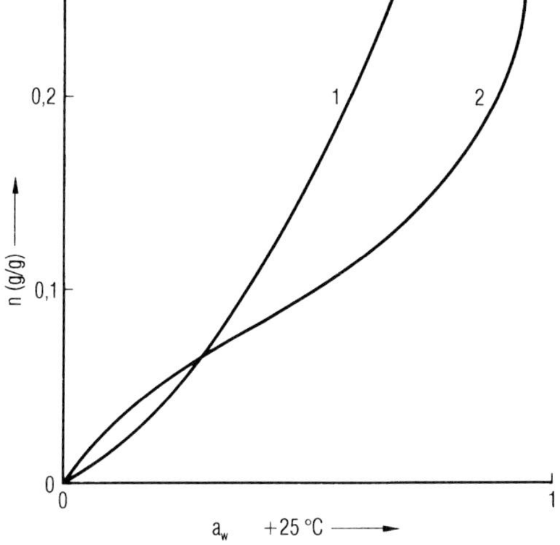

Fig. 4.2.1. Typical sorption isotherms of many food products.
n = mass of water/mass of dried product, a_W at approx. +25 °C.
1, Dried fruits; 2, wheat flower (Fig. 1 from [4.9]).

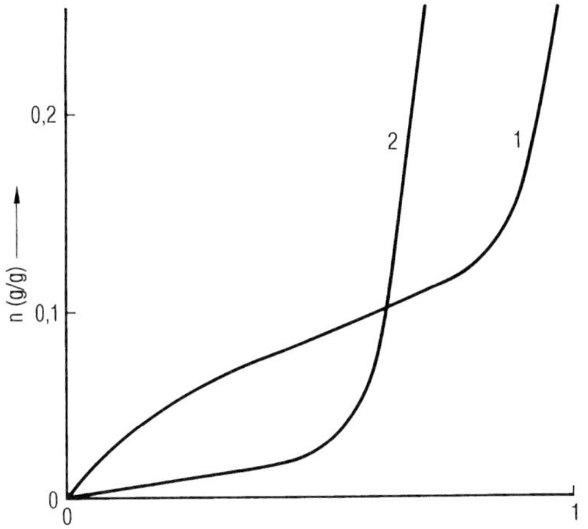

Fig. 4.2.2. Sorption isotherm for glucose. 1, at 30 °C; 2, at 80 °C (Fig. 2 from [4.9]).

and than decreases further (decreasing mobility, as local groups of molecules are formed). The auto-catalytic hydrolysis of fats is still active at low a_w and results in hydrolytic rancidity. The oxidation of fats increases at low $a_{w,}$ while oxidation of proteins decreases. The shelf life of freeze dried food therefore does not necessarily increases at low water activity values. There are several processes which work differently at low a_w. The optimum a_w of a product has to be determined for a maximum storage temperature and its various components. The same applies to the gas mixture in the packing, and specially for the O_2 content.

Poulsen [4.10] proposed removing the first 60 % of water by freeze drying and the remaining 40 % in a dryer with circulating air. In this process, a long rehydration time is accepted. The dry product has a higher density, is less brittle, and is less sensitive to O_2.

The cost advantage was expected to be approx. 20 %. Authors note: The possibility of achieving a comparable throughput per invested capital and the higher density of the dried product by freeze drying at a higher pressure, e. g. 3 mbar is not discussed. At 1.8 mbar TM or approx. 1.2 mbar CA T_{ice} has been approx. –15 °C. At 2 mbar; T_{ice} may be between –10 and –11 °C. The product may shrink to a certain extent as in air drying, but the approx. 70 % higher pressure would increase the heat transfer and reduce the volume of water vapor to be transported. In addition, the change from one plant to another could be avoided, while the cost of water removal from circulating air compared with freezing water depends on the temperature allowed during the removal of the 40 % of water.

4.1 Vegetables, Potatoes, Fruits and Juices

Spiess [4.11] has tested the aptitude of 16 types of vegetables and fruits for freeze drying. Some of the results are shown in Fig. 4.3. With strawberries, carrots, beans and peppers (Fig. 4.4) the type selected decrees (with the exception of peppers) the color, taste and consistency of the freeze dried product. Quick freezing in an air flow of –40 °C is more advantageous than a slow freezing in non moving air of –30 °C. A chamber pressure of more than approx. 1.3 mbar (TM), corresponding to approx. 0.85 mbar CA is, depending on the product, more or less detrimental to the taste, consistency and smell. The assumption, that p_{ch} corresponds approx. to p_S of the ice is not justified, since no water vapor can be transported under that condition from the sublimation front into the chamber. BTM measurements have shown, that the differences between p_s and p_{ch} can be e. g. 0.65 mbar and 0.5 mbar depending on the structure and the thickness of the already dried product. This relation changes when all ice has been sublimated. From the measurements in [4.1] one can conclude, that T_{ice} should be below –20 °C for the studied products. A shelf temperature above +60 °C is only slightly detrimental with peas and mushrooms, but strongly so with strawberries and raspberries.

Fig. 4.3. Survey of freeze dried vegetables and fruits. Blanching time between 2 and 4 minutes (part of Table 1 from [4.11]).

Product	Type	Size	Blanched at 100 °C	Freezing temp (°C)	T_{sh} (°C)	p_{ch} (mbar)	RM (%)
Cauliflower	Unknown	Rosebud	+	–24	60	1.0	2.8
Beans	Wade	1.5 cm	+	–30	70	1.3	2.0
Mushroom	Cultivated	0.3 cm	+	–30	60	1.3	2.2
Peas	Signet	Size III	+	–40	70	1.0	2.8
Pepper	Red, green	1 cm stripes	+	–24	80	1.3	2.4
Strawberries	Senga-S	Halves	–	–27	60	1.0	2.1
aspberry	Unknown	Whole	–	–40	60	1.0	2.1
Currant, red	Unknown	Whole	–	–40	60	1.0	2.0

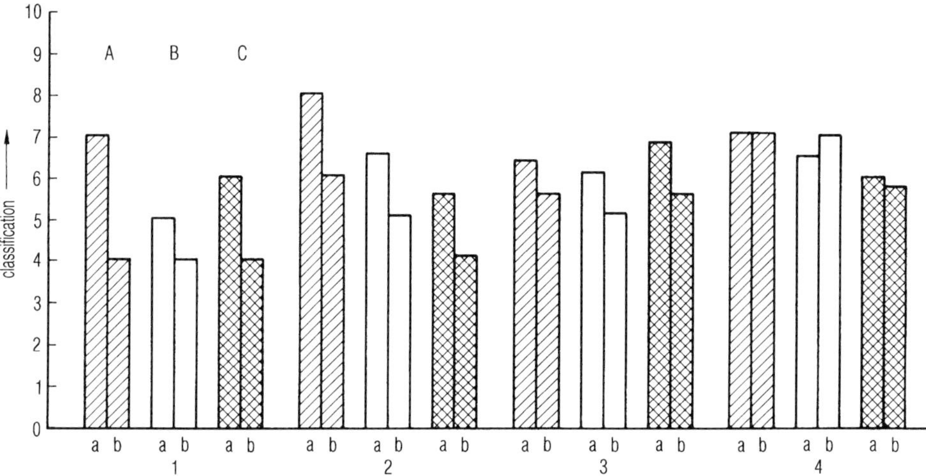

Fig. 4.4. Influence of fruit and vegetable type on color, taste and consistency of freeze dried products.

A Color	B Taste	C Consistency	
1 Strawberries	2 Carrots	3 Beans	4 Paprika
a Senga Sengana	a Nantaiser	a Wade	a Green
b Hummi Trisca	b Dutch early	b Unknown	b Red

classification: 10, excellent; 9, very good; 8, good; 7, rather good; 6, satisfactory; 5, mediocre; 4, small defects (Fig. 2 from [4.11]).

Kapsalis et al. [4.12] showed, that the residual moisture content RM of peas should neither be to small nor too high. During 84 days-storage at +43 °C and RM below 5 %, the thiamine content was barely reduced, but the carotene content fell to approx. 36 %. On the other hand, the thiamine content fell at RM 33 % to 81 %, while 50 % of the carotene content was preserved. These and other reasons not discussed here led Kapsalis to the conclusion, that an optimum RM does generally not exist, but only a desirable RM for one type of product under given storage conditions.

Medas [4.13] and Sauvageot and Simatose [4.14] arrived at similar conclusions for strawberries and orange juice. The freezing rate changed the RM and the rehydration differently for different types. Also, the retention of aroma depends not only on the freezing rate but also on the layer thickness of the juice, the original concentration, and the operation pressure. All these data are different for different types of one product. The retention of five main aroma components of mushrooms has been studied by Kompany and Rene [4.21]. They recommend for maximum retention, that during the first drying stage (until approx. 50 % of water has sublimated) to use a high T_{sh} (+90 °C) and a low p_{ch} ($5 \cdot 10^{-2}$ mbar). The temperature should then be decreased to +60 °C and the pressure increased to 0.5 mbar.

Lime [4.15] described the freeze drying of avocado salad (88.7 % avocado-meat, 4.6 % lemon juice, 0.7 % onion powder, 1.43 % NaCl, and 5.0 % cracker powder) ending with a

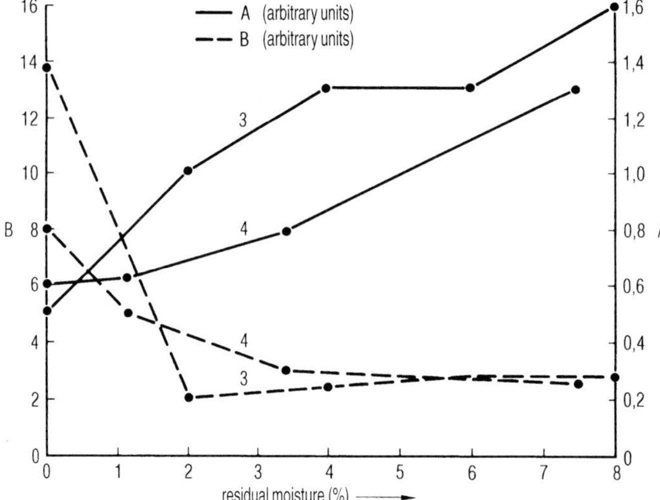

Fig. 4.5. Peroxide and free fatty acids data as a function the residual moisture content after storage of the dried product for 10 days at +37 °C. A, free fatty acids; B, peroxides. Two runs have been carried out, marked as 3 and 4 (Fig. 1 from [4.15]).

RM of less than 1 % up to 8 %. T_{sh} was +38 °C. The dried products were placed in cans at 20 °C room temperature with a maximum RM of 25 %. The product in the cans was under vacuum or air or N_2-stored for 48 weeks at –18 °C, +5 °C, +20 °C and +38 °C. Figure 4.5 shows the opposite effect of RM on the formation of peroxides and free fatty acids. For this product, the RM of 2–3 % would be optimum. In 39 samples it was 3.25 % and 2.51 %, average RM 2.8 %, standard deviation 0.2 %. The taste of the product packed in air, and stored at 20 °C was not more acceptable after 8 weeks. The avocado salad packed in N_2 or under N_2 was considered acceptable after 16–24 weeks. Storage for 48 weeks was only possible at +5 °C.

4.2 Coffee

The freeze drying of coffee extracts has been, and still is, the most frequent application of this process in the food industry. For economic reasons it is best to start with 40 % solid content in the extract. The final product is judged by the following criteria:

Color, bulk density, distribution of grain size, resistance against abrasion, aroma and taste. The first four qualities are mostly influenced by the freezing- and granulation process. Aroma and taste depend largely on the quality of the raw material and the process data during MD. However this statement applies only with certain restrictions. The color can also be influenced by the operation pressure during main drying, accepting a limited softening of the granulate. Partially molten surfaces reflect the light, shine, and have a darker color because the surfaces are more compact.

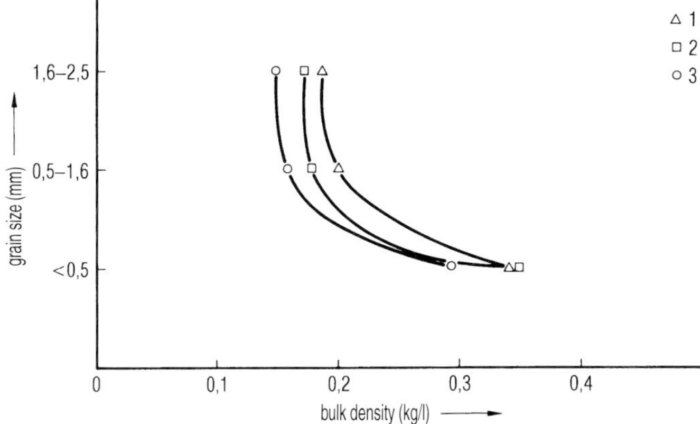

Fig. 4.6. Distribution of grain size as a function of bulk density of freeze dried coffee extract, which has been gassed with different pressures of CO_2 before freezing.
1, without CO_2; 2, 2 bar CO_2; 3, 12 bar CO_2.

Figure 4.6 shows the influence of CO_2 injection into the extract at approx. 0 °C. The CO_2- treated extract has been frozen in 20 min to –40 °C in a plate freezer and then ground. The bulk density can be reduced by the freezing process by approx. 25 %, while the color remains approx. the same. Figure 4.7 [4.16] compares the bulk densities of different grain size distributions by different freezing methods. In a precooler, the ice scraped from a cooled surface is mixed with liquid extract producing a pulp at e. g. –3 °C, which afterwards is frozen in a flow of cold air. The influence of the freezing rate can be seen clearly

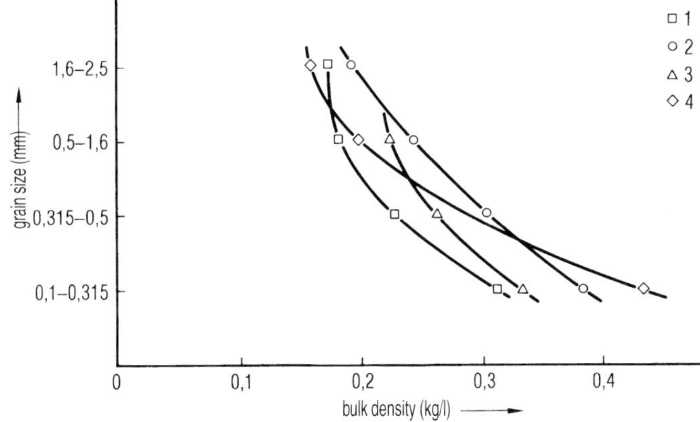

Fig. 4.7. Distribution of grain size as a function of bulk density of freeze dried coffee extract, which has been frozen by different methods.
1, In a plate freezer; 2, on a belt in a flow of cold air; 3, in trays in cold air; 4, partially frozen in a soft-ice machine (continuously scraped cold surface) and finally frozen as in 2.

if the extracts are frozen in cold air flows. Extracts frozen in trays do not show a measurable amount of particles above 1.6 mm, while the size distribution of extracts frozen on belts may be as follows:

2.5	to 1.6 mm	18.5 %
1.6	to 0.5 mm	30.5 %
0.5	to 0.315 mm	17.2 %
0.315	to 0.1 mm	28.8 %
	< 0.1 mm	4.9 %

The grain size distribution can also be changed by the use of different grinding equipment, e. g. by a series of different mills.

Flink [4.17] described the relative retention of volatile components in coffee extracts as a function of the freezing speed and the operation pressure:

Freezing	Relative retention rates (%) [1)] Drying at a chamber pressure (mbar)						
	0.26	0.4	0.53	0.66	0.8	0.92	1.06
Very slowly [2)]	92	96	78	77	66	67	34
Slowly [3)]	100	99	88	82	91	82	35
Slowly and foaming [4)]	67	61	49	53	57	44	63
Fast [5)]	47	53	38	38	44	35	36
Fast and foaming [6)]	48	–	42	42	43	32	29

1) Highest retention rate = 100 %
2) 15 mm layer in 24 h down to –10 °C, in 48 h down to –25 °C, in 72 h down to –40 °C
3) 15 mm layer in aluminum trays in resting air down to –40 °C
4) Coffee extract foam in a 'soft-ice-machine', and then frozen in 15 layers in resting air down to –40 °C
5) Coffee extract sprayed on a drum freezer at –52 °C with a layer thickness of 3 mm
6) Coffee extract with CO_2 foamed at +20 °C and than sprayed as in 5).

The frozen extract was ground and only the grain sizes between 1.2 to 2.7 have been filled in aluminum trays with a layer of 15 mm. During freeze drying the trays were placed between two shelves, the temperature of which were controlled in such a way that the surface never exceeded +40 °C. The pressure was measured by an Alphatron (ionization of the gas by α-radiation) and a second instrument calibrated with water vapor. The table shows that slow freezing and drying up to a pressure of 0.92 mbar results in the highest retention rates, while quickly frozen foam, and quick freezing in general, provide less favorable data. The gas mixture of the reconstituted freeze dried extract was analyzed at 80 °C by gas chromatography. The surface area below the 'peaks' in the diagram was used as measure of the retained volatile components.

Downey and Dublin [4.22] used near-IR reflectance spectroscopy to identify the origin of the coffee beans, which have been used for the freeze dried extracts: Pure Arabica, pure

Robusta or blends of these both. When reconstituted freeze dried extracts were analyzed, 56 test samples out of 65 were classified correctly.

The selected examples demonstrate that there is no ideal process for the freeze drying of coffee extract. One must compromise in the various steps of the whole process, from bean selection, roasting, extraction, freezing, freeze drying and packing to achieve the desired quality and cost of the freeze dried coffee.

4.3 Eggs and Rice

Pyle [4.18] described the freeze drying of whole eggs which have been frozen in ribbed aluminum trays in a cold air flow and freeze dried in two tunnels. The dry product had less than 1.5 % RM and was packed under N_2. The egg powder could be stored for 12 months at room temperature and has, after reconstitution with water, its original working qualities.

Mitkov et al. [4.19] produced RM data and DTA measurements of egg white, yolk and whole egg of fresh eggs, from which no different process data for the three products are deduced. It was shown that freezing at 6 °C/min led to a more homogeneous end product than freezing at 0.2 °C/min. The conclusions of the authors, that egg products have eutectic zones between –29 °C and –34 °C and can be dried at approx. –20 °C cannot be understood from the published data. The course of drying showed that T_{pr} during the first third of the run was approx. –30 °C; –20 °C was only used after 50 % of the drying time had been completed.

Shibata [4.20] described the freeze drying of boiled rice as a method to produce convenient food (fast food). The paper is quoted as the advantages of food freeze drying are summarized within it: Structure and taste of the boiled food can be restored in approx. 3 min by adding hot water of +80 °C, and the dried product can be stored well.

References for Chapter 4

[4.1] The Wall Street Journal, 14 July 1960

[4.2] Tchigeov, G. B., Leningrad Technological Institute of Refrigeration Industry, USSR: Progress in refrigeration and technology, Vol. III, 1960

[4.3] The Times, 20 September 1960

[4.4] Oetjen, G. W.: Die Entwicklung der Gefriertrocknung von Nahrungsmitteln. Forschungskreis der Ernährungsindustrie e. V. Bonn, October 1962

[4.5] Oetjen, G. W.: Economical aspects of industrial freeze-drying. Le Vide, No. 102, p. 531–540, 1962

[4.6] Spiess, W.: Verfahrensgrundlagen der Trocknung bei niedrigen Temperaturen. VDI-Bildungswerk, BW 2229, 1974

[4.7] Wolf, W., Jung, G., Spiess, W.: Technologische und technische Fragen bei der Gefriertrocknung von Lebensmitteln. Energiewirtschaft/Lebensmitteltechnik 7, p. 454, 1972

[4.8] Karel, M.: Stability of low and intermediate moisture foods. 6. International Course of Freeze-Drying and Advanced Food Technology, Bürgenstock (Switzerland), 1973

[4.9] Loncin, M.: Basic principles of moisture equilibria. 6. International Course of Freeze-Drying and Advanced Food Technology, Bürgenstock (Switzerland), 1973

[4.10] Poulsen, M. P.: Economy of combined freeze and air drying. International Institute of Refrigeration <IIR> Paper 314, (Conference Montreal 1991)

[4.11] Spiess, W.: Qualitätsveränderungen bei der Gefriertrocknung von Gemüse und Obst. Kältetechnik 16, p. 349–358, 1964. C. F. Müller Verlag, Hüthing GmbH, Heidelberg

[4.12] Kapsalis, J. G., Wolf, M., Driser, M., Walker, J. E.: The effect of moisture on the flavor content and texture stability of dehydrated foods. Ashrae Journal, 13, p. 93–99, 1971

[4.13] Medas, M., Simatose, D.: Freeze-drying and reconstitution of raspberries, influence of the chemical content and variety. International Institute of Refrigeration <IIR> (XIII Congress, p. 605–610, Washington, 1971

[4.14] Sauvageot, F., Simatose, D.: Some experimental data on the behavior of fruit juice of volatile components during freeze-drying. International Institute of Refrigeration <IIR> (Comm. X, Paris, 1969)

[4.15] Lime, B. J.: Preparation and storage studies of freeze-dried avocado salad. Food Technology V. 23, p. 43–46, 1969. Copyright © 1969 Institute of Food Technologists, Chicago IL, USA

[4.16] Unpublished data of the author

[4.17] Flink, J. M.: The influence of freezing conditions on the properties of freeze dried coffee. International Course on Freeze-Drying and Advanced Food Technology, Bürgenstock, Switzerland, 1973

[4.18] Pyle, H. A. A.: Egg pulp freeze-dried with absorption refrigeration. Australian Refrigeration, Airconditioning and Heating, 1969

[4.19] Mitkov, S., Bakalivanov St., Nikolova, T., Vatinov, T.: Determination of optimal parameters in lyophilization of egg white, egg yolk and a mixture of both. International Institute of Refrigeration <IIR> (XIII th. Congress, p. 739–748, Washinton, 1973

[4.20] Shibata, T.: Freeze-drying of cooked rice. Jpn. Kokai Tokyo Koho, 1991

[4.21] Kompany, E., Rene, F.: Optimal conditions for freeze drying of cultured mushrooms *(Agaricus bisporus)* – study of aroma retention. Recent Prog. Genie Procedes, 7 (30) Etudes et Conception d'Equipements, p. 267–272, 1993

[4.22] Downey, G., Dublin, I.: Authentication of coffee bean variety by near-infrared reflectance spectroscopy of dried extracts. J. Sci. Food Agric. 71 (1), p. 41–49, 1996

5 Metal Oxides, Ceramic-powders

Dogan and Hausner [5.1] presented a survey of the applications of freeze drying in ceramic powder processing, the three main objectivesof which have been pursued:

- Freezing and freeze drying of metal salt solutions, to obtain a homogeneous mixture of different components.
- Freezing and freeze drying of precipitates to minimize agglomeration during drying.
- Injection molding of powder-liquid mixtures at temperatures below the freezing point of the liquid and freeze drying of the frozen parts.

The metal solutions are sprayed into cold liquids for rapid freezing, after which the droplets are freeze dried and decomposed to metal oxides. Due to the homogeneous distribution of the components, the reactions in the solid state occur at lower temperatures compared with conventionally produced powders.

The drying of the precipitates conventionally leads to hard agglomerates, which densify still more during the calcination to oxides. In the freeze dried precipitates only soft agglomerates are formed with fine pores, as shown in Fig. 5.1. During sintering of this product, a high relative density can be achieved at substantially lower temperatures (300–400 °C), as shown in Fig. 5.2.

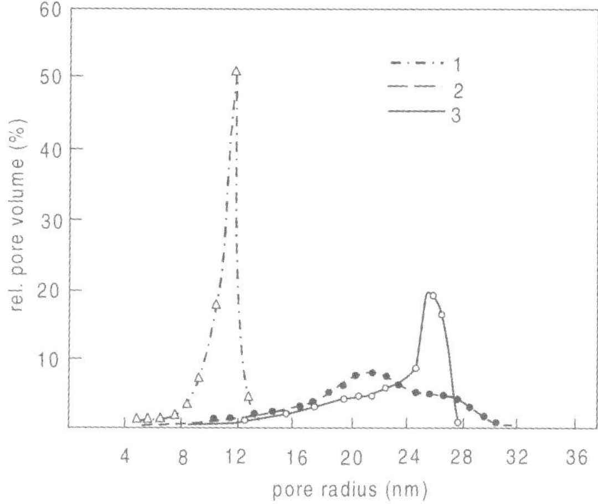

Fig. 5.1. Pore-size distribution of MgO-compacts.
1, freeze dried after precipitation and washing; 2, chamber drying at 120 °C after precipitation and washing; 3, commercially available product (Fig. 1 from [5.1]).

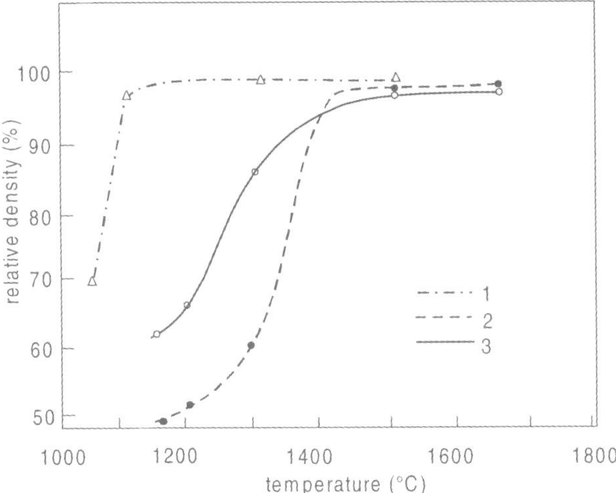

Fig. 5.2. Relative density of sintered MgO-compact. 1, 2, 3 as in Fig. 5.1 (Fig. 2 from [5.10]).

Conventionally, a large amount of organic binders must be used for the injection molding of ceramic parts. By injecting a mixture of water and organic solvents with the ceramic powder into the cold mold, small amounts of organic binders are necessary, allowing high heating rates during firing.

Reetz and Haase [5.2] used different freezing rates to freeze ZrO_2, and found slow freezing of this product to result in better technological qualities, e. g. free-flow and sinter ability than quick freezing. On the other hand, complex Zn- solutions can only be frozen quickly to arrive at a product homogeneous in chemical structure and grain size distribution.

Nagai and Nishino [5.3] froze solutions of K_2CO_3, Na_2CO_3, $MgSO_4$ and $Al_2(SO_4)_3 \cdot 18$ H_2O by spraying them into LN_2, and freeze dried the frozen droplets at 0.13 mbar and 100 °C for 2–3 days. The dry product consisted of hollow and porous balls with a diameter from 150 to 200 μm, and had to be milled for 16 h in a ball mill to make it workable. The calcinated product had a spherical (0.1 μm diameter) or plate-like form of 0.2 to 2 μm.

Torikai et al. [5.4] sprayed aqueous solutions of Mn^{2+}, Co^{2+}, and Ni^{2+} sulfate into LN_2 and produced uniform, spherical particles, which were freeze dried in a bottle at –80 °C. This was connected to two cold traps (LN_2), a diffusion pump, and a backing pump. Freeze drying of approx. 10 g took 2–3 days. The dry product in the form of $Mn_3CO_2Ni\ (SO_4)_6 \cdot$ 15–16 H_2 O could be transformed into a fine spinelle powder at 900–1000 °C during 1 hour.

Milnes and Mostaghaci [5.5] compared the consequences of different drying methods on the density, the sinter rate and micro structures of sublimated TiO_2 suspensions. Evaporation of water in a micro-oven and by radiation heating leds to strongly bound agglomerates, while freeze drying resulted in softly bound secondary clusters. The freeze dried powder reached in 2 h of sintering 98 % of the theoretical density, while differently dried powders needed twice as much time and had a less fine microstructure.

Kimura et al. [5.6] and Ito et al. [5.7] produced superconducting $YBa_2Cu_3O_7$-delta and $YBa_2Cu_3O_7$-x ceramic plates respectively from freeze dried carbonate- and nitrate solutions respec6tively of ytterbium, barium and copper. After calcination and sintering small plates of approx. 2 · 5 · 20 μm with high packing density were formed without pressing. [5.6] having a superconductivity equal to that of monocrystaline $YBa_2Cu_3O_7$-x [5.7].

Lacour et al. [5.8] produced the starting material for YBaCuO from freeze dried acetate solutions. The strong dependence of the electrical qualities on the salt concentrations was considered as surprising. (Authors note: The salt concentration strongly influences the structure of any freezing product.)

Kimura et al. [5.9] produced spherical $YBa_2Cu_3O_7$-delta material with a diameter between 20–30 μm from freeze dried nitrate solutions and its thermal decomposition.

Kondou et al. [5.10] compared the production of $Pb(Zr_xTi_{1-x})O_2$ (PZT) by solid-state reaction between TiO_2, ZrO_2PbO and the freeze drying of the nitrate salt solution. The solid state reaction requires 1100 °C, but the transformation of the freeze dried nitrates only 580 °C. Furthermore, the freeze dried product could be sintered better and showed at the Curie-temperature a two-fold larger dielectric constant than the PZT produced by solid-state reaction.

References for Chapter 5

[5.1] Dogan, F., Hausner, H.: The role of freeze-drying in ceramic powder processing. Ceram. Trans., 1 (Ceram. Powder Sci., PT.A), p. 127–134, 1988. Reprinted with permission of the American Ceramic Society, Post Office Box 6136, Westerville, Ohio 43086-6136. Copyright 1988 by The American Ceramic Society. All rights reserved.

[5.2] Reetz, T., Haase, I.: The Influence of freezing process on the properties of freeze-dried powders. Ceram. Powder Process. Sci., Proc. Int. Conf., p. 641–648, 1988. Edited by: Hausner, H., Messing, G. L., Hirano S.

[5.3] Nagai, M., Nishino, T.: II Aluminia ceramics fabricated by the spray-froze/freeze-drying method. International Institute of Refrigeration <IIR> (Com. C1, p. 186–190, Tokyo, 1985

[5.4] Torikai, N., Mejuro, T., Nakayama, H., Yokogama, Y., Sasamoto, T, Abe, Y.: Preparation of fine particles of spinel-type Mn-Co-Ni-Oxides by freeze-drying. International Institute of Refrigeration <IIR> (Com. C1, p. 177–183, Tokyo, 1985

[5.5] Milne, S. J., Mostaghaci, H.: The Influence of different drying conditions on powder properties and processing characteristics. Mater. Sci. Eng., A, A130 (2), p. 263–271, 1990

[5.6] Kimura, Y., Ito, T., Yoshikawa, H., Tachiwaki, T., Hiraki, A.: Growth and characterization of homogeneous yttrium-barium-copper-oxide ($YBa_2Cu_3O_7$-delta) powders prepared by freeze-drying method. Jpn. J. Appl. Phys., Part 2, 29 (8), L. 1409 – L 1411, 1990

[5.7] Ito, T., Kimura, Y., Hiraki, A.: High-quality yttrium barium copper oxide ($YBa_2Cu_3O_7$-x) ceramics prepared from freeze-dried nitrates. Jpn. J. Appl. Phys., Part 2, 30 (7B), L 1253 to 1255, 1991

[5.8] Lacour, C., Laher-Lacour, F., Dubon, A., Lagues, M., Mocaer, P.: Freeze-drying preparation of yttrium barium copper oxide. Correlations between electrical and microstructural properties. Physica C (Amsterdam), 167 (3–4), p. 287–290, 1990

[5.9] Kimura Y., Ito, T., Yoshikawa, H., Hiraki, A.: Superconducting Yttrium Barium, Copper Oxide ($YBa_2Cu_3O_{7-}$delta) particles prepared from freeze-dried nitrates. Jpn. J. Appl. Phys., Part 2, 30 (5A), L 798 to L 801, 1991

[5.10] Kondou, S., Kakojawo, K., Sasaki, Y.: Synthesis of Pb $(Zr_xTi_{1-x})O_3$ by freeze drying method. Nippon Kagaku Kaishi, (7), p. 753–758, 1990

6 Trouble Shooting and Regulatory Issues

6.1 Trouble Shooting

In Section 2.2.8, possible failures during the freeze drying process are classified in four categories, and the preventions and necessary actions briefly discussed. In this chapter, some unexpected or undesirable events are studied which, by experience may happen. The problems listed here are selected from the course of the freeze drying process. A breakdown of single components, for example pumps, compressors or valves are not included in this chapter. The list will be incomplete, but an attempt has been made to mention some of the more frequent events. The problems with leaks and their hunting is discussed in Section 2.2.8.

6.1.1 Prolonged Evacuation Time

The evacuation of the plant takes longer than calculated from the volume and the pumping capacity, in spite of the regular leak check. The event is most likely related to the formation of ice, which is condensed during loading on the precooled shelves. The extent of this depends on the shelf temperature during loading, and the moisture content of the gas in the chamber. One of the possible course of actions could be as follows: Check the condenser temperature in relation to the pressure reached. If the condenser temperature e. g. is –42 °C, $p_s \approx 0.1$ mbar, but the lowest pressure reached in the chamber is only ≈ 0.5 mbar, then the condenser capacity is not the bottle neck, but the slow sublimation of some ice. Raising T_{sh} from e. g. –40 °C could be started very slowly in such a way that the pressure in the chamber does not exceed p_s of the maximum tolerable T_{ice}, e. g. –22 °C ($p_s \approx 0.85$ mbar). This may take one or more hours, depending on the heat transfer to the undesired ice. When all excess ice is removed, the pressure will drop quickly to the value close to p_s of the ice on the condenser and the normal cycle can be started. If, in another example, the condenser temperature is the limiting factor, the condenser is at capacity limit and one has to delay the heating until the condenser temperature falls. All this can be avoided, if the door of the chamber is only partially opened during loading (Section 2.4) and a small pressure of dry gas is kept in the chamber.

6.1.2 Sublimation Front Temperature too High

After evacuation and at the start of heating, the trend of T_{ice} gets to close to the maximum tolerable T_{ice}. A reduction in operation control pressure (p_c) will immediately stop this trend (Fig. 2.37.1). (If the desired T_{ice} is not reached p_c could be raised.) As shown in Fig. 2.37.2 the slope of the function $T_{ice} = f(p_c)$ can become unfavorably steep. In this case, the shelf temperature should also be lowered or raised, but the effect of this change will take time before an equilibrium state is reached again (e. g. 0.5–1 h) depending on the heat transfer conditions and the heat capacity of the shelves and the heat transfer medium.

6.1.3 Sublimation Front Temperature Irregular

T_{ice} data fluctuate and the standard deviation of the measurements becomes larger than approx. 0.3–0.5 °C. The monitored p_c, T_{sh} and T_{co} are stable, but the pressure rise measurements change and vary as shown in Fig. 2.18.2. Most likely the valve between chamber and condenser does not close reproducibly. If the drive of the valve has been checked externally as far as possible, the run cannot be finalized automatically and hand operation should be used, if the time course of the program has been already established.

6.1.4 Slow Pressure Increase in the Chamber During Main Drying

In spite of a constant or decreasing condenser temperature, the pressure in the chamber rises after main drying has started correctly. This may be due to the development of a leak, though this unlikely in a plant leak-tested before the run, and while a newly developed leak may alter the pressure slightly. In addition a leak which worsens constantly during the run is possible, though even unlikely. Among other reasons, rising pressure may occur due to a constant increase in the pressure of permanent gases. This will increasingly hinder transport of water vapor and reduce the condensing effectiveness of the condenser (Fig. 2.17). The increase may occur for two reasons: After the start of MD, air dissolved or included in the frozen ice is freed as the ice is sublimated. The amount of included gas can vary by a factor of almost 100, depending on the material used and its fabrication history (see Section 1.2.1). If the capacity of the vacuum pumps is smaller than the amount of freed gas at the desired pressure, the condenser is increasinly filled by permanent gases. The other possibility is a misplaced suction pipe of the vacuum pump at the condenser. The common permanent gases have a higher density than water vapor. If the vacuum connection is not in the lowest area of the condenser, the gas will slowly fill the condenser chamber from the bottom up to the suction level of the pump, thus reducing condenser efficiency to a greater or lesser degree. The suction pipe should be connected as shown in Fig. 2.18.

6.1.5 Stoppers 'Pop Out' or Slide into the Vials

The 'pop out' of stoppers occurs mostly during evacuation. If the product is not completely frozen and contains highly concentrated, but not solidified inclusions, their water content may evaporate explosively when the vacuum is applied. This abrupt evaporation will also blow some product particles to the walls of the vial, as can be seen in the vials after drying.

Stoppers may also slide into the vials and cause their virtual closure. This can happen during freezing of the product, especially if the shelf temperatures are very low. The dimensions of the stoppers need to be tested not only for the pressure to close them after drying, but also for shrinking at low temperatures.

6.1.6 Traces of Highly Volatile Solvents (Acetone, Ethanol)

The presence of such traces always requires some special steps, depending on the amount of the volatile component. If the traces cannot be removed before freezing and drying, they can: (i). influence the structure during freezing; (ii). disturb the condensation of ice; and (iii) contaminate the oil in the vacuum pump. As shown in Fig. 6.1, the vapor pressure of ethanol and acetone are at –70 °C approx. $2 \cdot 10^{-2}$ and $3 \cdot 10^{-1}$ mbar (ice $3 \cdot 10^{-3}$ mbar) respectively. Ethanol melts at –114 °C and acetone at –95.5 °C. If the amounts are small, a very quick freezing might help to distribute the solvents very uniformly in the ice. If the amounts are large enough, they will form a veil on the condenser surface, disturbing the efficiency of the condenser, drip as a liquid from the condenser surface, and evaporate from the warmer condenser wall. Furthermore, gauges with hot wires (thermal conductivity gauges) should not be used and at last the oil of the vacuum pumps must be recycled and cleaned or it will be contaminated by the solvents and the pumps will no longer discharge the solvent vapors. These problems may be reduced by placing a LN_2-cooled trap between the condenser and vacuum pump set. Of course some ice will also condense in the trap, but this amount will be small, as the water vapor pressure of ice in the condenser is very small compared with the vapor pressure of the solvent at the condenser temperature. However this trap solves the problems in the condenser, mentioned before, only if it is designed also for this task. A substantial effort is justified to remove such solvents before freezing starts.

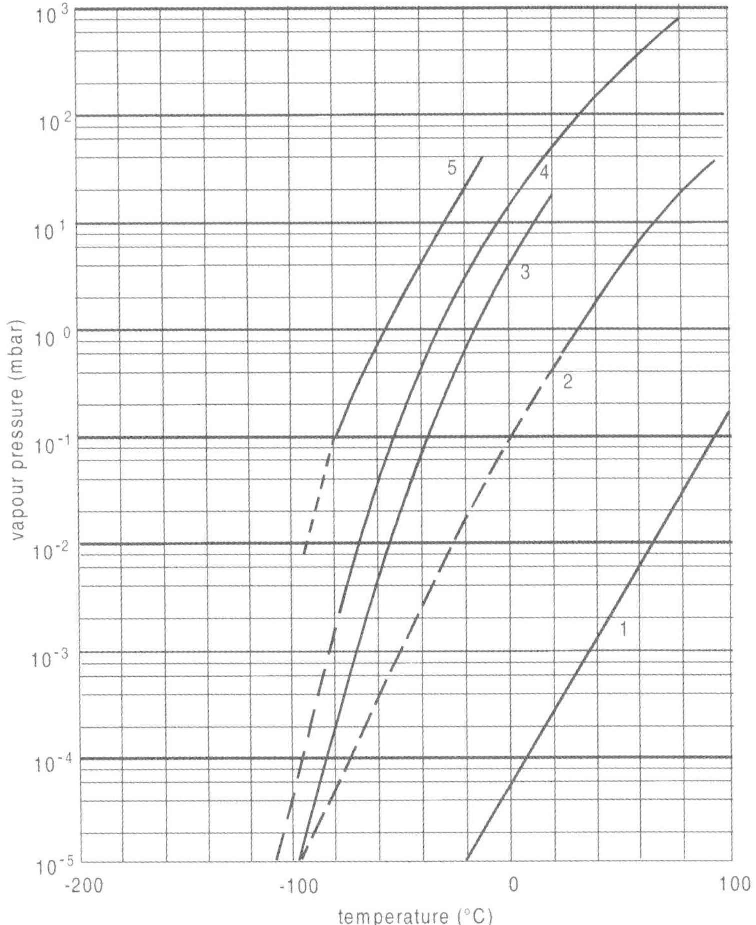

Fig. 6.1. Vapor pressures of solvents as a function of temperature.
1 glycerin, 2 DMSO ($C_3H_8O_3$), 3 water, 4 ethanol, 5 aceton.

6.1.7 Different Structure of the Dried Product in the Center and Border of a Shelf

The outer vials are influenced (if the shelf temperature is uniform) by a different temperature of the walls and door of the chamber. If the chamber walls and the door are not kept at shelf temperature, the outer vials must be shielded or they may be too warm during freezing (e. g. freezing differently) or too cold during secondary drying (see Fig. 1.68), and this may lead to a different residual moisture content, from that in inner vials.

6.2 Qualification and Validation of Processes and Installations

In the Validation Documentation Inspection Guide, US Department of Health and Human Services, Food and Drug Administration, 1993 the process validation is defined as:

- Establishing documented evidence, which provides a high degree of assurance that a specific process will consistently produce a product meeting its pre-determined specifications and quality attributes.
- The Guide to Inspections of Lyophilization of Parenterals, published by the US Food and Drug Administration, July 1993, contains among others the chapters 'Lyophilization Cycle and Controls', 'Cycle Validation' and 'Lyophilizer Sterilization/Design'.

In the European Union, the directive 91/356 EEC provides the principles and guidelines of Good Manufacturing Practice (GMP). In a series of Annexes, supplementary guidelines are covered, but until 1996 only 'Annex 1: Manufacture of sterile medical products' has been revised. In spite of all these guidelines and annexes, Monger [6.1] summarized the situation for the user of freeze drying processes and installations as follows: 'It might be expected that some substantial guidance would be provided. Regrettably, this is not so.'

Powell-Evans [6.2] provided a range of advice on how to 'streamline validation', which he calls 'one of the most time-consuming and costly exercises faced by pharmacetical manufactures'.

In following text an attempt is made to give an example of the types of documentation that must be prepared within the validation master plant for the installation qualification (IQ) and the operational qualification (OP). The examples given are restricted to the process and installation engineering aspects, and exclude the many other aspects, e. g. sterility, biological or chemical requirements, corporate policies or the production environment.

The information described are organized in six categories:

6.2.1 Quality of the product to be manufactured
6.2.2 Description of the process developed for the manufacturing of the product
6.2.3 Description of the production installations and their handling
6.2.4 Equipment performance tests
6.2.5 Qualification of the installations to document the ability of the equipment to operate the process described in Section 6.2.2
6.2.6 Documentation of the quality of the manufactured product and its comparison with Section 6.2.1.

6.2.1 Quality of the Product

The qualities of the product and their tolerances must be deduced from the protocols of the development. The methods to measure the qualities must be described as documented in the development protocols. The directives (a) to (t) are not detailed here. They must be written in accordance with company policies and the Quality Systems Manual for each individual production. The tolerances are given in brackets {e. g. ±1 °C}.

6.2.1.1 The product to be frozen can be stored and handled at maximum XX °C {±YY °C} for a period of maximum XX hours {+YY hours}. If the temperature limits and/or time limits are exceeded, directive (a) must be applied.

6.2.1.2 The product subcools by 4 to 8 °C, if cooled at a cooling rate of 1 °C/min {±0.2 °C/min}. Warning: If subcooling does not occur, the frozen product may have an undesirable structure, e. g. highly concentrated, highly viscous inclusions, which might dry slowly and will raise T_c. If the subcooling cannot be confirmed by measuring T_{pr} during freezing (e. g. due to automatic loading), the method given in directive (b) is to be applied. If a subcooling between 4 and 8 °C cannot be proven, directive (c) is to be applied.

6.2.1.3 After cooling the product to minimum –40 °C {+0 °C, –5 °C} the product has a uniform, fine structure. This product could be well but slowly dried. To induce some growing of the ice crystals the product must be kept for XX min {–0 min, +YY min} at a temperature of e. g. –30 °C {–3 °C, +0 °C}. With this tempered structure, t_{md} = XX hours {±0.5 h}. If t_{md} is shorter than or exceeds this time, directive (d) is to be applied.

6.2.1.4 The collapse temperature of the product frozen, treated as described in Section 6.2.1.3 T_c = –28.0 °C {±1,5 °C}.

6.2.1.5 The product temperature during secondary drying must reach $T_{pr,sd}$ +35 °C {+0 °C, –4 °C}. If $T_{pr,sd}$ is not measured directly, directive (e) must be applied.

6.2.1.6 At the end of secondary drying, dW of the product must be 0.6 % {±0.3 %} at +35 °C. If 0.6 % is not achieved, follow directive (f). The residual moisture content (RM) of the product in the closed vials measured by the methods given in directive (g) must be 1 % {standard deviation of all vials measured 0.4 °C or smaller}.

6.2.2 Description of the Process Developed for Manufacture of the Product

The process has been developed in a pilot plant with the following specification: Volume of the drying chamber V, shelf area A, shelf temperature controlled between –50 °C to +70 °C, maximum cooling rate of the empty shelves down to –40 °C is 2.5 °C/min, heating rate from 0 °C to +40 °C is 3 °C/min, temperature variation on each shelf < ±1 °C, roughness of the shelf surface in RA-Standard 1.5 (1.0–1.5 µm), condenser T_{end} –65 °C, surface of the condenser coils XX m^2, vapor flow density (s$^\#$) from the chamber to the condenser at XX $p_{ch,H2O}$ = XX (g/s cm^2) and at XX $p_{ch,H2O}/10$ = XX/30 (g/s cm^2) (*Note*: If the pressure decreases by a factor of ten, the vapor flow densiy decreases more e. g. by a factor of 30; see Fig. 1.89), vacuum pumpset two-stage, with gasballast at XX p_{ch} S = XX (m^3/h). All vacuum gauges measure the changes of a capacitance, type CA, T_{ice} by automatic BTM, DR by automatic pressure rise measurement, dW calculated by computer from DR data, automatic operation pressure control operates from 0.02 to 0.5 mbar, leak rate of chamber and condenser < XX (mbar L/s).

Filling of vials: Vials of type (B) of the manufacturer (C) have been selected as documented in file (D) and used exclusively during the process development. The required subcooling and the cooling rate are only ensured if the filing height in the vials B is XX mm {–0 %, +2 %} The time for filling and loading of XX vials in the pilot plant has been XX hours {±10 %}. For each development run XX number of vials have been used, filling 50 % of the available shelf area.

The chosen T_{ice} has been reached in the pilot plant at p_c mbar {±5 %} and at T_{sh} = –10 °C. *Note*: In a freeze drying plant with different dimensions and different number of vials p_c will be slightly different to achieve the desired T_{ice}. Therefore p_c must bee modified accordingly. In the pilot plant, a change of p_c to 1.1 p_c {±5 %} increases T_{ice} by 0.5 °C. In the pilot plant, p_c has been controlled by closing the valve between the condenser and the vacuum pump set.

After XX h of main drying, T_{ice} decreases in two measurements in succession by more than 2 °C from the maximum of all average T_{ice} calculated (called max. $T_{ice,ave}$) and the controlled operation pressure p_c can no longer be kept constant. The control of the operation pressure is terminated automatically. The chamber pressure decreases during several minutes to less than $p_c/8$. At that time, T_{sh} is automatically increased to +33 °C {±1 °C} and the DR – dW measurement switches on. After XX hours, DR shows a value of 0,05 %/h and dW was calculated as 0.8 %. Since the preset dW must be smaller than 0,8 % but larger than 0,6 % the secondary drying has been continued until dW = 0.71 has been reached. At that time the plant has been vented with gas G (specification in directive (h) to atmospheric pressure, and the vials have been closed. The data of three repeated runs are documented in file (E) and the data of p_{ch}, p_{co}, T_{ice}, T_{pr}, T_{sh}, DR and dW as a function of drying time shown as graphs.

In three additional test runs (each repeated three times), the following has been demonstrated:

1. p_c· has been increased until T_{ice} reached –28 °C {standard deviation 0.34 °C}. This happens, if p_c is increased by a factor 1.25 (e. g. from 0.10 mbar to 0.125 mbar). This higher T_{ice} decreases t_{md}, but cannot be accepted since T_c = –28 °C ± 1.5 °C. For safety reasons p_c + 0.5 % must remain the maximum tolerance (in this example).
2. The main drying has been terminated, when T_{ice} had decreased for the first time from max. $T_{ice,ave}$ by 1.5 °C followed by a secondary drying as described. The desired dW was not reached measurably early than by the change of 2 °C in two successive measurements.
   The decision remains, to change after two successive measurements showing T_{ice} reduced by 2 °C.
3. T_{sh} has been increased to +40 °C, but only until T_{pr} reached +25 °C, at which time T_{sh} was reduced to +30 °C. By this step the total drying time is reduced by 15 %.
   The visual inspection of the dried product from XX vials (cut open) did not show signs of collapsed occlusions. The solubility of the product in XX tested vials satisfied the directive (i).

Based on the product quality (Section 6.2.1) and the process development above, the following process has been adopted:

1. The filling height in the vials type B must be XX mm {–0 %,+2 %}
2. The filling and transfer times of the vials as carried out in the development is well below the specification in Section 6.2.1.1 and the temperature during this time did not exceed the specified maximum as proved in file (E).
3. The subcooling in the documented runs was between 4 and 6 °C, (average 5.3 °C). If the subcooling of 4–8 °C is not achieved, follow directive (c). If the subcooling cannot be measured during production, directive (b) must be applied.
4. The cooling rate down to below –40 °C must be between 0.8 and 1.2 °C/min. If it is outside this range, directive (c) must be followed.
5. After T_{pr} is below –40 °C it has to be increased to –30 °C {+0 °C, –3 °C} for XX min {–0 min, +YY min}.
6. The controlled operation pressure during the main drying is p_c mbar {–0 %, +10 %}, T_{sh} = –10 °C. With these data T_{ice} has been –29.5 °C {standard deviation 0.38 °C) in the pilot plant. The production department must check the function T_{ice} = f (p_c) for the production plant, and modify p_c if necessary to achieve T_{ice} = –29.5 °C{standard deviation < 0.4 °C}
7. The main drying time has to be t_{md} = XX hours {±0.5 h} If t_{md} is shorter or longer, directive (d) has to be applied.
8. The main drying is to be terminated automatically if the measured T_{ice} becomes (by two successive measurements) 2 °C smaller than max. $T_{ice,ave}$. At that time, the automatic pressure control is terminated and T_{sh} raised to +35 °C {+0 °C, –1 °C}. After the pressure control has been terminated, the pressure must drop to $p_{c/}8$ within 10 min. If the pressure remains higher, or it takes a longer time to reach this level, directive (j) must be followed.

9. During the secondary drying, $T_{pr,sd}$ must reach +33 °C {+2 °C, –3 °C}.
10. If $T_{pr,sc}$ is not measured directly, directive (e) has to be applied.
11. The end of SD is reached, when dW is 0.6 % {±0.3 %). If the dW is not reached in XX hours {±0.5 h) the drying can be prolonged by YY h. If dW is not reached at that time, follow directive (f).
12. After the drying is terminated, the valve between chamber and condenser is closed and the chamber vented with gas (G), specification in directive (k) to atmospheric pressure. Thereafter, the vials are closed with the stoppers, which have been treated as specified in (l).
13. The residual moisture content of the product in the vials is checked as specified in directive (m). The product in all vials measured has to have an RM 1 % {standard deviation smaller than 0.4 °C}. If this is not the case, follow directive (n).
14. If, during stoppering, one or more vials break, directive (o) must be applied.
15. The visual inspection of XX vials cut open does not show any sign of skin on the cake surface and of collapsed inclusion near the bottom of the vial. The solubility test, described in directive (p) shall be satisfactory for the dried product in all vials tested.

6.2.3 Description of Production Installations and their Handling

6.2.3.1 The following documents of the manufacturer can be used as part of this qualification: Description and instruction manual, installation drawings and – instructions, test and take over-certificates, and soft ware documentation. It is recommended to list and specify all necessary documents and their format in the purchase order, as this reduces time and effort for the buyer as well for the seller.

The installations have been taken over from the seller at (date and time). Before the take-over, the seller has successfully corrected complaints listed in appendix 1 of the take-over protocol. The freezing and freeze drying plant is registered as GF 50/95/1 and the loading and unloading system as GF 50/95/2.

6.2.3.2 The electrical power of both plants (XX kW) is supplied by the main power supply of building (F). An emergency diesel power supply of YY kW is installed, which switches on 1 min after a power break. (If YY is not equal to XX: The sequence of components to be connected to the emergency set is listed in directive (q), deposited at the control center). The control- and calculating system of both plants is connected by shielded cables to an independent power supply system, called computer power, that excludes influences of power variation or breakdown.

6.2.3.3 The cooling water (maximum XX m^3/h, maximum temperature +28 °C) is taken from the central supply system, which is protected against power failure and has an ample reservoir. No compressed air is used by GF50/95/1. All drives are electric or hydraulic.

6.2.3.4 The training of the personal on the installations GF 50/95/1 and 2 is organized as laid down in directive (r).

6.2.3.5 The calibration of all instruments of the installations is carried out by the calibraion service of the quality department, following the rules of that service. The computer programs used in the two installations are permanently tested by a program tested and accepted by the quality department. Errors or deviations are recorded and must to be evaluated by the quality department as given in directive (s).

6.2.3.6 Changes in the program must be carried out as described in directive (t).
If one or more of the following data are not supplied by the manufacturer, it is recommended to measure and to include them in the qualification document of the plants:

- The calculated evacuation time for the chamber- and the condenser volume by the described pump set: e. g. XX min down to 1 mbar, YY min down to 0.1 mbar, and ZZ min down to 0.03 mbar.
- The leak rate of chamber and condenser measured as proposed in Section 2.2.6.
- The flow of water vapor between chamber and condenser at least at two different pressures, as proposed in Section 1.2.4: e. g. $p_{ch,H2O} = 0.5$ and 0.05 mbar.
- The temperature differences between the product of 10 different vials filled with specified product and the specified filing height, e. g. five of these vials placed in the center, and five at the border of the shelf. It is recommended to measure these differences during freezing and secondary drying with and without the shielding described in Section 1.2.1, (see Fig. 1.68). The temperatures during freezing and secondary drying should be as specified in the process qualification. The results will give the information, whether the applied shielding method is satisfactory for the process or indicates necessary improvements.

6.2.4 Equipment Performance Tests

6.2.4.1 The leak test, described in directive (u) has been carried out twice in a row and on a third occasion, 24 h later. The leak rates $2.0 \cdot 10^{-3}$, $2.3 \cdot 10^{-3}$ and $2.5 \cdot 10^{-3}$ mbar L/s are below the requirements in directive (u). At the same time, the evacuation time has been measured. The average of the three measurements have been: 1.1 *a* min (±0.2 min) to 1 mbar; 1.3 *b* min (±0.2 min) to 0.1 mbar; and 2.2 *c* min (±0,4 min) to 0.03 mbar (*a*, *b*, *c* are the calculated evacuation times). The results are acceptable (the degassing of the walls starts below 1 mbar). The final pressure after XX h of pumping has been between 0.015 and 0.025 mbar, which is below the manufacturer's specification.

6.2.4.2 The roughness of the shelves and the maximum pressure at the stoppering device has been documented in the take-over protocol. The maximum cooling rate of the empty shelves has been 2.2 °C/min from 0 °C to –40 °C. The maximum heating rate from 0 °C to +40 °C is 2.5 °C/min. Both data meet the specification. The temperature differences on all shelves are smaller than ±1 °C in the temperature range between –40 °C and +40 °C, which corresponds with the specification.

6.2.4.3 The water vapor flow has been measured at two different pressures by using the instructions given in 6.2.3.6. In the calculated data, the time for evacuation is not taken into account; therefore the data may have with an error of ±10 %. The vapor flow has been at a pressure $p_{ch} = 0.9\ S_m^{\#}$ (g/h) and at 0.1 $p_{ch} = 0{,}8\ S_m^{\#}$ (g/h). $S_m^{\#}$ is the maximum possible flow at the tested pressure for a ratio $l/d = 2.5$ given in Fig. 1.89. (The data given in the figure have to be multiplied by free surface area of the connection between chamber and condenser.) The factors 0.9 and 0.8 show, that the accuracy of such a test cannot be only a few percent, but the fact that both values are clearly below 1.0 indicate the ratio of l/d is larger than 2.5. In summary, the vapor flow performance of the plant can be accepted. At the end of both tests the condenser surface has been covered by an ice layer of approx. 1.5 cm (by calculation). The visual inspection confirmed this as far as possible. Before both tests, and after the second test, the condenser temperature has been T_{co} with changes smaller than ±1.5 °C. The defrosting time for the condenser by steam has been approx. XX min. The condenser performance is within the specifications.

6.2.4.4 The warning of undesirable trends during the process has been tested by providing undesirable trend data by hand for the following trend measurements:

- cooling speed during freezing
- T_{ice} at constant p_c
- T_{ice} changes of more than 0.6 °C between two measurements after 2 h of MD until 8 h of MD. (*Note*: In the first 2 h the thermodynamic equilibrium is not reached and after 8 h MD comes to the end at which T_{ice} decreases anyway.)
- the pressure decrease after SD has started does not reach 0.15 p_c in 10 min.

In all cases the trend warning has been activated, indicating the maximum tolerance of the trend.

6.2.4.5 The alarm system has been tested by applying by hand data outside the preset tolerances for the following process data:

- T_{pr} at the end of freezing
- t_3 during which T_{pr} must be reached
- T_{ice}
- $T_{sh,md}$
- p_c
- T_{co}
- t_{md}

- $T_{sh,sd}$
- $p_{ch,sd}$
- t_{sd}

All alarm signals showed the correct identification and the acoustic alarm warned about the event.

All tests within the activities of Section 6.2.4 are recorded and documented in file (F).

A similar test protocol for the loading and unloading installation GT50/95/2 must be established. However no example can be given, since this kind of equipment can be based on very different design layouts.

6.2.5 Quality of Installation to Document the Ability of Equipment to Operate Processes (described in Section 6.2.2)

In the following table the important process steps (Proc. no.), the process description (Process quantity) (measure), the related target data (Target data) and their tolerances (tolerances) are listed and compared with average data measured in three runs (Ave. act. data) and the minimum and maximum data measured in the three runs (min./max.). The last two data have to be taken from the protocols and to be listed. In the last column the identification number of the runs, in which the two extreme data are measured is listed (Ident. no.) The last two column are not given, with the exception of proc. nr. 1.1 as an example. The table is a proposal of how the comparison could be made. The list may not be complete in all possible cases and is concentrated on the time-, pressure- and temperature data. Other methods may be preferred to make the ability of the equipment transparent.

Proc. no.	Process quantity (measure)	Target data (tolerances)	Ave. act. data (min./max.)	Ident. no.
1.1	Time between product being finished and start of freezing (h)	t (+10 %)	0.93 t (0.85 t/1.01 t)	001/003
1.2	Temperature during 1.1 (°C)	T (±3 °C)		
1.3	Filling height in vials (mm)	XX (0 %, +2 %)		
2.1	Leak rate of chamber and condenser (mbar L/s)	< XX (+0)		
2.2	Subcooling (°C)	4–8 (±0)		

2.3	Cooling rate down to –40 °C	0.8 to 1.2 (°C/min) (±0)
2.4	Resting time at –30 °C (min)	XX (–0, +YY min)
2.5	Temperature during 2.4 (°C)	30 (–3, +0 °C)
3.1	Condenser temperature during MD	< –45 °C (+0)
3.2	Time of evacuation to p_c (min)	XX (+30 %)
3.3	Controlled pressure during MD (mbar)	p_c (–0, +10 %)
3.4	T_{sh} after 3.2 completed (°C)	–10 (±1.5 °C)
3.5	T_{ice} 1 h after 3.2 (°C)	–30 °C (+0/–1 °C)
3.6	Standard deviation of 3.5	< 0.4 °C
3.7	Duration of MD from 3.2 to 4.1	t_{md} (±0.5 h)
4.1	Change from MD to SD	T_{ice} = max. $T_{ice,ave}$ –2 °C
4.2	p_{ch} 10 min after 4.1 (mbar)	$p_c/8$ ($< p_c/7$)
4.3	T_{St} during SD (°C)	+35 (+0, –1 °C)
4.4	T_{pr} during SD to reach	33 (–3, +2 °C)
4.6	T_{co} during SD	< –55 (+0)
4.7	DR after 1,5 h of SD (%/h)	3 (1 to 5 %/h)
4.8	dW at the end of MD (%)	0.6 (±0.3 %)
4.9	Time of SD (h)	t_{sd} (±0.5 h)
5.1	Time of venting with gas (G) (min)	XXX (–0, +20 %)

Such (or a similar) comparison not only proves that the process in the installations can be completed reproducible within the given tolerances, but also that the information may be helpful:

- to assess the target data and their tolerances. If the average data of all three runs are approximately in the middle of the tolerances, the target data and their tolerances are an optimum.

- to assess the reproducibility of the total process. If the maximum and minimum actual data relate mostly to the same run, reproducibility is just within the tolerances. If one run is not referred to at all, this run would be the best, and one can try to find the reasons why this has happened.
- to select such process numbers whose average actual data are very close to the target data and study, whether the tolerances can be reduced or enlarged if the maximum and minimum data are close to the tolerances in all three runs.
- to calculate the standard deviations of the average actual data and use these in future runs as a trend warning when the standard deviation with new data added changes measurably.

6.2.6 Documentation of the Quality of the Products Manufactured (in comparison with 6.2.1)

The product development department and/or quality assurance department prove and document that the products manufactured in the runs described above have the specified quality.

References for Chapter 6

[6.1] Monger, P.: Freeze Dryer Validation, p. 157, Good Pharmaceutical Freeze-Drying Practice, Editor: Cameron. P., Interpharm Press, Inc., Buffalo Grove, Il., USA 1997

[6.2] Powell-Evans, K.: Streamlining Validation, Pharmaceitical Technology Europe, Vol. 10 (12), p. 48–52, 1998, Advanstar Communication, Chester, CH1 4RN, UK

Appendix

Abbreviations, Symbols and Unit of Measure

symbol	meaning	unit of measurer
Å	Angstrom unit	10 nm
α	heat transfer coefficient	kJ/(°C m^2 h), J/(°C cm^2 s)
a_w	water activity	p/p_s
BTM	Barometric Temperature Measurement	
b/μ	permeability of water vapor through the dried product	kg/(h m mbar)
CA	capacitive vacuum gauge	
CPA	cryo protective agent	
c_f	specific heat of solids	kJ/(kg °C)
c_g	concentration of solids at T_g	% (w/w)
$c_{g'}$	concentration of solids at $T_{g'}$	% (w/w), g/g
c_{ice}	specific heat of ice	kJ/(kg °C)
c_p	specific heat of gas (constant pressure)	kJ/kg °C)
c_w	specific heat of water	kJ/(kg °C)
DR	Desorption Rate, desorbed water in % of solids per hour	%/h
DSC	Differential Scanning Calorimetry	
DTA	Differential Thermal Analysis	
d	thickness of layer, diameter	cm, m
dp	pressure rise	mbar
dt	time of dp	s, h
dW	desorbable water in % of solids	%
Δp	$p_s - p_{H2O,ch}$	mbar
E	enthalpy	kJ/kg
ER	electrical resistance	Ω
ε	radiant efficiency, radiance of the emitter/ radiance of a black body	

F	area	m^2, cm^2
F_{sh}	shelf area in a drying chamber	m^2, cm^2
f	frequency	Hz
h	hour(s)	h
Ic	ice, cubic	
Ih	ice, hexagonal	
J_{nmr}	coupling constant, measure of the splitting of lines in NMR measurements	
J^*	nucleation rate	nuclei/(volume time)
K	thermodynamic temperature	K
K_{su}	heat transfer coefficient from the heating medium to the freezing zone	kJ/(m^2 h °C)
K_{tot}	heat transfer coefficient from the heating medium to the sublimation front s	kJ/(m^2 h °C)
LN_2	liquid nitrogen	
LR	leak rate of a plant	mbar L/s
LR_{max}	maximum tolerable leak rate of a plant	mbar L/s
LS	sublimation energy of water	kJ/kg
LW	evaporation energy of water	kJ/kg
l	length	m, cm, µm, nm
Λ	wave length of light	nm
λ	thermal conductivity	kJ/(m h °C)
λ_g	thermal conductivity of frozen product	kJ/(cm s °C)
λ_{tr}	thermal conductivity of dried product	kJ/(cm s °C)
M	mol	Mr g
MD	main drying, sublimation drying	
m	mass	k
m_{H2O}	mass of water, sublimate during MD	kg
m_{ice}	mass of water frozen to ice	kg
m_{fest}	mass of solids	kg
Δm	part of frozen water	kg/kg
min	minute(s)	min
NMR	Nuclear Magnetic Resonance	
OPC	Operation Pressure Control	
p	pressure	bar, mbar
p_c	controlled operation pressure during MD	mbar
p_{ch}	pressure in the drying chamber	mbar
p_{co}	pressure in the condenser	mbar
p_{end}	end pressure	mbar
$p_{H2O,ch}$	partial water vapor pressure in the chamber	mbar
p_{ice}	p_s of ice	mbar
p_{md}	pressure during MD	mbar
p_{pg}	pressure of permanent gases	mbar

p_s	saturation vapor pressure	mbar
p_t	total pressure	mbar
Δp	$p_s - p_{H2O,ch}$	mbar
Q	quantity of heat	J, kJ, W s
Q_e	melting energy of ice	kJ/kg
Q_{tot}	total quantity of heat	kJ
q	density of heat flow	W/m^2
qL	leak rate	mbar L/s
qpv	suction capacity	mbar L/s
RM	residual moisture content	% (water in % solids)
RTD	resistance temperature gauge	
r	radius	m, cm
ρ_g	mass density of a frozen product	kg/m^3, g/cm^3
S	suction speed	m^3/h, L/s
S^*	spin	
S^{**}	magnetic field strength	G
$S^{\#}$	gas- or vapor stream	g/h, g/s
SD	secondary drying	
SEM	scanning electron microscope	
s	second(s)	s
$s^{\#}$	gas- or vapor stream density	$g/(cm^2\ s)$
σ	unit conductance	$kJ/m^2\ h\ K^{-4}$
T	temperature	°C
T_0	freezing temperature of water	°C
T_1	temperature of water at process start	°C
T_{am}	"antemelting" temperature	°C
T_c	collapse temperature	°C
T_{co}	condenser temperature	°C
T_d	devitrification temperature	°C
T_e	eutectic temperature	°C
T_g	glass transition temperature	°C
$T_{g'}$	glass transition temperature if all freezable water is crystallized	°C
T_{tot}	time weighted average of temperature difference ($T_{sh} - T_{ice}$) during MD	°C
T_h	crystallization temperature of hexagonal ice	°C
T_{hc}	temperature of homogeneous crystallization	°C
T_{ice}	temperature of ice at the sublimation front	°C
$T_{ice/n}$	sum of all n T_{ice} measurements divided by n	°C
T_{im}	"incipient melting" temperature	°C
T_k	crystallization temperature of hexagonal ice	°C
T_m	melting temperature	°C
$T_{m'}$	temperature of bilayer transition	°C

T_{md}	temperature during MD	°C
T_{me}	temperature of the heating and cooing medium	°C
T_{pr}	product temperature	°C
T_r	recrystallization temperature	°C
T_{sc}	temperature to which a product subcools	°C
T_{sh}	shelf temperature	°C
T_{str}	temperature of a radiant surface	°C
T_{tb}	tray bottom temperature	°C
Th	thermocouple	
TM	heat conductivity vacuum gauge	
t	time (space of –)	h, min, s
t_e	freezing time	h, min
t_{md}	main drying time	h
t_{sd}	secondary drying time	h
t_{SGR}	spin-lattice-relaxation	s
t_{SR}	spin-spin-relaxation	s
t_{to}	total drying time	h
UFW	"unfreezable water"	% (of total water)
V	volume	m^3, L, cm^3
V_{ch}	volume of a drying chamber	m^3, L
v	speed	m/s
v_c	growing speed of nuclei	cm/s
v_f	cooling rate	°C/s
x_f	part of solids	kg/kg, %
x_w	part of water above 0 °C	kg/kg
$x_{w'}$	part of ice	kg/kg
ξ_w	part of water in the start-up product	kg/kg

Index

S

T